Basic Concepts in Environmental Biotechnology

Basic Concepts in Environmental Biotechnology

Edited by

Neetu Sharma
Abhinashi Singh Sodhi
Navneet Batra

CRC Press
Taylor & Francis Group
Boca Raton London New York

CRC Press is an imprint of the
Taylor & Francis Group, an **informa** business

First edition published 2022
by CRC Press
2 Park Square, Milton Park, Abingdon, Oxon, OX14 4RN

and by CRC Press
6000 Broken Sound Parkway NW, Suite 300, Boca Raton, FL 33487–2742

CRC Press is an imprint of Taylor & Francis Group, LLC

ISBN: 978-0-367-65259-3 (hbk)
ISBN: 978-0-367-67469-4 (pbk)
ISBN: 978-1-003-13142-7 (ebk)

DOI: 10.1201/9781003131427

Typeset in Times
by Apex CoVantage, LLC

Contents

Foreword

Environmental biotechnology and innovative applications that have emerged from this multidisciplinary area, in the last decade alone, have a far-reaching implication in today's global society; therefore, it gives me immense pleasure to witness this book come to fruition. It is now increasingly accepted that research stemming from the broad field of environmental biotechnology can address several sustainable development goals (SDGs) set forth by the United Nations. SDGs can be grouped under 17 goals as a blueprint for the long-term stewardship and environmental resilience of our planet. The SDGs that can be addressed using environmental biotechnology and the use of emergent techniques of modern biotechnology are as follows: clean water, clean energy, no poverty, zero hunger, good health and overall sustainability of our planet. Microbial life forming the basis of environmental biotechnology is beginning to be much better understood through the application of emergent techniques of modern molecular biology, such as metagenomics, metatranscriptomics and metaproteomics. Therefore, this book will significantly enhance our understanding of various forms of environmental biotechnologies for environmental stewardship. For example, chapters on water, soil and air pollution will provide the current state of these environmental systems and how scientists are mitigating pollution from the environment. Biodegradation of organic, inorganic and solid waste management of contaminants for environmental conservation as well as impact assessment and monitoring using microbial technologies will help readers better understand the application of these modern tools for environmental applications. The topics included in this book will not only benefit students and academicians but also environmental resource managers and professionals alike and, therefore, I applaud the entire team for bringing this resourceful book to fruition.

Ashvini Chauhan, PhD
Professor of Environmental Biotechnology
School of the Environment
Florida Agricultural and Mechanical University
Tallahassee, Florida, USA

Preface

Environmental biotechnology implies principles of multidisciplinary sciences of environmental engineering, metabolic engineering, rDNA technology and omics to study the role of microbes and plants in tackling environmental issues. With the rapid industrialization and development of new innovations, there is a rising pressure on natural resources. The dream of sustainable development is remote without the contribution of environment biotechnology.

Basic Concepts in Environmental Biotechnology presents a comprehensive review of environmental topics in a compilation of 16 chapters related to environmental problems, control and use of biotechnology to develop innovative products. Each chapter provides a detailed context of problems and solutions of different topics with diagrammatic illustrations and tables for students, researchers and other professionals. The topics include the basics of environmental biotechnology, different types of pollution and their control measures, solid waste management and exploiting the role of microbes and plants to decipher problems such as the detoxification of marine and soil pollutants. Topics also cover alternative approaches in the form of a microbial route for the production of biofuels, biosensors, bioplastics and bioleaching and content related to risk assessment and environmental management systems.

This book has been updated to include recent studies and innovations made in this field and incorporates case studies that will be helpful for undergraduate and postgraduate students studying modules in the field of environmental studies, pollution control strategies, environmental biotechnology and more.

Acknowledgments

We would like to express deep and sincere gratitude to the Principal and the Management of GGDSD College, Sector-32-C, Chandigarh, India, for providing us with the necessary facilities. We are extremely grateful to our colleagues Mr. Arvind Behal and Dr. Sonu Bhatia for their extensive support and advice. Our thanks are also due to all the contributing authors and their host institutions, and we convey our special appreciation to them. Our special thanks to CRC Press/Taylor & Francis Group for publishing this book on Environmental Biotechnology, which surely will benefit the biological and environmental sciences professionals and enthusiasts.

Finally, we would like to express our sincere thanks to our families and friends for their continuous support, valuable prayers and love.

Contributors

Dr. Varenyam Achal, School of Ecological and Environmental Sciences, East China Normal University, China

Dr. Wamik Azmi, Department of Biotechnology, Himachal Pradesh University, Summer-Hill, Shimla (H.P.), India

Dr. Navneet Batra, Department of Biotechnology, GGDSD College, Chandigarh, India

Dr. Sonu Bhatia, Department of Biotechnology, GGDSD College, Chandigarh, India

Dr. Xueyan Chen, School of Ecological and Environmental Sciences, East China Normal University, China

Ms. Rajni Kashyap, Department of Biotechnology, Himachal Pradesh University, Summer-Hill, Shimla (H.P.), India

Dr. Saurabh Gupta, Mata Gujri College, Fatehgarh Sahib, Punjab, India

Dr. Jing He, School of Ecological and Environmental Sciences, East China Normal University, China

Dr. Mahdieh Houshani, University of Tabriz, Tabriz, Iran

Dr. Hanuman Singh Jatav, Sri Karan Narendra Agriculture University, India

Dr. Alok Jha, Department of Biotechnology, Institute of Bio-Medical Education & Research, Managalayatan University, Beswan, Aligarh (U.P.), India

Dr. Amit Joshi, Department of Biotechnology, SGGS College, Chandigarh, India

Dr. Malarvizhi Kaliyaperumal, Centre for Advanced Studies in Botany, University of Madras, Guindy Campus, Chennai, India

Ms. Damanjeet Kaur, Mata Gujri College, Fatehgarh Sahib, Punjab, India

Dr. Vishnu D. Rajput, Southern Federal University, Russia

Ms. Alka Rani, Department of Biotechnology, Himachal Pradesh University, Summer-Hill, Shimla (H.P.), India

Dr. Yogesh Rawal, Department of Zoology, Panjab University, Chandigarh, India

Dr. Ajay Sharma, Department of Chemistry, Career Point University, Tikker-kharwarian, Hamirpur, Himachal Pradesh, India

Dr. Anil Kumar Sharma, Department of Biotechnology, Maharishi Markandeshwar University, Mullana, Ambala, India

Dr. Himanshu Sharma, Department of Biotechnology, GGDSD College, Chandigarh, India

Dr. Indu Sharma, Department of Biotechnology, Maharishi Markandeshwar University, Mullana, Ambala, Haryana, India

Dr. Neetu Sharma, Department of Biotechnology, GGDSD College, Chandigarh, India

Mr. Tarun Sharma, Department of Biotechnology, GGDSD College, Chandigarh, India

Dr. Jagtar Singh, Department of Biotechnology, Panjab University, Chandigarh, India

Mr. Abhinashi Singh Sodhi, Department of Biotechnology, GGDSD College, Chandigarh, India

Dr. Meenakshi Suhag, Department of Environmental Science, Kurukshetra University, Haryana, India

Dr. Sarieh Tarigholizadeh, University of Tabriz, Tabriz, Iran

Dr. Abhinay Thakur, Department of Zoology, DAV College, Jalandhar, Punjab, India

Ms. Mony Thakur, Department of Microbiology, Central University of Haryana, Mahendragarh, Haryana, India

Dr. Anoop Verma, School of Energy and Environment, Thapar Institute of Engineering and Technology, Patiala (Punjab), India

Dr. Meenu Walia, Department of Biotechnology, SD College, Ambala Cantt, Haryana, India

Dr. Vinod Yadav, Department of Microbiology, Central University of Haryana, Mahendragarh, Haryana, India

Abbreviations

AFLP	amplified fragment length polymorphism
ALM	Advanced Locality Management
AMD	acid mine drainage
ASTM	American Society for Testing and Materials
BASI	Brewster angle straddle interferometry
BAT	best available technology
BATNEEC	best available technology not entailing high cost
bio-PE	bio-based polyethylene
bio-PET	bio-based polyethylene terephthalate
bio-PVC	bio-based polyvinyl chloride
BTEX	benzene, toluene, ethylbenzene, xylene
CAX	cation/proton exchanger
CEN	European Committee for Standardization
CPCB	Central Pollution Control Board
CRISPR	clustered regularly interspaced short palindromic repeats
DENL	day-evening-night level
DGGE	denaturing gradient gel electrophoresis
DoE	Department of Environment
EC	environmental clearance
EFC	evanescent-field coupled
EIA	environmental impact assessment
EK	electrokinetic
ELISA	enzyme-linked immunosorbent assay
EMS	environmental management system
EPA	Environmental Protection Agency
EPS	extracellular polymeric substances
FET	field-effect transistor
FHWA	Federal Highway Administration
GABA	gamma-aminobutyric acid
GMOs	genetically modified organisms
HDPE	high-density polyethylene
HMs	heavy metals
IPCC	Intergovernmental Panel on Climate Change
ISO	International Organization for Standardization
JnNURM	Jawaharlal Nehru National Urban Renewal Mission
LDPE	low-density polyethylene
LFG	landfill gas
LLDPE	linear low-density polyethylene
LOD	limit of detection
MCGM	Municipal Corporation of Greater Mumbai
MES	microbial electrochemical system
MFC	microbial fuel cell

MICP	microbial-induced calcium carbonate precipitation
MitBASE	mitochondrial DNA database coordination
MoEF	Ministry of Environment and Forests
MRSA	methicillin-resistant *Staphylococcus aureus*
MT	metallothioneins
MTP	metal tolerance protein
MWCNT	multi-walled carbon nanotubes
NEPA	National Environmental Policy Act
NGS	next-generation sequencing
NRAMPs	natural resistance-associated macrophage proteins
PA	polyamides
PAHs	polycyclic aromatic hydrocarbons
PBAT	polybutylene adipate-*co*-terephthalate
PBS	polybutylene succinate
PBSA	polybutylene succinate-*co*-adipate
PBT	polybutylene terephthalate
PC	polycarbonate
PCL	polycaprolactone
PCR	polymerase chain reaction
PCS	phytochelatin synthase
PDLA	poly(D-lactic acid)
PDLLA	poly(DL-lactic acid)
PET	polyethylene terephthalate
PGPRs	plant-growth promoting rhizobacteria
PHA	polyhydroxyalkanoates
PHB	poly-β-hydroxybutyrate
PLA	polylactic acid
PLLA	poly(L-lactic acid)
PMC	Pune Municipal Corporation
POPs	persistent organic pollutants
PP	polypropylene
PS	polystyrene
PSA	prostate-specific antigen
PTT	polytrimethylene terethalate
PUR	polyurethane
PVC	polyvinyl chloride
QCM	quartz crystal microbalance
RAPD	randomly amplified polymorphic DNA
RPM	respirable particulate matter
SELEX	systematic evolution of ligands by exponential enrichment
SERS	surface-enhanced Raman spectrum
SMS	Stree Mukti Sanghatan
SOLiD	sequencing by oligonucleotide ligation and detection
SPCB	State Pollution Control Board
SPCDE	surface plasmon-coupled directional emission
SPM	suspended particulate matter

SPME-GC/MS	solid-phase microextraction-gas chromatography/mass spectrometry
SPMEs	secondary plant metabolites
SPR	surface plasmon resonance
SWaCH	solid waste collection and handling
SWCNT	single-walled carbon nanotube
TGGE	temperature gradient gel electrophoresis
TNT	trinitrotoluene
UM-BBD	University of Minnesota Biocatalysis/Biodegradation Database
USEPA	United States Environmental Protection Agency
UXO	unexploded ordnance
WBCSD	World Business Council for Sustainable Development
WHO	World Health Organization

1 Environmental Biotechnology
An Overview

Sonu Bhatia and Jagtar Singh

CONTENTS

1.1 INTRODUCTION

Large-scale urbanization and industrialization has led to environmental pollution and deterioration. Over the decades, different types of conservation strategies have been employed to preserve global resources, but no fundamental solution has been attained. In this context, governmental agencies and the scientific community have been exploring the applications of biotechnology to address the unresolved environmental issues. Biotechnology is paving the way for environmental protection by becoming a significant part of the governmental policies aimed at achieving the goals of sustainable development. Environmental biotechnology plays a pivotal role in unfolding the primary structural and functional link between the environment and its inhabitants. Information obtained from different approaches, including bioremediation, rDNA technology and omics, is being applied widely to redefine the microbial abilities for environmental betterment.

Thus, environmental biotechnology is defined as a scientific discipline using biological processes to control or resolve environment-related issues, including

DOI: 10.1201/9781003131427-1

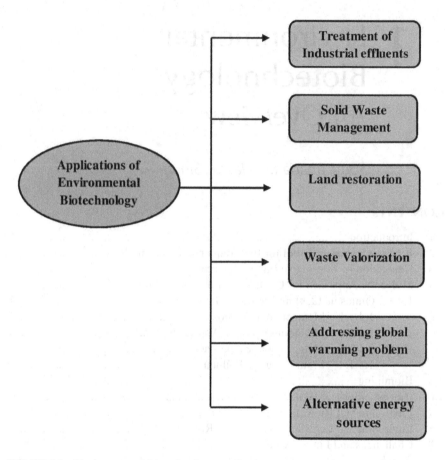

FIGURE 1.1   Environmental biotechnology applications.

remediation of contaminated sites, pollution monitoring, renewable and eco-friendly energy source generation, and so on (Figure 1.1).

The Following Are the Common Environmental Issues Being Faced by the World

1. Industrial effluents
2. Air pollution
3. Soil contamination
4. Greenhouse gas emission
5. Radioactive waste
6. Land degradation

1.2   BIOREMEDIATION AS A TOOL FOR ENVIRONMENTAL BIOTECHNOLOGY

Bioremediation is a promising approach that makes use of microorganisms and their processes to curb different environmental issues. Conventional methods for waste treatment are either less efficient or involve the use of chemicals, thus not an

TABLE 1.1
List of Some Selected Pollutant-Degrading Microorganisms

Sr. No.	Microorganism	Targeted Pollutants	References
1.	*Arthrobacter nicotinovorans*	Pesticides	Aislabie et al., 2005
2.	*Aspergillus flavus*	Heavy metal	Akhtar et al., 2013
3.	*Aspergillus niger*	Oil	Damisa et al., 2013
4.	*Bacillus subtilis*	PAHs	Kotoky et al., 2017
5.	*Cladosporium* sp.	Aromatic hydrocarbon	Birolli et al., 2017
6.	*Doratomyces nanus*	Polychlorinated biphenyls	Mouhamadou et al., 2013
7.	*Enterobacter* sp.	Pesticides	Singh et al., 2004
8.	*Flavodon flavus*	Bleaching effluent	Raghukumar, 2004
9.	*Geobacillus thermodenitrifican*	Aliphatic hydrocarbon	Abbasian et al., 2015
10.	*Gordonia sihwensis*	Aliphatic hydrocarbon	Brown et al., 2016
11.	*Marasmillius* sp.	PAHs	Vieira et al., 2018
12.	*Microbacterium* sp.	Heavy metals	Coretto et al., 2015
13.	*Mucor racemosus*	PAHs	Passarini et al., 2011
14.	*Mycobacterium* sp.	PAHs	Chauhan et al., 2008
15.	*Phanerochaete chrysosporium*	Azo dyes	Cripps et al., 1990
16.	*Pseudomonas putida*	PAHs	Lee et al., 2018
17.	*Sphingomonas* sp.	PAHs	Ghosal et al., 2016
18.	*Tintoporia barbonica*	Paper and pulp industry effluent	Fukuzumi et al., 1977
19.	*Trichoderma* sp.	Dye	Jebapriya and Gnanadoss, 2013

environment-friendly approach. Microorganism-based methods are an eco-friendly approach due to the use of biological entities. Bacteria, fungi, yeast and algae are used as a whole cell, or their metabolites are applied for the treatment of various environmental contaminants. Microbial enzymes are being used to treat various industrial effluents from, textile industry, pharmaceutical industry, chemical industry, petroleum refineries and radioactive waste.

Bioremediation is a cost-effective and noninvasive method to restore the environment. Microorganisms can be used to degrade the complex organic and inorganic pollutants into simpler and less toxic forms or altogether remove them from the site. A list of some selected microorganisms are mentioned in Table 1.1.

1.3 GENERATION OF ALTERNATIVE ENERGY SOURCES

For ages, humans have remained depended on conventional fossil fuels to fulfill their energy demands. But fossil fuels, due to their limited stock, will be exhausted shortly. Therefore, the search for clean alternative energy fuels with similar or improved energy output has shifted the world's focus on environmental biotechnology. Microorganisms have been studied for their role in the synthesis of clean fuel commonly known as biofuels. The conventional biofuels include bioethanol, biodiesel and biohydrogen fuels.

1.4   MOLECULAR APPROACH FOR CONTROLLING ENVIRONMENTAL POLLUTION

The gradual development of molecular tools has significantly enhanced our ability to understand different microbe-based bioremediation mechanisms. Tools and techniques, including randomly amplified polymorphic DNA (RAPD), amplified fragment length polymorphism (AFLP), polymerase chain reaction (PCR) and electrophoresis, have been employed to derive precise information related to metabolism and the biogeochemical cycle of microbes in the environment. PCR-DGGE- (polymerase chain reaction denaturing gradient gel electrophoresis) and PCR-TGGE (polymerase chain reaction temperature gradient gel electrophoresis) have been employed to assess degradation pathways of bacterial communities in a bioremediation process. Pesticide-contaminated sites in soil have been assessed for microbial diversity by using RAPD, which has also been used to identify viruses and to study their diversity.

1.4.1   Omics in Tackling Environmental Pollution

The detailed study of microbial habitat can be helpful in understanding the natural biodegradation processes and improving their ability to decontaminate the site. In recent times, molecular tools like next-generation sequencing (NGS) are being applied extensively to curb environmental pollution by monitoring and studying the microbial community composition of the contaminated site. The efficient biodegradation mechanism involves knowledge about the indigenous microbial community and pollutant type and concentration. There are many omics-based approaches, including metagenomics, transcriptomics and proteomics, and they can be employed for understanding the complex biodegradation mechanisms. These approaches, in combination with bioinformatics, provide detailed information related to genome, growth profile, metabolism, protein profile and environmental abilities of unculturable microorganisms (Table 1.2).

1.4.1.1   Metagenomic Approach

Metagenomic analysis involves direct DNA extraction from environmental samples and subjecting them to sequencing techniques like NGS and SOLiD (sequencing by oligonucleotide ligation and detection) for further assessment of microbial ecology and environmental genomics research. NGS high-throughput sequencing technology includes illumina sequencing, Roche 454 sequencing and ion torrent. The SOLiD technique involves a ligation reaction utilizing the probe recognition method, which is less error-prone. This approach has revolutionized the human ability to have insight into the genome and related concepts of unculturable microorganisms. High-throughput microarrays, for example GeoChip microarrays, have also been used to perform metagenomic analyses. These are useful in monitoring biogeochemical processes and assessing microbial communities for their structure and dynamics. A microbial ecology and biodegradation

TABLE 1.2
Common Techniques in Environmental Biotechnology for Microbial Community Studies

Molecular Technique	Applications
RISA/ARISA	• To characterize and compare microbial communities in different environments
	• To identify organic and inorganic pollutants degrading microbes
	• To identify functional genes involved in biodegradation
DGGE/TGGE	• To characterize microbial communities and their functional genes
	• To track the shift in microbial community structure in various contaminated sites
RFLP	• To study biogeographical patterns in microbial communities
	• To assess microbial communities in wastewater systems
RAPD	• To profile microbial community in sites contaminated with pesticides, fertilizers, etc.
DNA microarrays	• To detect novel prokaryotic taxa
	• To annotate genes
Metagenomics	• To characterize uncultivable microbes in various environments
	• To determine phylogenetically relevant genes
Metatranscriptomics	• To profile the whole genome expression of microbial communities
Metaproteomics	• To identify proteins produced by complex microbial communities
Metabolomics	• To identify and quantify metabolites
	• To illustrate quorum sensing

profile of contaminants can be studied by using the PhyloChip microarray based on high-throughput phylogenetic analyses.

Metagenomic analysis has been placed in two categories:

1. **Sequence-based analysis**

   In this analysis, DNA samples are processed for sequencing for retrieving gene-related data. The sequence-based approach employs sequence analysis for further functional prediction of the gene. The method is beneficial in identifying new genes providing microbes with their bioremediation ability.

2. **Function-based analysis**

   Function-based metagenomic techniques are based on studying the proteins encoded by environmental DNA. Screening for enzymatic activity can be done from the cloned DNA fragments expressed in a host. This approach can be useful in the finding of novel genes and their role in metabolism, which is helpful in the exploration of unknown environmental sites. Complete metabolic pathways can be elucidated by comparing organisms from various communities by employing sequence databases.

1.4.1.2   Metatranscriptomic Approach

This deals with the analysis of RNA content of the environmentally essential microorganisms, thus providing insights into the complex gene expression and regulation system required for bioremediation. Specific genes can be identified and quantified for further analysis.

1.4.1.3   Metaproteomic Approach

This involves the study and characterization of the protein profile of microorganisms from environmental samples. The procedure is very convenient for identifying different sets of proteins being expressed by genes for the bioremediation of pollutants in the environment. Further, protein profiling is useful for knowing the changes a microbe undergoes during the bioremediation process.

1.4.1.4   Monitoring the Pollutants

Real-time information about the concentration of environmental pollutants and contaminants is an essential feature in selecting and designing mechanisms for controlling, treating and completely removing the contaminants. Conventional pollution indicators rely upon frequent sampling and extensive and costly laboratory protocols. Therefore, for in recent years, the development of biosensors has provided an alternative to this. Biosensors offer several advantages over conventional methods.

1. Specific, sensitive and stable detection
2. Availability of portable systems
3. Rapid detection

A biosensor is an integrated device used to detect the presence of an analyte, such as a biomolecule or a microorganism present in the environment. It consists of three parts (Figure 1.2):

1. Biocomponent that recognizes the analyte and generates a signal
2. Signal transducer
3. Signal reading device

Biosensors have been applied for the detection of different kind of pollutants, including organic matter, heavy metals and microorganisms.

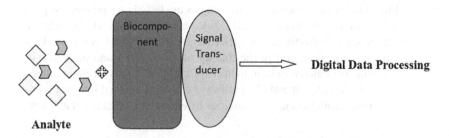

FIGURE 1.2   Schematic diagram representing the basic mechanism of biosensors.

Types of Biosensors

1. Optical biosensors
2. Electrochemical biosensors
3. Thermal biosensors
4. Whole-cell biosensors
5. Immunological biosensors

1.5 BIOMINING

Biomining is a microorganism-based economical alternative for the treatment of specific mineral ores to extract minerals such as copper, uranium and gold. This technique has a significant advantage as it is inexpensive, can work with lower concentrations and can attain extraction yields of over 90%. A range of mineral-oxidizing bacteria (*Acidithiobacillus ferrooxidans*, *Acidithiobacillus thiooxidans*, *Acidithiobacillus caldus*, *Leptospirillum ferrooxidans*, etc.) can effortlessly oxidize iron- and sulfur-containing minerals. Fungal strains have also been looked upon for biomining; for example, *Aspergillus niger* and *Penicillium simplicissimum* helped in the mobilization of Cu/Sn by 65% and Al, Ni, Pb and Zn by more than 95%. Conventional mining techniques can have ill effects on the environment; thus, microbial-based metal recovery processes provide a competitive advantage.

1.6 BIOPESTICIDES AND BIOFERTILIZERS

Microbial pesticides or biopesticides control pests using nontoxic mechanisms. They are generally naturally less toxic, affect only the target pest, show effect in minute quantities and are biodegradable. They are used as a part of integrated pest management programs to control pests and minimize the impact of chemical-based pesticides on the environment. Biofertilizers constitute a significant part of integrated nutrient management because of their use as a renewable resource for sustainable agriculture. Efficient strains of microorganisms can be employed to encourage crop plants' uptake of nutrients by their rhizospheric interactions. They hasten some microbial processes in the soil, which results in augmenting the availability of nutrients in a form that is easily assimilated by the plants. Several microorganisms and their association with different plants are being explored for the production of biofertilizers, for example, nitrogen fixers, phosphate solubilizers and phosphate mobilizers. Biofertilizers have a longer shelf life, are cost effective, cause no harm to the ecosystem and show enhanced survival on seeds and soil.

1.7 ENVIRONMENTAL BIOTECHNOLOGY INFORMATICS

With the advent of systems like bioremediation, environmental risk assessment and microbial ecology, the use of computers and information technology is being employed in the study of biological systems—this is known as environmental biotechnology informatics.

A variety of *in silico* algorithms, software and web resources are utilized to infer or associate the molecular and omics data. Experimental information can be stored or retrieved from numerous databases like the MEROPS peptidase database (structure of proteolytic enzymes) and MitBASE (mitochondrial DNA database coordination). Microbial biocatalytic reactions and xenobiotic biodegradation pathways are enlisted in the University of Minnesota Biocatalysis/Biodegradation Database (UM-BBD). It contains 187 pathways, 1287 reactions, 1195 compounds, 833 enzymes, 491 microorganism entries and 259 biotransformation rules (http://umbbd.msi.umn. edu/). MetaRouter is used for maintaining diverse information correlated with bioremediation and biodegradation. Based on the stored data, mathematical models of biological systems can be derived to plan further experimentation.

1.8 ENVIRONMENTAL LEGISLATIONS AND THEIR ROLE IN EMS

To reduce industrial pollution, restrain the use of hazardous materials and reduce waste, environmental laws are being practiced worldwide. Several legislations target "point-source" emissions to avoid pollution from particular discharge points like chimneys and waste pipes. Other prohibitions deal with pollution from the start of an industrial process. Significant roles in an environmental regulatory process are played by governments, environmental lawyers, regulatory agents, industrial managers and green activists involving constant interactions to address various issues.

The first formal integration of the environmental impact assessment (EIA) process in a legislative figure was introduced by the National Environmental Policy Act (NEPA) in the United States. Numerous countries such as Australia, Canada, Sweden and New Zealand adopted such legislation soon after that. Over the last two decades, EIA has progressed in most of the UN Environment Programme member nations due to increasing global environmental issues, including climate change, loss of biodiversity and degradation in water quality.

Environmental impact assessment comprises basic methodology, including screening, scoping, broad study, impact prediction progress statement, assessment, public participation, monitoring and follow-up measures. EIA was introduced in India in 1994 by the Ministry of Environment and Forests (MoEF) making EIA compulsory for 32 major polluting activities. Several amendments have been made over the past few years to improve the quality of the environment. This involves the development of a stringent decision-making process related to the environmental clearance (EC) process. The current EC process is based on a two-tier system concerning both central and state authorities. MoEF's Impact Assessment division (IA), MoEF's regional offices and the Central Pollution Control Board (CPCB) are involved at the level of the central government. SPCBs (State Pollution Control Boards) and state Departments of Environment (DoE) work at the state level.

A systematic and transparent process known as environmental management system (EMS) has been established industry-wide for proper implementation of environmental goals while taking into account legislative policies and individual responsibilities, followed by regular auditing of its elements. EMS international standards such as ISO 14001 is widely implemented and supported by companies worldwide. EMS consists of three key areas—identifying the sources of pollution, managing emergency conditions and building up preventive solutions. The overall goal of EMS can be achieved by focusing on the technical aspects such as impact assessments, measurement of contaminant discharge, audits and inspections, which will lead to the development of "green" solutions with the implementation of cleaner technology. The administrative aspects are based on current environmental regulations, standards and so on and the social aspects consist of training, communication, coordination and more.

The best available technology (BAT) principle has been adopted by various industries to represent their commitment in the direction of environmental protection. Available techniques are referred to as those that allow industrial sector execution under economically and technically feasible conditions, simultaneously considering the costs and advantages of them.

Location-specific circumstances influence the costs of meeting the goals of environmental protection. BAT takes into consideration marginal costs/benefits, which in turn are location dependent. A technology considered "best" in one country might not be regarded that way in another country. It can be assessed by cost-benefit analysis; for example, a place with low levels of pollution and high environmental quality would require less costly remediation methods considered at BAT. Biotechnological processes can be considered as BAT/BATNEEC, especially in the areas of wastewater treatment, solid waste management and soil remediation.

1.9 CHALLENGES AND FUTURE ASPECTS

The greatest challenge that environmental biotechnology faces is to strike a balance between the technical and economic parameters considered during monitoring, manufacturing and waste management. Environmental technologies are competitively assessed in terms of remediation abilities and pollution control, as well as market viability. Operational sustainable development from a business viewpoint, known as eco-efficiency, was formed by the World Business Council for Sustainable Development (WBCSD). Eco-efficient technologies provide the best opportunities for saving costs with sustainable production and consumption. Risk assessment has become mandatory for studying the impact of new technologies, including biotechnology. Genetically engineered plant varieties with increased pest resistance, yield or nutrition should undergo stringent testing and evaluation before its environmental release. It becomes essential for national regulatory agencies to follow principles of a precautionary approach to ensure the safe release of genetically modified organisms into the environment. With progress in biotechnological interventions like enzyme engineering, process engineering and strain improvement, biodegradation abilities of microbes can be further enhanced. The development of new biomembrane technologies can enhance wastewater treatments. Also, biowaste can be used for valorizing and developing renewable energy sources.

1.10 CONCLUSION

Environmental biotechnology focuses on the sustainability of modern human society by digging out crucial materials from renewable resources and by decreasing dependence on nonrenewable resources. Water and energy issues are critically addressed by environmental biotechnology by treating poor-quality water sources, thereby eliminating public health risks while shifting the future fuel utilization trend from fossil fuels to renewable ones. Environmental biotechnology has proved its potential in pollutant detection, monitoring, remediation and development of cleaner technologies. The need of the hour heightens the implementation of environmental biotechnology to safeguard environmental quality, along with economic benefits by capturing valuable resources. Modern biotechnology should be evaluated for its consequences, opportunities and challenges by policymakers to provide its economically viable use.

2 Air Pollution
Causes, Effects and Control of Pollutants

Malarvizhi Kaliyaperumal and Indu Sharma

CONTENTS

2.1 INTRODUCTION

Air pollution is a complex mixture of chemical, physical and biological air pollutants that alter the natural environment. Earth's atmosphere is supposed to be balanced with essential gases such as nitrogen (78%), oxygen (21%) and other trace gases (such as argon and carbon dioxide) that are essential to life. Pollutants of many natural and anthropogenic origins can have an intense and destructive effect on the ecosystem. However, the latter is of great concern to developed and/or developing countries and a difficult task for developers, policy makers and environmental organizations to address. The variations of pollutant and weather changes modify the concentration of pollutants in time and space.

People moving from the countryside to cities made the air condition worse because of factories and the domestic heating. Urban air quality and health has been deteriorating over time and compares unfavorably to that of the countryside or rural areas. Countries in southwestern Asia, along with other eastern Asian countries, experience the highest rates of mortality from air pollution (~59% of total global deaths) in the world. Most southwestern Asian countries exceed ambient air quality standards recommended by the World Health Organization (WHO) because of desert dust storms, power plants, petrochemical activity and traffic emissions. The more common pollutants produced in high-density urban areas are carbon monoxide (CO), nitrogen oxides (NO_x), sulfur oxides (SO_x), ozone (O_3), particulate matter (PM) and benzene (C_6H_6). The inventory includes anthropogenic sources, such as transport (road, rail, ship and aviation), large-scale power generation (from coal,

DOI: 10.1201/9781003131427-2

diesel and gas power plants), small-scale power generation (from diesel generator sets for household use, commercial use and agricultural water pumping), small- and medium-scale industries, dust (road resuspension and construction), domestic (cooking, heating and lighting), open waste burning and open fire, and non-anthropogenic sources, such as sea salt, dust storms, biogenic substances and lightning. It includes aerosols, smoke, fumes, dust, ash and pollen.

Particulate matter consists of a combination of both solid particles and liquid droplets found in the air. The size of the PM may vary, for example, PM_{nano}, PM_1, $PM_{2.5}$ and PM_{10}, and the different sizes have been associated with varying health effects (Figure 2.1). Total PM fraction includes both PM_{10} and $PM_{2.5}$, while PM_{10} includes $PM_{2.5}$ and vice versa. These fractions are generated either through physical motion or through gaseous chemical reactions in the atmosphere. The composition of PM is highly variable in relation to countries, industries, seasons and weather conditions. Fine particulate matter, that is, PM of 2.5 microns in diameter and less, is of prime public health concern. It is also known as $PM_{2.5}$ or respirable particle because it penetrates the respiratory system more deeply than larger particles do. In highly polluted countries, $PM_{2.5}$ is largely made up of sulfate and nitrate particles, elemental and organic carbon and soil. In urban settings, the overall average of $PM_{2.5}/PM_{10}$ is about 0.42, which is higher when there is low precipitation, dust storm activities, enhancement of sea/land breezes and traffic activities. Depending on the duration

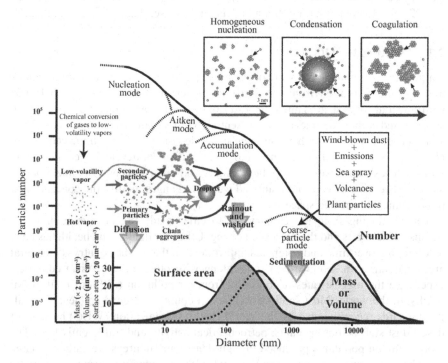

FIGURE 2.1 Origin of common particulate matter. (From: http://radontestingdallas.com/indoor-air-quality-testing-composition-dust-particles-particulate-matter/#respond.)

of exposure to $PM_{2.5}$, health issues vary from mild to adverse. Many studies from Europe and elsewhere lend strong support to the wide array of mortality and morbidity outcomes associated with $PM_{2.5}$. In addition to epidemiological evidence linking PM pollution to a diversity of health outcomes, there is also extensive research that aims to determine the underlying mechanistic pathways. Most of the time, however, it remains unmeasured and unnoticed, so WHO termed it a "silent killer".

The global burden of disease assessments estimated 0.74 million and 1.1 million premature deaths in 1990 and 2016 due to outdoor $PM_{2.5}$ and ozone pollution and 0.99 million and 0.78 million premature deaths in 1990 and 2016 due to household (indoor) $PM_{2.5}$ pollution. Some particles, such as dust, dirt, soot and smoke, are large or dark enough to be seen with the naked eye. Others are so small they can only be detected using an electron microscope. Irrespective of the origin, classical air pollutants can take many forms, including PM, CO, CO_2, SO_2, ground-level O_3, nitrogen dioxide (NO_2), methane, radioactive decay and certain toxic chemicals that are extensively studied for their health impacts (Table 2.1).

Air pollution severely damages the lungs and airways. It is a potential contributor to many diseases that are harmful to vital organs. Mechanistic evidence

TABLE 2.1
Classical Pollutants in Air

Criteria Pollutants	Sources	Environmental Effects	Health Effects
Ozone (O_3)	i. Industrial emissions ii. Motor vehicle emissions iii. Sunlight-mediated chemical reaction of NO_x and VOCs	i. Damages plant tissues, decreases crop yields ii. Creates cracks in rubber iii. Fades paint and fabric dyes	Affects the respiratory system, eyes and throat and causes chest pain
Sulfur dioxide (SO_2)	i. Burning of fossil fuels ii. Industrial processes iii. Petroleum refining	i. Damages animals and plants; sometimes leads to death ii. Corrodes paint and metals iii. Fades dyes and paint iv. Causes acid rain	Severely damages the lungs
Nitrogen oxides (NO_x)	i. Automobiles ii. Burning of fossil fuels	Affects the plant community, especially crops	i. Irritates eyes, lungs, nose and throat ii. Increases vulnerability to disease
Carbon monoxide (CO)	Only 10% from fuel combustion and fires	None	i. Extremely toxic at low concentrations ii. Birth defects iii. Nausea and dizziness iv. Shortness of breath

TABLE 2.1
(Continued) Classical Pollutants in Air

Criteria Pollutants	Sources	Environmental Effects	Health Effects
Particular matter (PM)	i. Agriculture ii. Windstorms iii. Burning of fossil fuels iv. Industrial processes	i. Corrodes metal ii. Damages crops and trees iii. Causes acid rain	i. Respiratory disease ii. Throat, lungs and eye irritation
Lead (Pb)	i. Gasoline ii. Paint iii. Smelting of metals	i. Leads to soil toxicity ii. Alters plant metabolism	i. Brain damage in children ii. Behavior problems iii. Nerve disorders iv. Anemia and kidney damage

Source: Adapted from Botkin and Keller, 2007.

indicates that $PM_{2.5}$ is associated with oxidative stress, systemic inflammation and alteration of the cardiac electrical processes. Animal studies, complemented by experimental studies in humans, provide robust evidence of vasoconstriction and systemic pro-inflammatory responses. Scientific conclusions about the evidence for a causal link between $PM_{2.5}$ and adverse health outcomes have been confirmed. According to WHO (2016), 91% of the world population were living in a polluted environment where air quality was not within WHO guidelines. Because of outdoor air pollution, 4.2 million premature deaths were recorded in cities and rural areas of the world. In the monitoring database of the 100 countries released by the WHO in 2018, India has 14 of the top 15 cities with the with the worst effects from $PM_{2.5}$ pollution.

2.2 CAUSES OF AIR POLLUTION

The increase of world population and the overexploitation of natural sources have negative impacts on both nonliving and living systems. Industrialization and the migration of populations have led to clearing of vegetation in the cities. In developing countries, these lead to inadequate living space and promote more slums in the outskirts of cities. Accordingly, people are acclimatized to live with poorly built sanitary and drainage systems. The main causes for air pollution may be due to natural or anthropogenic sources. Diverse anthropogenic activities lead to indoor and outdoor air pollution. Both are due to the depletion of fossil fuels such as petroleum substances, the burning of coal, wood and dry grass and construction activities. Furthermore, toxic gases like carbon monoxide, hydrocarbons (HCs) and nitrogen oxides (NOs) released from automobiles and industries lead to air pollution. Agricultural activities, constructions activities (residential, commercial, industrial, road etc.) and mining operations are the other human-made causes.

2.2.1 Natural Pollutants

Pollution occurring from natural phenomena is caused by periodic activities of physical, chemical or biological substances not associated with human activities. Naturally occurring particulate matter includes dust from the earth's surface (crustal material), spores, pollen, plant and/or animal debris, toxic gases, smoke, volcanic activity, ozone and sea salt in coastal areas. The levels of air pollutants produced by different natural factors are likely to vary under prevailing conditions. Volcanic emissions are considered to be the major cause of natural air pollution because it is also an important source of atmospheric gases, aerosols and ash. An active volcano produces enormous amount of gases such as hydrochloric acid (HCl), SO_2, hydrogen sulphide (H_2S), and hydrogen fluoride (HF) and ash, which can be dispersed by wind over a large area. Sulphate-based aerosols are the secondary products and mainly consist of volcanic plume and cloud. The total sulfur emission from active volcanoes is only ~14% of the anthropogenic and natural emission. Kilauea, one of Hawaii's active volcanoes, in 1983 emitted 300 metric tons of SO_2 in the beginning and 30,000 tons during vigorous eruption. Even prior to 1983, SO_2 emitted by Kilauea was ranging from 50,000 to 100,000 tons/year. A survey among Hawaiian children revealed that volcanic eruption with high concentrations of sulfur dioxide and sulfate particulates ("vog") is associated with respiratory symptoms, asthma or bronchitis, and reduced lung function.

The cyclic process of deterioration of trachyte in urban/rural sectors and industries resulted in different degrees of air pollution. In Padua, Italy, trachyte alteration significantly contributed to air pollution, in addition to domestic combustion and automobiles. Other natural sources of air pollution include thunderbolts, which produce significant quantities of NO_x; algae on the surface of the oceans, which emit H_2S; wind erosion, which introduces particles into the atmosphere; and humid zones, such as swamps, peat bogs and little deep lakes, which produce methane (CH_4). Low concentrations of O_3 occur naturally at ground level and are formed in the presence of sunlight by reactions between NO_x and volatile organic compounds (VOCs). Methane released during animal digestion and VOCs emitted by vegetation reacts with NO_x, SO_2 and few carbon compounds to produce hazes.

2.3   INTERRELATIONSHIP OF INDOOR AND OUTDOOR AIR POLLUTION

The issue of indoor air pollution (IAP) also piqued the interest of many scientists, as people spend more time indoors (>80%) than outdoors. Outdoor to indoor air movement is either dependent on mechanical or natural ventilation or infiltration. IAP originates from the outdoors and is affected by deposition rate, air exchange rate and penetration efficiency.

Globally, 41% of the world population is exposed to household air pollution resulting from cooking with polluting fuels and technologies. IAP has also been coupled with combustion processes, household activities, smoking etc. In Korean urban areas, VOCs are transported from the outdoor environment to the indoor, suggesting the quality of indoor air is affected by outdoor air pollution. Experimental data from different ventilation conditions with an indoor to outdoor ratio and black smoke showed the highest indoor/outdoor ratio under a ventilated environment; however, a low I/O ratio was observed in closed window circumstances. In the central part of India, an I/O relationship of PM is ≤ 2.5 µm, but an I/O ratio above 1 is shown on road sides and in urban areas. If the I/O ratio is larger, transport and deposition of fine PM indoors is relatively higher. I/O relationship studies revealed that the I/O ratio depends on indoor human activities and ventilation. Climate changes also affect indoor air quality besides the pollutant attributes, building structure and human activities. Warmer weather, relative humidity and precipitation are most likely to elevate indoor fungal spores, which could cause allergic reaction in children. Approximately 0.78 million premature deaths have been reported from exposure to $PM_{2.5}$ in the indoors. Smoke released from biomaterials may contain PM, CO and a few organic compounds, which are transported to the indoor environment. Cooking-related activities may elevate the PM indoors by a factor of 1.5 to 2.7.

Besides residential areas, indoor air quality in hospitals is a major concern to be monitored and assessed periodically since it is a key factor in protecting workers, patients and attendants against occupational diseases. Many VOCs, PMs, polycyclic and monocyclic aromatic hydrocarbons, CO, CO_2, BTEX and NO_2 in hospital rooms are a result of outdoor air entering through ventilation systems. In inpatient wards of French hospitals, more hundreds of commercial products and substances have been used as disinfectants, which may affect health since they influence indoor air quality. While disinfectant and cleaning activity lowers the microbial contaminations within hospitals, overall biochemical aerosols, PMs and dust concentration usually increases. Vacuum bags accumulate bioaerosols like bacteria, molds, mites, allergens and mycotoxins. In well-aerated indoor hospital environments, the level of microbial aerosols may be comparatively low if sustainable facilities are considered. Natural ventilation in the childcare units results in a higher I/O ratio of PM and ozone than does mechanical ventilation.

Many anthropogenic sources like oil paints, fragrant decorations, wooden construction materials and indoor plants emit different concentration of VOCs. Poultry farm buildings emit bioaerosols in the atmosphere that may affect humans. In China, attention has been given to household air pollution in rural areas; furthermore, air pollution policies must include household air pollution in urban settings, where an increase of air pollution by 1% has resulted in a 2.9% rise in house hold health expenditure. Tobacco smoke containing PMs is another indoor life-threatening air pollutant to passive smokers and nonsmokers. In many countries, radon, a carcinogenic radioactive gas, causes lung cancer when houses are poorly ventilated.

The following are the major types of indoor air pollutants, which can be more dangerous than outdoor air pollutants.

- Carbon monoxide is an invisible and odorless gas. It's formed by the incomplete combustion of fossil fuels. Carbon monoxide stops oxygen supply for the normal activity of the human body, causing the person to experience sleepiness, dizziness, headaches, increase heart rate, nausea and/or confusion. An increased level of carbon monoxide is lethal to the biotic components.

- Radon is a colorless, odorless gas that is found ubiquitously in low levels. It is present naturally from the uranium present in earth. Being exposed to prominent levels of radon may increases the risk of getting lung cancer.
- Nitrogen dioxide is a common oxide of nitrogen. It is a toxic and corrosive gas. NO_2 is different from nitrous oxide (N_2O), an oxide of nitrogen that is therapeutically useful when administered by trained dental professionals. NO_2 irritates the eyes, nose, throat and respiratory tract. High-level exposure of NO_2, such as at the site of a building fire, can lead to pulmonary edema or lung injury. Moderate exposure can lead to acute or chronic bronchitis. Low-level exposure can impair lung function for people with asthma and chronic obstructive lung disease.
- Secondhand smoke, also known as environmental tobacco smoke, comes from incompletely burned tobacco products. According to the Environmental Assistance and Protection Department of Forsyth County, secondhand smoke contains more than 4,700 harmful chemical ingredients. Exposure to secondhand smoke can cause eye, nose and throat irritation and lung damage. In the long term, exposure can cause many of the same health problems as smoking does, such as wheezing, pneumonia, bronchitis and lung cancer. Asthma attacks may be produced by secondhand smoke exposure.
- Lead is soft metal that is very toxic if consumed at a high level. Lead was widely used in house paint. Lead particles and dust can become airborne, leading to hazardous indoor air pollution. Exposure to lead can damage to the brain, nervous system, kidneys and liver. If children are exposed, they may develop behavioral problems, short attention spans, lower IQ levels and delayed growth.
- Asbestos is the group of minerals found naturally all over the world. The US Environmental Protection Agency declared asbestos unsafe in 1971, listing it as a hazardous air pollutant. While asbestos is not hazardous when intact, troubling asbestos fibers causes them to become airborne, where they could enter the lungs. In the long term, exposure to asbestos can lead to lung cancer and asbestosis. Asbestosis is an inflammatory condition of the lungs that causes coughing, trouble breathing and permanent lung damage.
- Molds are types of fungi that grow indoors and outdoors. Some molds are harmless, while others are dangerous. Mold can trigger an allergic reaction, nasal stuffiness, eye or throat irritation, swelling, coughing or wheezing, headaches and skin irritation and asthma attacks. Severe reactions can lead to breathing problems.

2.4   SAMPLING OF AIR POLLUTANTS

- Air quality can be monitored by frequent air sampling and analyses. A wide variety of simple to complex techniques are available to predict IAP and OAP. The US Environmental Protection Agency (EPA) employs a complex reference method for assessment of suspended particulate matter (SPM) in air, consisting of solid and low vapor pressure liquid particles with particle sizes ranging from 0.01 to 100 μm or larger. Three major types of sampling methods are available for measuring air pollutants. (1) Source or stack sampling can be done at the location of pollution discharges like exhaust gases and automobiles. This will reveal the data based on comparison of the obtained results with emission standards recommended by EPA; (2) ambient sampling pertains to measuring outdoor air pollutants; and (3) indoor air sampling from industry, hospitals, universities and residential places is done to protect the health of workers or residents.

- A number of general principles and factors are involved in selecting effective sampling devices for air pollution. Concurrently, the size, nature, settling property, inertial impaction and effects of thermal and electrical forces of pollutants are additional factors of sampling method. The EPA Air Quality System (AQS) has essential tools for monitoring in compliance with ambient air pollution standards. Sampling can be categorized depending on prevailing weather conditions such as dry (only dry deposition), wet (only during rain) and bulk (wet and dry deposition are collected together). Mosses and lichens are called "biomonitors" or "pollution indicators" that could be an alternative indirect sampling method to evaluate the air pollutants. Numerous studies have shown that no sampling device could collect all particulate matter efficiently.

- Inertial sampling is of two main types: impaction and impingement. Impaction devices can collect and retain particles from an aerosol stream to a collecting surface. The monitoring devices include two-stage impactors (large and smaller particles), Andersen cascade impactor (size range of 0.3–11 μm), high-volume Andersen cascade impactor (size range of 7.5–8 μm), multiple-slit high-volume cascade impactor (large sample size), automated cascade impactor (pollutant concentration with size range of 0.5–35 μm) and $PM_{2.5}$ for inertial particle size separator. EPA recommends a well impact or ninety-six (WINS) or a very sharp cut cyclone (VSCC) with polytetrafluoroethylene (PTFE) or Teflonâ filter to collect $PM_{2.5}$ and particles less than 2.5 μm more efficiently. In virtual impactors such as dichotomous sampler, the larger particles impacting onto a collection surface will be slowly pumped through a filter. The dichotomous sampler has a lower slow flow rate but efficiently separates particles into two size ranges: fine particles (less than 2.5 μm aerodynamic diameter) and coarse particles (2.5 to 10 μm). Hence, dichotomous samplers have more advantages than conventional impactors. They are widely used in collecting dusts, mists, fumes, soluble gases and particulate matter. A Greenburg-Smith impinger is able to collect insoluble particles >2 μm. Various centrifugal separation devices are employed to collect the particles from the air stream by centrifugal force. Cyclone samplers are efficient collectors for particles less of than 2 μm. The collection efficiency depends on particle density and flow rate. A VSCC sampler is a centrifugal separation device, recommended by EPA to collect $PM_{2.5}$. A variety of size-selective inlets and devices are available for sampling of particulate matter.

- Filter-based sampling methods are a combination of filtration and impaction, which is cost effective and can be used for simple and/or complex analyses of PM. Filters capture air pollutants by different mechanisms like inertial impaction, direct interception, diffusional deposition, electrical attraction and gravitational attraction. Efficiency of the mechanism depends on the flow rate, the composition and nature of the particles, particle size and the type of filter media. For PM sampling alone, four filters—cellulose fiber, glass fiber, mixed fiber, and membrane filters—are available.

- By gravitational sampling method, the amount of solid or liquid air pollutants reaching the ground over a stated period can be determined. Here, the duration may be from 24 hours to more than 30 days. The sampling periods reported for total solids are 24 hours to as much as 30 days. In electrostatic precipitation devices, the particles are electrically charged and accelerated toward an oppositely charged electrode. A number of mechanisms for charging particles have been used, including friction with solid material, flame ionization, radioactive charging and high-voltage corona discharge. The most widely used mechanism is the corona discharge because of its efficiency, speed and controllability. This sampler is widely used for mass concentration analysis, sampling for radioactive particles and particle size analysis. Particles less than 5 to 0.005 μm in diameter are collected using a thermal precipitator sampler. A grab or whole-air sampler is handy and simple; any closed container or flask can be used to trap the air at any location. The sample of air drawn may be analyzed later. This method cannot be employed when small amounts of pollutants are present.

2.5 MEASUREMENT OF AIR POLLUTANTS

- Measurements of pollutants are carried out by collection followed by manual analysis, continuous or automatic analysis, concentration meters etc. In case of manual method, the pollutants can be analyzed widely using adsorption spectrophotometry, atomic absorption spectrophotometry and ICP (inductively coupled plasma mass spectrometry) emission method. The quantitative concentration range is determined based on the sensitivity of the analyzer and volume of the sample. The accuracy is directly proportional to collection efficiency.

- Gas chromatography represents a differential analytical method of gaseous and volatile air pollutants. The length of the column determines the degree of separation and detection efficiency. Many known and unknown volatile compound samples can be identified with the help of a mass spectrometer with respect to standard substances. In concentration meters, an air sample is introduced into a measuring chamber and the quantity of target substance is measured directly. But concentration meters have limitations in detecting co-existing gases. With a continuous analyzer, an automated device analyzes the supplied air samples and generates the results either at intervals or continuously. But limit of detection (a small amount of analyte reliably distinguished from zero) and limit of quantification (lowest amount of quantified detected analyte with accuracy) solely depends on measurement accuracy. According to IUPAC (International Union of Pure and Applied Chemistry), standard deviation of detection limit can be three-fold higher than blank values. Spectrophotometry-based measurement better fulfilled most of the recommendations made by IUPAC than chromatography methods did. Moreover, environmental concentrations have no objectives, so effective mean value of two digits is reported to be sufficient.

2.6 EFFECTS OF AIR POLLUTION

Pollution of indoor and outdoor environments poses a major problem for human health. WHO's first report indicates that air pollutants could damage health by causing mild irritation, premature death, asthma, cardiovascular or lung disease, cancer etc. It has emphasized the strategies for adapting air quality standards without formulating any guidelines. WHO estimated 3.7 million deaths in 2012, and the count has increased to 4.2 million in 2016. People with a low social and economic status, living in or near heavy traffic areas, industrial areas, power plants and poor housing are more vulnerable to air pollution–related health problems. Age group, epigenetic variation, smoking, obesity, genetic variation, pregnancy and occupation are the crucial susceptible factors of the ill effects of air pollution.

2.6.1 IMPACT OF AIR POLLUTION ON HUMAN DISEASE

Of the 20 most polluted cities in the world, 14 of them are in India and China alone. Common health problems among people are associated with sick building syndrome (SBS), which results in irritation of nose, eyes and throat, skin ailments and allergies. This can be reduced when the living or working area is properly ventilated. Airborne allergens decrease sleeping efficiencies. Several birth cohort investigations revealed that DNA methylation occurred in mitochondrial genes when exposed to NO_2. Of all organs, the lungs undergo severe damage by filtering PM, which is accumulated as "soot" in the lungs and ultimately leads to early-generation respiratory bronchioles, possibly resulting in lung carcinoma. Exposure to air pollution is associated with impaired trachea-bronchial tree development, prenatally and postnatally, and childhood asthma. Cord blood samples from several birth cohort studies showed that prenatal NO_2 exposure was associated with DNA methylation in several mitochondria-related genes, as well as in several genes involved in antioxidant defense pathways. Higher level of ambient PM_{10} is reported to be responsible for reduced marathon performance of athletes. Lower sperm counts and quality have been reported among peoples living in highly polluted areas. Likewise, reduced fertility rates and high risk of miscarriage is also associated with air pollution. Air pollution exposures mainly include systemic inflammation, increased stress, epigenetic modifications induced by exposures and immune response caused by airway damage. Dampness often attracts pollutants, such as molds and mites that cause respiratory problems and rhino conjunctives if the houses have ventilation ≤0.5 air changes per hours (ACH), and children may have bronchial obstructions. Potential air pollutants for lung cancer is radon, polycyclic aromatic hydrocarbons, asbestos, benzene, and environmental tobacco smoke (ETS). Gaseous pollutants such as anesthetic gas are harmful to pregnant women and damage the liver and kidneys.

An elevated level of ozone causes aggressive behavior and negative emotions. Bad odor is associated with behavioral disorders and cognitive deficiencies. The toxicity of PM is size dependent; fine particles reach alveoli and are transferred into the bloodstream. Women (25–45 years) may become vulnerable to uterine fibroids when exposed to $PM_{2.5}$ or O_3. $PM_{2.5}$ or NO_2 exposure is associated with elevated levels of maternal hypothroxinemia in pregnant women.

Air pollution also affects other biotic components of the ecosystem. Adverse effects of air pollution on ecosystems and biota have been investigated to a great extent throughout the world. Various scientific analyses revealed that air pollution is a limiting factor for phenotype, plant survivorship and productivity. Gaseous form of air pollutants, especially SO_2, ozone and NO_x, affects the growth patterns and various physiological processes of plants. This will have a negative effect on plant-environment relationships, resulting in reduced leaf and petiole length, leaf injury, leaf morphology reducing the photosynthetic pigments, flowering phonology and floral morphology, excess of reactive oxygen species and depletion of lipid peroxidation. In urban areas, plants on roadsides are affected by automobile exhaust and showed significant changes in their metabolism and other morphological variations. High relative humidity and soil moisture may intensify the SO_2 injury to plants.

Apart from living organisms, even monuments and works of art can be damaged by pollution, especially in city centers. Higher level of SO_2, gases (burning of fossil fuels), NO_x and PM are primary pollutants and cause either decay or blackening of monuments. Dry pollutants deposited on buildings are activated by rain, gradually change into sulfurous acid and finally into sulfuric acid, which results in acid rain. The stones of monuments act as repositories for pollutants accumulated on the surfaces, which are not washed off but frequently soaked by rainwater. In urban settings, SO_2 oxidized into sulfate reacts with $CaCO_3$ and produces calcium sulfate on the surface of monuments, which subsequently results in darkening of exterior surfaces. The European Commission published a review on the impact of air pollution on buildings and monuments. Air pollution damaging the cathedrals of Mechelen (Belgium) and Seville (Spain), the sculptures of the Palais du Louvre in Paris (France), York Minster (England) and the Trajan's column in Rome (Italy), with different environmental and climatological conditions, has been reported. The Taj Mahal, one of the seven wonders of the world and the historical monument of the 17th century, has been affected by total suspended particulate matter (TSPM), respirable suspended particulate matter (RSPM) and NO_2 and SO_2. Other historical buildings such as Charminar, Jama Masjid, Mecca Masjid and Badeshali Ashoorkhana are also affected by dust particulates. Recently, researchers found that air pollution may have an effect on the intensity of received signal in 3G/4G mobile terminals. An experiment performed in both indoor and outdoor environments showed that a certain amount of attenuation and cross-polarization occurs in electromagnetic waves when encountering suspended particles in the atmosphere. The dispersion and absorption of the waves are the two causal mechanisms that strongly depend on the dielectric constant and the size of particles, so there could be a possibility of conversion of electromagnetic energy into other forms of energy.

WHO recommends framing new policies that could support cleaner transport, energy-efficient homes, power generation, industry and better municipal waste management that would reduce key sources of outdoor air pollution. Therefore, air pollution policy should be drafted that quantifies and spatially maps out pollution and assesses the impact of sources at the local scale. As the economic status of individuals improves, the usage of liquified petroleum gas kerosene and electric-based cooktops are higher. This will have a certain impact on controlling indoor air pollution. The United Nations should recommend policies to control both indoor and outdoor air pollution.

3 Soil Pollution
Sources, Control and Treatment Measures

Meenu Walia, Amit Joshi and Navneet Batra

CONTENTS

3.1 INTRODUCTION

A thin layer over earth's rocky surface, consisting of organic and inorganic compounds, constitutes the soil. Dark uppermost layers made up of decayed plant and animal remain an organic material, while a rocky base made up of chemical and physical weathering of bedrock is an inorganic component. The presence of unwanted chemicals or materials and the presence of chemicals beyond permissible concentration in the soil, which affects non-targeted microbes, is known as soil pollution. These contaminants may occur naturally in the soil at toxic concentrations, but most of them originate in human activity. Dumping solid/liquid wastes, mining, industrialization, intensive agricultural practices and the uncontrolled use of pesticides/insecticides has led to extensive damage to soil quality and its functioning. Soil pollution, as compared to water or air pollution, has a property of concealment and latency, and its effect is visible with a change in the growth of plants and the health status of human or animal. Mostly, soil pollution is irreversible, especially due to heavy metals, and has long-term effects.

It is considered one of the most important threats throughout the world, including Europe, Asia and America, followed by sub-Saharan Africa and Latin America. As the soil gets polluted, it will affect biodiversity, biological cycles and water cycles.

Soil pollution has been classified as the following:

Point source pollution: Release of pollutants into the soil because of certain specific events within a particular area lead to point source pollution. Various anthropogenic activities, including improper waste and wastewater disposal, uncontrolled landfills, extensive agrochemicals application in fields of abundant factory sites, mining and smelting are sources of contamination.

Further, the release of aromatic hydrocarbons and toxic metals from oil products and leakage from oil refinery storage tanks and batteries or radioactive waste add to point-source pollution.

Diffuse pollution: Pollution spread over vast areas and accumulating in the soil is known as diffuse pollution. This type of pollution does not have a single identified source. Diffuse pollution involves uncontrolled waste disposal and contaminated effluents, landfilling of sewage sludge, agricultural use of pesticides and fertilizers, sources from nuclear power, weapons activities, persistent organic pollutants and flood events.

3.2   SOURCES OF SOIL POLLUTION

Natural and anthropogenic sources contribute to soil pollution. These have been classified as the following.

3.2.1   Sewage

Domestic sewage and wastewater from industries, especially the chemical industry, contain various compounds such as nitrogen, phosphorus and potassium as well as heavy metals, phenols, cyanide and many more toxic and harmful chemicals. As per the World Bank report (2012), approximately 1.3 billion tons of municipal solid waste was generated every year worldwide and was expected to rise to 2.2 billion tons by 2025. Mining and manufacturing industries add a number of contaminants as waste. When untreated sewage of industrial wastewater is used for irrigation, these harmful materials will enter the agricultural soil and contaminate it.

3.2.2   Solid Waste

Municipal and industrial solid waste add solid pollutants to the soil. Besides this, much solid waste material is used in agricultural practices, such as plastic films in greenhouses, garbage, paper cardboard, food waste, battery metals, glass, packaging material etc. These materials do not readily decompose and remain in the soil for a long time. In Indian cities, 50–80,000 metric tons of solid waste are produced every day, which results in soil pollution and clogging of drains, barriers in water movement, foul smell and increased microbial activities.

3.2.3 PESTICIDES AND CHEMICAL FERTILIZERS

To improve the production level of crops and to overcome deficiencies of major nutrients (like nitrogen, phosphorus, potassium, calcium, magnesium, sulphur etc.) in soil, extensive use of chemical fertilizers is carried out, which leads to soil pollution. Excessive use of nitrogen fertilizers results in soil compaction and alteration in biological properties, thereby affecting crop yields and quality. Rock phosphate used for phosphate fertilizers contains traces of lead (Pb), arsenic (As) and cadmium (Cd), which do not degrade and accumulate in the soil above toxic levels. The presence of potassium beyond a certain limit decreases the vitamin C and carotene content in fruits and vegetables. In addition to this, pesticides, herbicides and insecticides sprayed on crops are ultimately transferred to soil. Chemicals such as dichlorodiphenyltrichloroethane are fat-soluble and enter our food chain, affecting both humans and animals (Table 3.1).

3.2.4 ATMOSPHERIC DEPOSITION

Due to process of deposition and precipitation, numerous chemicals are added in soil from the industrial waste gases formed from solid particles (soot, dust, smoke) or liquid particles (mist). Acid rain caused by the mixing of pollutants with air and rain results in contamination of soil.

3.2.5 OIL POLLUTION

Due to rapid population growth and industrial development, use of oil materials and its derivatives have increased.

TABLE 3.1
Agriculture Source of Soil Pollution

Sr. No.	Sources
1	Use of pesticides
2	Solid waste
3	Livestock waste
4	Poor-quality water for irrigation
5	Overfertilization
6	Irrigation with untreated water

TABLE 3.2
Types of Pollutants on Soil and Its Subsurface

Sr. No.	Source Type	Example
1	Industrial	Heavy metals, solvents, effluents
2	Agrochemicals	Pesticides, herbicides, insecticides, fertilizers
3	Land disposal	Sewage water, sludge
4	Irrigation	Saline water, treated wastewater
5	Atmosphere	Acid rain, contaminated dust
6	Domestic or municipal waste	Surfactants, pharmaceuticals, salts
7	Sea and surface waste	Dissolved chemical pollutants, salt

Source: Modified from Rodríguez-Eugenio, McLaughlin and Pennock, 2018.

Various types of pollutants in the soil are included in Table 3.2.

3.3   TYPES OF POLLUTANTS IN SOIL

As discussed earlier, most of the soil pollutants are due to human activities. Based on chemical characteristics, these are classified into organic and inorganic compounds (Table 3.3).

3.3.1   ORGANIC POLLUTION

Both natural and synthetic organic compounds (dioxins, PAH, DTT and organic solvents) contribute to soil pollution and lead to an adverse effect on human and plant health. A polycyclic aromatic hydrocarbon (PAs) is hydrophobic and resistant material formed during crude oil formation, gas work, forest burning and fuel combustion. Several PAH group compounds are known to be carcinogenic. Polychlorinated biphenyls (PCBs) and their derivatives originating from industry are accumulated in fat tissue after entering the body and attach to the nervous system.

TABLE 3.3
Chemical-Based Classification of Soil Pollutants

Inorganic	Metal/metalloid	Cd, Pb, Cu, Zn, As
	Non-metals	Cyanide, Ammonium, Sulphates
Organic	Chlorinated	Dioxins, PAH, DTT, PCBs
	Non-chlorinated	Ethane, benzene, ethylbenzene, xylene, toluene

Source: Modified from Swartjes, 2011.

3.3.2 INORGANIC POLLUTION

Natural processes (like rock weathering, volcanic eruption and crustal changes) and human activities (like mining, building, smelting, chemical production) add inorganic pollutants, which include metals, semimetals, heavy metal compounds, acids, alkali and salts. These occur in soil either in the liquid phase (as a hydrated ion, soluble complexes or as a colloidal ingredient) or in the solid phase (in precipitates as insoluble or as crystal lattices of silicate). Heavy metal compounds are generally insoluble but are mobilized by acidification and cause serious environmental damage (Table 3.4). Heavy metals are metals and semimetals with a density higher than 5 g/cm^3. Essential heavy metals (Mn, Fe, Cu, Zn) are required in low concentration in living organisms, whereas Cd, Pb, Hg and others constitute nonessential heavy metals. Heavy metals like Mn, Fe, Co, Ni, Mo and Zn act as micronutrients and play a role in synthesis and the functioning of various macromolecules like carbohydrates, chlorophyll and nucleic acids, secondary metabolites, however, these metals are toxic. These heavy metals remain in the environment and have bioaccumulative properties; however, their effect depends on the dose and duration of exposure. Heavy metals are added to soil because of natural (geogenic/lithogenic) and anthropogenic activities. Weathering of metal-bearing rocks and volcanic eruptions constitute natural sources, whereas industrialization and urbanization are anthropogenic sources that have increased heavy metals in soil.

3.3.3 MICROBIAL POLLUTION

Many harmful microbes can invade the soil and multiply, thereby destroying the soil ecosystem and human health. These unwanted microbes have their origin from untreated faeces, municipal waste, livestock manure, garbage and waste from slaughterhouses. Contact with contaminated soil is a major source of food contamination.

TABLE 3.4
Permissible Limits for Heavy Metals in Soil

Element	Desirable Maximum Levels of Elements in Unpolluted Soils (mg/kg)
Cd	0.8
Zn	50
Cu	36
Cr	100
Pb	35
Ni	35

Source: Modified from Ogundele, Adio and Oludele, 2015.

TABLE 3.5
Soil-Borne Infectious Diseases (bold) and Their Causative Agents (italics) Split into Two Groups: Edaphic Pathogenic Organisms (EPOs) and Soil-Transmitted Pathogens (STPs)

Edaphic Pathogenic Organisms	Soil-Transmitted Pathogens
Actinomycetoma: (e.g. *Actinomyces israelii*)	**Poliovirus**
Anthrax: *Bacillus anthracis*	**Hantavirus**
Botulism: *Clostridium botulinium*	**Q fever**: *Coxiella burnetii*
Campylobacteriosis: e.g. *Campylobacter jejuni*	**Lyme disease**: *Borrelia* sp.
Leptospirosis: e.g. *Leptospira interrogans*	**Ascariasis**: *Ascaris lumbricoides*
Listeriosis: *Listeria monocytogenes*	**Hookworm**: e.g. *Ancylostoma duodenale*
Tetanus: *Clostridium tetani*	**Enterobiasis** (pinworm)
Tularemia: *Francisella tularensis*	**Strongyloidiasis**: e.g. *Strongyloides stercoralis*
Gas gangrene: *Clostridium perferingens*	**Trichuriasis** (whipworm): *Trichuris trichiura*
Yersiniosis: *Yersinia enterocolitica*	**Echinococcosis**: e.g. *Echinococcus multicularis*
Aspergillosis: *Aspergillus* sp.	**Trichinellosis**: *Trichinella spiralis*
Blastomycosis: e.g. *Blastomyces dermatitidis*	**Amoebiasis**: *Entamoeba histolytica*
Coccidioidomycosis: e.g. *Coccidiodes immitis*	**Balantidiasis**: *Balantidium coli*
Histoplasmosis: *Histoplasma capsulatum*	**Cryptosporidiosis**: e.g. *Cryptosporidium parvum*
Sporotrichosis: *Sporothrix schenckii*	**Cyclosporiasis**: *Cyclospora cayetanensis*
Mucormycosis: e.g. *Rhizopus* sp.	**Giardiasis**: *Giardia lambila*
Mycetoma: e.g. *Nocardia* sp.	**Isosporiasis**: *Isospora belli*
Strongyloidiasis: e.g. *Strongyloides stercoralis*	**Toxoplasmosis**: *Toxoplasma gondii*
	Shigellosis: e.g. *Shigella dyseneriae, Pseudomonas aeruginosa, Eschericia coli*
	Salmonellosis: e.g. *Salmonella enteric*

Source: Adapted from Jeffery and Van der Putten, 2011.

The pathogens in soil responsible for causing human diseases can be divided into two groups depending on the closeness with soil. True soil organisms, also known as edaphic pathogenic organisms (EPOs), have their usual habitat in soil; for example, bacterial pathogens and all fungal pathogens. The other group consists of soil-transmitted pathogens (STPs) consisting of soil-transmitted pathogens. These are obligate pathogens and must infect a host to complete their life cycles (Table 3.5).

3.3.4 RADIOACTIVE POLLUTION

Entry of radioactive material into the soil through radioactive wastewater to the ground, dumping of radioactive solid waste or emission accident leads to soil pollution. These radioactive materials produce alpha, beta and gamma rays upon decay and damage cells and tissues. In addition, entry of these into the human body through the food chain causes internal damages.

3.4 IMPACT ON HEALTH

Soil contaminants affect health as they move in the environment. Due to leaching, they might end up in underlying groundwater. Some may volatilize into the air or tightly bind to soil particles. Soil texture, total organic matter, moisture level, pH, temperature and the presence of other chemicals are a few factors that affect the distribution of soil pollutants in plant or animal systems. Soil pollutants move through different pathways, as shown in Figure 3.1. In addition, these affect animals and the yield and quality of crops.

3.5 CONTROL OF SOIL POLLUTION

For maintaining a good environment, one has to adopt a reduce, reuse and recycle policy. Various biological and non-biological methods are used for controlling soil pollution (Table 3.6). Reducing the use of chemical fertilizers/pesticides and replacing them with biofertilizers and bio-insecticides will help in decreasing soil pollution. Reuse of solid waste such as plastic bags, paper and glass bottles help in to reduce soil pollutants. These can be recycled, and recovered material is used for the manufacture of fresh materials.

3.5.1 BIOLOGICAL METHODS

Contaminated soils can be treated with biological remediation procedures. Attenuation of contaminants during their transition to the vadose zone (region from earth's ground surface extending to the upper surface of major water-bearing structure) can be achieved using microbial degradation, resulting in decreased groundwater contamination. Three methods can be employed for *on site* bioremediation of polluted soils: *in situ*, prepared bed and bioreactor based techniques. Naturally occurring microbes treat the contaminated soil *in situ* under controlled conditions of pH, moisture control etc. or adapted microorganisms can also be introduced into the site to enhance the treatment. *Ex situ* treatment technology can involve solid-phase and slurry-phase bioremediation. In the solid-phase technique, contaminated soils can be physically removed to a newly prepared area where enhanced bioremediation procedures can be employed, whereas in a bioreactor based system, the soil is subjected to the treatment in an aqueous medium (slurry phase). Further, physical and chemical methods can be used to enhance treatment conditions. Bioremediation aims to detoxify a pollutant resulting in nonhazardous products that pose no harm to the environment. To achieve better results, degradation pathways of the soil contaminants should be well studied before implementing bioremediation procedures on site. The contaminant site should be well characterized, including soil features, subsurface hydrogeology and soil microbiota. Microbiological studies of such sites help to verify viable microbial communities that can achieve biodegradation of the pollutants. A range of soil contaminants, including amines and alcohols, halogenated aliphatic compounds, chlorinated and non-chlorinated phenols, PAHs, PCBs, pesticides, industrial manufacturing waste, sludge and municipal wastewaters have been treated efficiently using biological remediation.

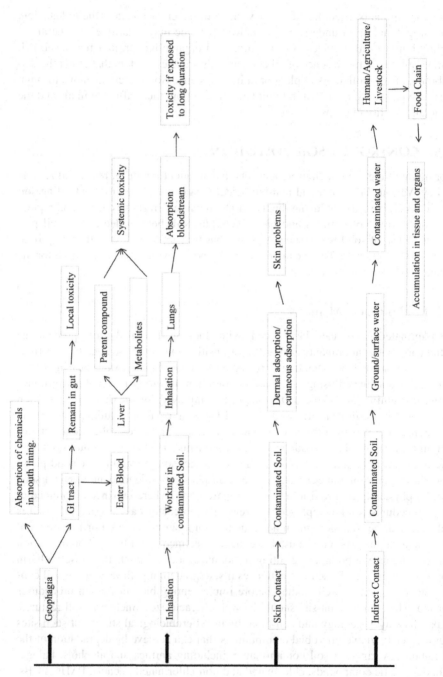

FIGURE 3.1 Pathways for contaminants from soil to humans.

TABLE 3.6
Biological and Non-Biological Methods of Soil Pollution Control

Treatment	Advantages	Disadvantages
Land farming	• Comparatively simple design and implementation • Duration of treatment is short (6 months to 2 years under optimal conditions)	• Difficult to achieve reductions in a concentration higher than 95% and lower than 0.1 ppm • Large area required • Air quality issues can be caused due to dust and vapor formation during aeration
Bioventing	• Equipment used is readily available and easy to install • Minimal disturbance to the treatment site • Costly off-gas treatments may not be required • Can be easily combined with other technologies (e.g., air sparging, groundwater extraction)	• Higher concentrations may be toxic to microorganisms • Cannot be used for site conditions with low soil permeability • Might require injection wells for nutrients and air • Only unsaturated zones of soils can be treated with this method, whereas saturated zones of soils and groundwater require different approaches
Natural attenuation	• Less remediation waste generation and lower environmental impact • Easily combined with other technologies	• High cost of site characterization • The possibility exists for continued migration
Phytoremediation	• Cost-effective when compared with conventional options	• Seasonal technology with low profundity
Biosparging	• Equipment is readily available • Inexpensive • Removal, treatment, storage or discharge of groundwater is not required	• Complex chemical, and physical and biological processes and their interactions are not well understood.
Bio-rehabilitation *in situ*	• Material dissolved in infiltrated and saturated zone can be degraded • Equipment is readily available	• Biomass or precipitation can result in obstruction of the hole • Requires continuous monitoring and maintenance
Vitrification (thermal)	• Well-developed technology • Reduced/eliminated mobility of contaminants.	• Rigorous energy and high temperatures are required • Total costs of the process increase due to the presence of water in the soil • Treatment of off-gases is required before release
Incineration (thermal)	• Contaminant toxicity and volume can be reduced • Wide commercial usage	• No destruction of metals • Screening is required with clay soils or rocky soils

TABLE 3.6
(Continued) Biological and Non-Biological Methods of Soil Pollution Control

Treatment	Advantages	Disadvantages
Soil washing (physical-chemical)	• Reduction in contaminant volume • Commercially used	• Contaminant volume is reduced, but its toxicity is unaffected • Less effective for soils with a high percentage of silt and clay
Soil vapor extraction (physical-chemical)	• Easy-to-install equipment • Site operations minimally disturbed • Treatment duration is less (6–48 months)	• Concentration can be reduced by up to 90% • Soil sites with low permeably are not preferred; only treats the unsaturated zone • Atmospheric discharge of the extracted vapor needs to be treated
Electrokinetic (others)	• *In situ* technology, which is used for the actual removal of metals rather than only stabilization	• Alkaline soils reduce effectiveness of the process • Soil moisture is required
Supercritical extraction	• Efficient procedure requiring shorter extraction time and less energy intake • Solvent residues are not left in the soils	• Critical pressure and temperature are required

Source: Modified from Pavel, Vasile and Gavrilescu, 2008.

3.5.2 Non-Biological Methods

Various physical and chemical methods are used for the treatment of polluted soil. Using solvent extraction, contaminated soil is mixed with solvents to remove various metals like cadmium, nickel, chromium, copper etc. Hydrocarbon and halogenated hydrocarbons are also excluded from soil using this technique. Supercritical extraction is another technique used for removal of contaminants from the soil. Further thermal methods are used for the destruction of pollutants by using either direct or indirect heating. Steam stripping involving the injection of steam into soil and evaporation of contaminant followed by their removal is also reported.

3.6 CONCLUSION

Soil pollution is a challenging area as it impacts different spheres of life directly or indirectly. The origin of different food chains directly depends on the soil, hence it is crucial to monitor soil health. The type of contaminants in soil is governed by many factors like the agricultural practices, industrial activities and waste treatment options used in a particular locality. To carry out remediation practices, it is important to have the complete data related to the origin, types and toxic potential of contaminants under study. Due to the increasing complexity and toxic behaviour of pollutants, one type of approach often results in incomplete removal or degradation of contaminants. The alternative approach involving a consortia of physical, chemical and biological methods can be employed to tackle this issue and attain better results in terms of contaminant removal. In the end, we conclude that reaching a sustainable solution is not possible without the active participation of government and citizens. Recycling, controlled use of chemicals and waste reduction at a source will go a long way to restrict the entry of contaminants in soil.

4 Water Pollution
Sources, Control and Treatment Measures

Anoop Verma

CONTENTS

4.1 INTRODUCTION

Water is one of the crucial aspects of sustainable development and an integral part of socioeconomic development, energy generation, human survival and a balanced environment. Water also plays an important role in adapting to climatic changes and thus serves as the pivotal link between the environment and humanity. In July 2010, the United Nations General Assembly perceived the privilege of each individual to have adequate water for personal and residential uses. It was announced as a human rights issue. Therefore, as the worldwide population increases, there is an alarming need to adjust the majority of commercial requirements for water resources to meet the goal of human networks having sufficient water assets for their needs.

DOI: 10.1201/9781003131427-4

Water is a basic asset in the lives of individuals who are profited by its utilization and are hurt by its exploitation and uncertainty (salinity, drought, acidity, flooding and polluted quality). It is a limited resource with no substitute, and consumption of contaminated water puts humans at risk. There are numerous ways that water designed for human use can be contaminated, including waste from industries like food processing, mining and construction, radioactive wastes from power-producing industries, agricultural and domestic waste and by different microbiological organisms. Presently, water is being decontaminated using different strategies, but researchers are seeking for more progressive and less expensive techniques. These include technologies such as UASB (upflow anaerobic sludge blanket), nanotechnology and biotechnology, utilization of specific microbial species and more.

Approximately 80% of the world's untreated wastewater is disposed of into natural reservoirs, thus polluting streams, lakes and seas. This unrestricted contamination of water is imperiling human health and killing more people every year than does war or any other kind of violence. Also, drinkable water sources are limited—less than or equal to 1% of the earth's freshwater is available for human use. If prevention measures are not taken, the difficulties will increase by 2050, when the worldwide requirement for freshwater is estimated to be one-third more than it is presently. For a better understanding of the problem and what can be done, the chapter provides a detailed discussion of what water pollution is about, what its causes are, and how preventive and treatment measures can be taken.

4.2   WATER POLLUTION

Water is considered polluted if some harmful components or conditions are present to such an extent that it can't be utilized for a particular use. Olaniran (1995) explained water pollution as the presence of hazardous substances (pollutants) in water to such an extent that it is no longer appropriate for drinking, cooking, washing or other applications. According to Water (Prevention and Control of Pollution) Act, 1974, in India, water pollution means such defilement of water or such modification of the natural properties of water or disposing of sewage or trading contaminated outlet or any liquid, solid or gaseous substances into natural water reservoirs, which is likely to render a nuisance in receiving water, making it hazardous or destructive to public well-being or security, to commercial, residential, agricultural or industrial areas, or to animals, plants or aquatic organisms.

4.2.1   Sources of Water Pollution

This section gives information about the most general sources of water pollution: direct and indirect pollutant sources.

Direct water polluting sources incorporate effluent discharge from various types of industries, factories, refineries that discharge a fluid of varying quantity and quality straight into water resources.

Indirect water polluting sources incorporate pollutants that penetrate into the water supply from the groundwater system or soil and also from the air through rainwater. Groundwater and soils consist of residues from agricultural practices, for example, pesticides, fertilizers, etc. and inappropriate discarding of industrial outlets. Air contaminants consist of gaseous discharges from factories, plants, bakeries and automobiles.

If contamination of water originates from a single location, for example a release pipe connected to a processing plant, it is known as point-source contamination. If contamination originates from a single source but from a wide range of dispersed locations, it is known as nonpoint-source contamination (Figure 4.1).

4.2.2 WATER POLLUTANTS

Water pollutants are the substances that cause water contamination. Based on the source, water pollutants can be categorized as domestic wastes and sewage, surface runoff, industrial effluents, thermal pollution or marine pollution.

4.2.2.1 Domestic Wastes and Sewage

This is one of the prominent sources of surface and groundwater contamination in developing nations like India because of the huge gap between the quantity of sewage production and the resources to discharge it safely. The main concern isn't just the absence of resources but also the inability of existing resources and treatment plants to facilitate safe discharge.

4.2.2.2 Surface Runoff

Toxic components (coming from fertilizers and pesticides) present on the surface of soil penetrate into natural reservoirs during rainfalls. The outflow of manure-rich water into various watercourses causes the problem of eutrophication in them. Also, the presence of excessive pesticides in water affects aquatic flora and fauna.

4.2.2.3 Industrial Effluents

Industrial effluents containing a high amount of toxic substances like lead (Pb), arsenic (As), mercury (Hg), cadmium (Cd) and others discharged into different watercourses with no treatment are significant sources of water contamination. Various types of fluid effluents containing harmful synthetic compounds, heavy metals, acids and bases also are discharged into water streams, which not only affects human health but also aquatic flora and fauna. Instances of large-scale ejection of industrial effluents into water streams in a developing nation like India include Gomti (Lucknow), Hoogli (Kolkata), Yamuna (Okhla, Delhi) and more.

4.2.2.4 Thermal Pollution

Temperature over the typical range is known as thermal pollution or calefaction. It occurs because of the ejection of heated water from power plants and

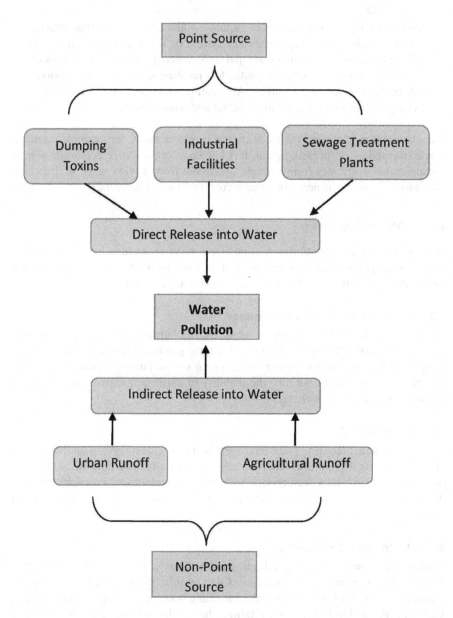

FIGURE 4.1 Water contamination sources. (From: Centre on Control of Pollution Water, Air and Noise.)

industries into natural streams. Various procedures associated with producing thermal pollution include water for cooling condensers, feeding boilers for steam generation, auxiliary plant cooling, ash handling and gas washing. The immediate impact of an incremental change in water temperature is the reduction of oxygen concentration in watercourses. Most aquatic organisms have a temperature tolerance range, out of which they either die or move to more congenial surroundings downstream.

4.2.2.5 Marine Pollution

Contamination of seas and ocean water by diverse sources is known as marine pollution. Almost all types of waste—industrial and commercial wastes, mine tailings, heavy metals, domestic sewage, agricultural wastes and surface run-offs, and even those which are non-biodegradable and radioactive in nature—is disposed of into the oceans. Occasionally, oil finds its way to the seas during a war period and thus becomes one of the critical reasons for killing a notable number of aquatic animals and birds. Oil is one of the most dangerous marine pollutants, especially when it drifts over the surfaces of seas. Point sources of marine pollution are industrial effluents and domestic sewage, and nonpoint sources are surface runoffs, including agricultural wastes, deforestation and desertification.

4.3   EFFECTS OF WATER POLLUTION

Water contamination is a critical issue as it influences all domains of ecosystem including human beings, plants, and animals.

1. Water contamination has an adverse impact on human health, and some of its issues are discussed as follows:

   Developing countries usually lack wastewater treatment facilities, thus exposing its *Homo sapiens* to different water-borne infections such as cholera and typhoid. In extreme cases, an epidemic of a disease might occur, for example, hepatitis A or E, malaria, schistosomiasis or other gastrointestinal problems. According to the World Health Organization (WHO) report, water-borne infections are responsible for killing about 3.4 million people each year in the world. The presence of nitrogen in excessive amounts in drinking water also poses some health risks, particularly to infants.

2. Environmental concerns related to water contamination are briefly explained next:

   Surface runoff from land with a high concentration of nutrients tends to contaminate large water streams. The eutrophication (high nutrient content) condition promotes the growth of algae and various different forms of aquatic life. However, excessive algae growth tends to influence aquatic animals, including fish, by increasing and diminishing their respective oxygen supplies and sometimes blocking their gills, which results in their death. Hence, the aquatic ecosystem is influenced in all aspects, antagonistically exasperating all the evolved ways of life.

3. Water pollution's effects on an ecosystem's flora and fauna are concisely discussed as follows:

   Animals, either aquatic or terrestrial, are susceptible to danger produced by wastewater. One recent example of marine contamination was an oil spill affecting over 16,000 miles of the US coastline, resulting in a tremendous loss of aquatic flora and fauna. Disposing solid waste directly into watercourses also endangers animals in numerous ways.

4. Water contamination also results in fiscal losses, particularly to the public sector. Some of them are elucidated next:

   It is evident that expanding water contamination will cause excessive burden to current treatment plants as well as to the foundation of new ones. Fishing industries are facing great economic loss because of the decreasing fish population. The recreational and travel industries are also influenced negatively as an enormous amount of money is needed, for example, to clear polluted water from algae.

4.4   PREVENTIVE MEASURES FOR CONTROLLING WATER POLLUTION

Water contamination in natural streams can be recognized and evaluated based on different parameters, for example, pH, amount of dissolved oxygen (DO), biochemical oxygen demand (BOD) and pathogen count (coliform). According to drinking water quality criteria, the level of DO should be ≥ 6 mg/L, BOD should be <2 mg/L and coliform should be ≤ 50 MPN/100mL (CPCB 2017). If the available water resource does not conform to these criteria, the water is unfit for drinking until it passes treatment process standards.

Diluting contaminated water in the natural reservoir is one of the effective measures for controlling water pollution. However, various strategies adopted for controlling water pollution can be abridged as follows:

1. Pretreatment of sewage is required before discharging it into natural reservoirs.
2. Water contamination because of organic pesticides and herbicides can be diminished by utilizing specialized and less stable chemical compounds in the production of herbicides or pesticides. Furthermore, application of bio-manure/composts should be encouraged.
3. Wastewater with a low concentration of radioactive materials can be treated using oxidation ponds.
4. Direct disposal of high-temperature water into the seas should be avoided, as it negatively influences aquatic fauna. Techniques like cooling ponds, wet cooling towers, simple cooling and dry cooling towers can be used for reducing thermal pollution.

5. A quality check of treated domestic and commercial wastewater should be done in treatment plants before releasing it into natural water systems.
6. Strong exercising of rules and regulations for wastewater treatment shall be implemented.
7. Highly toxic compounds should not be drained into natural water systems.
8. No human or animal corpses should be cast into water bodies.
9. Restrictions should be implemented on idol submersion and domestic chores such as bathing and laundry on riversides.

4.5 MODERN WASTEWATER TREATMENT TECHNOLOGIES

Some of the recent developments in the wastewater treatment sector are discussed in this section. With the main concern on controlling water contamination, new innovations like biotechnology, bioremediation and nanotechnology have reinforced the approaches in the field of wastewater research. While new technologies are replacing old ones, some of the anaerobic and aerobic wastewater treatment technologies are still in use.

4.5.1 Biotechnology

Biotechnology is a wide research area involving the application of organic processes, living cells or microorganisms to evolve some new innovations. New technologies and devices created by biotechnologists cultivate a high rate usage in agricultural and industrial practices, research and medical sectors. Some recent studies include the use of synthetic biological techniques hinged on carbon sources, for example, carbon monoxide, carbon dioxide, methane and methanol for the production of biofuels and chemicals including electro-biotechnology and biorefinery. Westbrook et al. (2018) audited the ongoing advancement made in biorefinery for the manufacturing of bio-polymers, biofuels, organic acids and specialist chemicals emerging from glycerol. They outlined the preeminent approach towards the microbial glycerol biorefinery by focusing on simulation and fermentation of glycerol using four specific bacteria—*Clostridium*, *Lactobacillus*, *Klebsiella*, and *Citrobacter*—concentrating the yield of 2,3-butanediol through biotechnological manufacturing. They also focused on strategies of involved metabolic and processing engineering. Bordoloi et al. (2019) assessed the ramifications of fundamental ecological parameters on the different carbon endpoints in different stages of degrading contaminants in biofiltration, with the aim of recovering carbon to better understand and enhance the performance of the biofilter model.

4.5.1.1 Bioremediation

Bioremediation is a delicate bioengineering approach to remedy polluted sites using microorganisms, earthworms and plants. Also, this technique can be used for stabilizing deteriorated lands and restraining soil erosion. The bioremediation system implemented using microorganisms is known as micro-remediation, whereas those carried out by plants are known as phytoremediation. Additionally, earthworms are known for executing the environmental disinfection process; this is called vermin-remediation. Presently, the bioremediation system is built upon microorganisms endemic to the polluted sites, thus driving them to enact by providing them with an optimum dosage of chemicals and nutrients crucial for their growth. Das et al. (2010) presented a revised review on the degradation of petroleum hydrocarbons emitted from the petrochemical industry by microbes under various biological surroundings. They observed that bioremediation works fundamentally as biodegradation, indicating absolute mineralization of organic pollutants into carbon dioxide, cell protein, water and inorganic components or conversion of obscure organic pollutants into simpler compounds by the action of microbes. Jain et al. (2011) discussed compelling bioremediation in treating petrochemical polluted water and soil based on efficiency, durability, administrative and economic factors. However, current bioremediation frameworks are constrained by the proficiency of the endemic microbes. Thus, analysts are exploring techniques to introduce polluted sites with foreign microbes that are hereditarily engineered to specifically degrade the pollutants of the targeted sites. Bisht and Sandeep (2015) investigated the biotechnical phytoremediation technique known as rhizoremediation, which involves both rhizospheric microorganisms and plants. Also, they assessed the remediation of polycyclic aromatic hydrocarbons (PAHs) using rhizoremediation along with certain ecological factors and demonstrated its significance over other bioremediation techniques like bioaugmentation.

4.5.2 NANOTECHNOLOGY

Nanotechnology is a technique that controls materials/particles on a nuclear or subatomic scale, particularly to construct infinitesimal devices like robots. It also offers numerous advantages for improving current environmental remedying technologies and developing new innovations that are superior to existing technologies. According to Yunus et al. (2012), nanotechnology consists of three principle qualities that can be practiced in the environmental field. These include disinfection and remediation, pollutant detection involving sensing and locating, and contamination prevention. Anjum et al. (2016) outlined and examined the usage of different categories of nano-materials for the treatment of wastewater. This incorporates four major categories. The first one, nano-adsorbents—carbon nanotubes, zinc oxide, ferric oxide, titanium oxide, activated carbon and magnesium oxide, which are generally used for heavy metal removal from wastewater. The second one, nano-catalysts—electro-catalyst, chemical oxidant, photo-catalyst and Fenton catalyst—are used potentially for evacuating both biodegradable and non-biodegradable pollutants. Third one, nano-membranes—carbon nanotube membranes, hybrid nano-membrane and electrospun nano-fibers—are utilized for adequate eradication of heavy metals, foulants and dyes. The fourth one, the assimilation of biological techniques with nanotechnology—anaerobic digestion and algal film bioreactor—is used for the purification process of wastewater. Nano-materials have various anatomical and morphological properties which make them quite efficient adsorbents and, thus, successful in solving numerous environmental complications, including wastewater treatment. Further, nano-materials with carbon or metal oxides or zero-valent metal and nanocomposites can be used extensively for the evacuation of heavy metals from wastewater.

4.5.3 Aerobic and Anaerobic Wastewater Treatment Technology

Biological wastewater treatment involves the utilization of organically existing microbes that convert complex organic compounds into simpler ones. This sort of treatment can be categorized into two broad types: aerobic wastewater treatment and anaerobic wastewater treatment. A decent beginning stage when choosing an on-location wastewater treatment technology is to examine the nutrient and biological content, wastewater temperature and flow rate. After that, factors like the plant's power utilization and operational expenditures could be considered.

4.5.3.1 Anaerobic Treatment

The anaerobic treatment process is a vital, proficient procedure under which microbes convert the organic material present in wastewater into biogas without oxygen. A gastight roof/coverage is provided over the anaerobic container to achieve oxygen-free conditions.

Anaerobic treatment is also called anaerobic digestion (AD). It is usually utilized for treating hot industrial effluents, and the treatment itself offers numerous advantages over other biological wastewater treatment processes (aerobic treatment), requiring lower vitality prerequisites, fewer synthetic substances and low sludge generation. The sludge produced during the process is quite steady and intact and can be used as soil fertilizer/manure. Biogas generated during this process is exclusively rich in methane and can be remedied and utilized as a sustainable source of energy, thus aiding on economic and environmental fronts.

An excellent example of AD is the upflow anaerobic sludge blanket reactor. This has been perceived as a significant wastewater treatment innovation among AD systems. The reactor is a methane (CH_4)-generating digester, which works under anaerobic conditions, operates by anaerobic microbes and forms a granular sludge blanket. It consists of a three-stage separator, thus empowering the reactor to segregate sludge, gas and water under high turmoil conditions. This favors for conservative, less expensive UASB designs. UASB was assessed in detail for treating domestic wastewater and determining the future scope of its innovation. The researchers found that UASB is superior to other wastewater treatment techniques and executes better performance, being environmental friendly, takes up less space, and has low economic and maintenance requirements.

4.5.3.2 Aerobic Treatment

Under aerobic treatment, microbes transform the organic material present in wastewater into CO_2 and biomass in the presence of oxygen. Proper oxygen-rich air is circulated within the aerobic tanks for the survival of the aerobic microorganisms and an air compressor or aerator is operated in the wastewater so that aerobic microbes can feed on the contaminants present in the water.

Aerobic frameworks act as solitary techniques or help in enhancing anaerobically treated effluent quality by its further biochemical oxygen demand and total suspended solids (TSS) removal. It is also known as the biological nutrient removal (BNR) technique because of its capability to remove nutrient content—both nitrogen and phosphorus—from the wastewater. Apart from having the disadvantages of higher vitality for air circulation and larger sludge production, the aerobic system plays a vital role in wastewater treatment processes. The system helps in meeting the stringent effluent discharge requirements and thus enables industries to release their effluents securely into outlet reservoirs.

One of the best examples of aerobic wastewater treatment processes is the activated sludge process (ASP). This is a biological process that accelerates the waste degradation of wastewater with the addition of activated sludge into it. The reactor's mixture is oxygenated and blends and mixes properly for a predefined time, thereby enabling the sludge to settle under gravity due to sedimentation. The settled activated sludge is either disposed of directly or returned back to the aeration tank. The system is known for producing a top-notch treated effluent with reasonable economical support. Innovations in the present technology are forthcoming, and one of them is the extended aeration sludge process (EASP).

4.6 CONCLUSION

Being a sustainable natural asset, water resources are under constant stress due to consistently expanding industrialization and urbanization. The stress of an expanding populace, deforestation, disposal of untreated wastewater into natural reservoirs, and the utilization of inorganic pesticides, herbicides, insecticides and chemical fertilizers are causing water contamination. This water contamination is rendering an ever-greater proportion of the accessible water and thus making it unsafe for drinking purposes. Also, various diseases, for example, cholera, typhoid fever and diarrhea, are being transmitted through polluted drinking water.

Currently, numerous technologies that can treat and disinfect wastewater, for example, nanotechnology and biotechnology, are available. However, various studies are being undertaken throughout the world to develop better innovations that can treat wastewater at a lower cost. The major agenda of all the water-purifying techniques is to make sure that in the distant future everybody has access to safe and clean water within reasonable expense.

5 Noise Pollution

*Himanshu Sharma, Yogesh Rawal
and Navneet Batra*

CONTENTS

5.1 INTRODUCTION

Pollution has been a significant concern for life in this decade. Air pollution was always considered as the primary pollution affecting the environment, health, and earth as a whole. Noise pollution is also racing up with air pollution. Noise is derived from the Latin word "nausea" meaning "unwanted, unpleasant or unexpected sound". Therefore, noise comprises all unwanted sounds in the surroundings produced by others, and it is imposed on us without our approval, frequently against our wills, at odd times, places, and volumes. Environmental noise pollution is a new menace to health and well-being. Noise pollution is more ruthless and extensive than ever before. It will continue to amplify in extent and severity because of urbanization and expansion in the sources of noise such as air traffic, rail, and highway. In the 21st century, there is virtually no escape from the human-made epidemic of noise. Urban planning aims to increase connectivity between inhabitants and their regular destinations, leading the dwellers to live closer to road traffic. Consequently, our homes,

DOI: 10.1201/9781003131427-5

streets, cars, theatres, restaurants, parks, fields, and other public places of relaxation and recreation have become nothing but noisy areas. Even though noise is considered as a slow and subtle killer, minimal efforts have been made to eliminate it. Disputes over noise, leading to violent behaviour causing severe injury or death, are being reported regularly by media. This dehumanization is commonly seen in the modern, crowded, and noisy urban environment.

5.2 MEASUREMENT OF SOUND EXPOSURE METRICS AND THRESHOLDS

A decibel (dB) is the standard unit for the analysis of sound-pressure intensities or loudness levels based on a logarithmic scale. The doubling of sound intensity would produce a 3 dB raise in sound pressure levels for diffuse sound fields. However, for human hearing, a 10 dB raise would be identified as an approximate doubling of loudness. Likewise, 60 dB sounds are recognized as roughly four times as loud as 40 dB sounds, in spite of having a pressure level 100 times higher. The WHO recommended continuous background noise in hospital rooms to be < 35 dB and an outside night time noise limit of 40 dB at night. High sound levels in hospitals and other areas are exceeding the permissible limits, and little attention has been paid towards the hazards of noise pollution. The lowest sound pressure which can be heard is zero on a decibel scale and is considered as the threshold of hearing. Soft voice scales to 20 dB, quiet office sound to 40 dB, normal conversation to 60 dB and physically agonizing sound level at 80 dB. These sounds are measured on a digital display type HTC sound level meter (3241—c type II data logger).

Sound pressure level is defined as a logarithmic measure of the effective pressure of a sound relative to a reference value. The reference sound pressure in air is 20 µPa (2×10^{-5} Pa), which is considered as the threshold for human hearing at a sound frequency of 1000 Hz.

A *decibel scale* is a logarithmic scale to measure the sound pressure level. A two-fold increase in sound energy (e.g., two identical jackhammers instead of one) will cause the sound pressure level to increase by 3 dB. A 10-fold increase in sound energy (10 jackhammers) will cause the sound pressure level to increase by 10 dB, which is perceived as about twice as loud.

L_{max}: The highest sound pressure level in a given period.

L_{eq}: The equivalent continuous sound level, which is an average level of sound pressure within a specific time period. If a filter is used for frequency-credence, the average level is indicated as L_{Aeq}. The filter and time period used are indicated in subscript, for example, L_{Aeq8h} or L_{night}. The commonly used signals for traffic noise are L10, L50, L90, and Leq, where L10 is the top end of the level range, although it can be substantially less than the occasional peak. L90 is the background noise in the absence of noise sources.

L_{DEN}: L_{DEN} (day-evening-night level), also referred to as DENL, is filtered average sound pressure level measured over a 24-hour period. A 10 dB penalty

TABLE 5.1

Ambient Noise Standards as Prescribed by the Central Pollution Control Board (CPCB), 2000 New Delhi, India, for Different Types of Zones

Sr. No.	Area	Leq dB (A)	
		Daytime	Nighttime
1.	Industrial area	75	70
2.	Commercial areas	65	55
3.	Residential area	55	45
4.	Silence zone	50	40

Source: www.cpcbenvis.nic.in.

TABLE 5.2

Ambient Noise Standards as Prescribed by the Central Pollution Control Board (CPCB) 2000, New Delhi, India, for Residential Areas in India

Sr. No.	Location	Acceptable Noise Level in dB
1.	Rural	25–35
2.	Suburban	30–40
3.	Residential	35–45
4.	Urban (residential and business)	40–45
5.	City	45–50

Source: www.cpcbenvis.nic.in.

is added to the night, a 5 dB penalty is added to the evening period, and no penalty is added to the average level in the daytime.

L_{DN}: L_{DN} (day-night level) measure is similar to the L_{DEN}, but it excludes the 5 dB penalty during the evening period. The penalties are imposed to indicate people's extra sensitivity to noise during the night and evening. Tables 5.1 and 5.2 illustrate the permissible limits of sound as set by the Central Pollution Control Board.

5.3 SOURCES OF NOISE POLLUTION

There are abundant sources of noise, among which are automobiles on roads, railway traffic, airway transportation, boilers, generators, air conditioners, fans, vacuum cleaners, construction equipment, loudspeakers, lawnmowers, manufacturing processes, garbage trucks, neighborhood and religious functions, and industrial and occupational activities, to name a few. Sources of traffic noise depend on the amount of traffic, the speed of the traffic, the type of vehicle, traffic flow, and the number of heavy vehicles in the traffic stream. Several studies in India reported the problem of noise pollution, and the noise intensity in metropolitan cities exceeds the standard limits, which are likely to increase the deafness among the inhabitants of these cities.

Usually, the intensity of traffic noise is increased by heavier traffic volume, fast-moving traffic, and large numbers of heavy vehicles. Vehicle noise is a blend of the sounds produced by the engines, exhaust, horns, and tires. The loudness of traffic noise is further amplified by defective mufflers or silencers or other faulty tools on vehicles. A reduction in noise can result from technological improvements in the engines of automobiles, railcars, and airplanes. Traffic noise levels are affected by distance, topography, natural vegetation, and human-made obstacles. Traffic noise is not typically a severe problem for those people who live more than 150 m or 30 to 60 m away from heavily travelled and lightly travelled roads, respectively.

5.4   NOISE TESTING SOFTWARE AND TOOLS

PTVVisum is the software tool used for macroscopic modelling of the transport, and it is used worldwide by leading organizations. It allows the analysis of peripheral traffic effects: emissions of harmful gases and noise level. This software offers two protocols of noise emission calculation, namely Noise-Emis-Rls90 and Noise-Emis-Nordic.

Otoacoustic emission testing is a diagnostic test of hearing loss induced by noise. Otoacoustic emissions are a discharge of acoustic energy from the cochlea that can be traced in the ear canal. This testing is specifically used to identify hearing defects in newborn babies and children, leading to the advancement in audiological assessment. It could specify the noise-induced changes in the inner ear that are not detected by audiometric tests. So it is twice as sensitive as audiometry tests and can be used for monitoring noise-induced hearing loss in the workplace of different organizations.

A noise barrier in Lucknow, India, has been proposed using the modified version of the Federal Highway Administration (FHWA) model, which uses traffic data such as traffic volume, speed, slope, and ground cover. The height of the noise barrier was evaluated as 4.2 meters above the floor level of the flyover with long continuous structures and 15 feet high. It is made of various materials to considerably reduce the noise level from 63.79 dB to 55 dB and also blocks the roadside view by creating massive barriers, thus changing the urban scenery. This FHWA model conducted for the Indian environment has shown excellent results.

5.4.1   NOISE POLLUTION BASED ON THE MEASUREMENT SCALE

Noise pollution arising out of any source is categorized as direct, indirect, cumulative, and post-impact.

- Sounds of a normal conversation between 40 and 60 dB are safe for the ear and have no health hazard.
- Sounds emerging from tape recorders have a sound intensity of nearly 70 dB, which is also safe for the ear.

- Sounds levels above 80 dB are related to both an increase in violent behaviour and a decrease in sympathetic response to others.
- Sounds of heavy traffic have a sound intensity of 90 dB, which is higher than the safe range. It can cause temporary hearing loss; if not treated immediately, it can cause permanent impairment.
- Sounds of drills and other machines have a sound intensity of 100 dB.
- Sounds of aircraft engines have a sound intensity of 100–200 dB. A higher noise level of 160 dB causes total deafness, rupturing eardrums and damaging the inner ear. It may also result in high blood pressure, stomach ulcers, palpitations, nervous problems, irritation, and anger and can also affect a pregnant woman's embryo.
- Sounds of rockets during takeoff have a sound intensity of 200 dB. It causes total deafness by damaging the inner ear and the eardrums. It also causes high blood pressure, stomach ulcers, palpitations, nervous problems, and irritation.

5.4.2 Noise Pollution Based on the Characteristic

- *Noise intrusions*: Their brief quality characterizes noise intrusions. Classic examples of sources of noise intrusions are aircraft, motorbikes, trucks, road drills, and sirens. Their noise interrupts personal activities such as study, sleep, conversation, relaxation, meditation, and entertainment. Noise intrusions are pointless and annoying. The noisy motorbike or a car operating with a defective silencer is a typical example of noise intrusion.
- *Property boundary noise*: This is the intrusion from neighbours, construction sites, industry, and commercial establishments. It interferes with concentration, sleep, and communication.
- *Impulsive noise*: These are more significant contributors to human infuriation than slower transitory sounds are even when both produce the same reading on a sound level meter. The more substantial annoyance is due to their upsetting effect and also because the human ear responds faster than does the circuitry in a sound level meter.
- *Noise in healthcare settings*: Regardless of attempts to normalize noise pollution, it has become an ill-fated fact of life globally and has even breached the boundaries of hospitals and healthcare systems. According to World Health Organization guidelines, the values for continuous background noise in hospital patient rooms are 35 and 30 dB during the day and night, respectively, with night time peaks not exceeding 40 dB.
- *Occupational noise*: The hearing loss induced by exposure to occupational sounds remains a problem and is the focus of standards for hearing protection. Many countries impose health and safety norms that specify maximum sound exposure, regular noise assessments, and audiometric testing along with protective action, equipment which is meant to protect both workers and the public from noise exposure. The standard of sound level in industrial settings varies internationally. For example, in the UK (Control of Noise at Work Regulations), set levels of L_{Aeq8h} at 80 dB and

85 dB is considered mandatory. However, in the United States, the Occupational Safety and Health Administration, (OSHA), has set the permissible limit at L_{Aeq8h} 90 dB. Although noise-induced hearing loss is a well-known problem in industrial settings, musicians and military workers also suffer from noise-induced hearing loss.

- *Social noise exposure*: The youth of today are exposed to relevant higher degrees of social noise due to personal music players, attending nightclubs and rock concerts. Safer products and public health campaigns at a large scale are needed to educate the youth regarding the risk of hearing loss. Noise-cancelling headphones are effective measures for reducing the hazards of personal music players.

5.5 HAZARDOUS EFFECT OF NOISE POLLUTION ON HEALTH

Noise pollution is not merely an annoyance; instead, it has a wide range of adverse health, social, and economic effects. The possible health effects of noise pollution are several, persistent, and medically noteworthy. Noise produces direct and collective undesirable effects that impair health and degrade residential, working, social, and learning environments. Noise pollution impairs task performance like reading attention at school and problem-solving and dealing with errors at work and decreases motivation and memory.

- Memory deficits adversely influenced by noise are a recall of subject content and recall of incidental details.
- Cognitive development is impaired when homes or schools are in a noisy area such as highways and airports; it may contribute to feelings of helplessness in children. Therefore, more attention needs to be paid to the effects of noise on the ability of children. Noise has been used as a harmful stimulus in a variety of studies since it produces the same kinds of effects as other stressors.
- Risk during pregnancy: Noise during pregnancy may increase the risk of hearing loss in the newborn, shortened gestation period, and intrauterine growth retardation. The level of noise in the neonatal intensive care unit may cause cochlear damage and congenital malformations in the premature infant.
- Sleep is a critical activity that offers mental, physical, and emotional restoration. Therefore, sleep disturbances are considered toxic to temper, performance, and health. Transportation noise has become the primary source of sleep disorder more during the night than during the day. It was the observation that humans respond to neutral sounds even during sleep. Sleep disorder is considered the most toxic non-auditory effect because of its immense impact on the quality of life and daytime performance. As per the meta-analysis, aircraft noise annoys the most, road noise to a lesser extent, and rail noise the least. Environmental noise, especially that caused by transportation means, is a growing problem in our modern cities. It is considered as a cause of exogenous sleep disturbances, after somatic problems and day tensions.

- Hormone imbalance: Slow-wave sleep, which is considered as the most restorative sleep stage, is associated with decreased blood pressure, heart rate, and sympathetic nervous activity as compared to sleeplessness. Throughout slow-wave sleep, growth hormone is released and the stress hormone cortisol is inhibited. Therefore, good sleep is an essential modulator of hormonal release, glucose regulation, and cardiovascular function.
- Glucose metabolism: Sleep limitation and limited quality sleep affect glucose metabolism, thereby reducing glucose tolerance and insulin sensitivity and leading to loss of appetite (lower levels of leptin and higher levels of ghrelin) as well as cortisol levels.
- Cardio metabolism: Sleep disturbances may play a significant role in the growth of the cardiometabolic disease. Cardiovascular diseases like coronary artery calcifications, atherosclerosis, and atherogenic lipid profiles have been associated with disturbed sleep.
- Psychiatric disorders: The relationship between disturbed sleep and mental disorders has been closely related. It has been reported that insomnia is not only a symptom of psychiatric disorders, but it also adds to the risk of relapse or the severity of psychiatric symptoms. This further increases the frequency of violence at the domestic level and at work as well.

5.6 TREATMENTS MEASURES

5.6.1 EDUCATION

Most people are unaware of the health hazards of noise, which can be passed on to the generations to come. Therefore, educating people through advertisements on radios, televisions, newspapers, and camps could be a significant step in making citizens aware of the hazards related to loud noise. Until and unless the health hazards of noise are made public, citizens will not abide by the rules and regulations strictly. Moreover, the level of noise varies from individual to individual.

5.6.2 INSTALLATION OF NOISE BARRIERS

Noise barriers should have relative effectiveness and be aesthetic to provide a striking outcome complementary to the surrounding community and providing a public image. They should be cost-effective to build and to maintain and to gain the tolerability and keen support of the surrounding community and decision-making authorities. Design aspects of noise barriers should be such that they reduce noise levels up to 15 dB at the receptor. The noise barrier should be dense and located close to the road. Such barriers can be approved for flyovers situated in residential or silence zones like hospitals and schools. The obstruction of barrier alteration placement between the highway and the observer will screen the observer and consequently lower the sound level. The noise decline caused by the barrier depends on the path difference of the sound wave that travels over the barrier compared with direct sound wave transmission to the receiver and also the frequency of the noise emission.

5.6.3 SOUND-ABSORBING TILES

The mounting of high-performance sound-absorbing tiles and panels in new and old constructions can improve concentration at work among staff.

5.6.4 ENVIRONMENTAL STRATEGIES TO REDUCE NOISE IN HOSPITALS

Environmental involvements may help reduce the noise levels in hospitals. Installing sound-absorbing ceiling tiles, reducing noise sources by adopting a noiseless paging system, and providing single-bed rooms rather than multiple bedrooms or dormitories could be key involvements.

5.6.5 THERAPEUTIC STRATEGIES

The interaction of genetic and environmental factors might explain the hearing loss after exposure to noise; therefore, identification of susceptibility genes might help to categorize the population with high risk and further improve the hearing capability in affected individuals. Stem cell therapy could be a potential restorative therapy to recover the damaged sensory circuitry in the cochlea. Collective exposure of high-level noise (MRI scanners) and cytotoxic drugs or cancer treatment drugs such as cisplatin increases the risk to the inner ear and the auditory nerves. Additionally, oxidative stress could lead to cochlear cell damage; the antioxidant glutathione and cytoprotective drug D-methionine have been successfully used in animals to prevent noise-induced hearing loss, but clinical trials for safety and efficacy in human beings is still in process.

5.6.6 SOURCE CONTROL THROUGH DIRECT MONITORING

Direct regulation to set strict rules for the maximum emission level for noise sources is the only strategy that assures noise reduction. Implementing this approach in the United States, the air transport noise exposures of >65 dB L_{DN} have seen a 90% reduction since 1981 to only 0.015% in 2007 in spite of a six-fold increase in travel by airways.

Initiatives have been taken by several countries to combat the noise menace. For instance, the USA has created sites where human-originated noise is not tolerated. Likewise, the Netherlands government forbids the building of residences in locations with 24-hour average noise levels exceeding 50 dB. In Great Britain, the Noise Act authorizes local authorities to impound equipment, instruments, and gadgets and impose a penalty on people who create excessive noise at night. Other countries are investing in "porous asphalt" technology, which can restrict traffic noise by up to 5 dB.

5.6.7 MAPPING

Geographic noise maps to adjust the environmental information are the ways to safeguard against noise and are based on accurate information. European Union governments have already prepared noise maps of roads, railways, and airports (Commission to the European Parliament and the Council, 2011). The US government has noise maps of the world's oceans to inspect the impact of noise on marine species (National Oceanic and Atmospheric Administration, 2012). Measurement and mapping of noise levels would assist in accessing sound information, integrating data systems, evaluating conditions, and further expanding the technology to facilitate in implementing protective measures.

5.6.8 STATE AND LOCAL ACTION EMPOWERING POLICE

State and local governments can pass regulations on sources of noise. Municipal regulation into noise ordinances to regulate the timing and intensity of noise may or may not be effective at reducing noise; therefore, empowering police to impound the source of noise should be considered.

The cost-effective legal involvements at state and local levels can be through carefully spending money and by changing the built environment. Municipal noise sources (construction equipment, emergency sirens, garbage and street maintenance equipment, and transit vehicles) may be reduced by careful purchasing of the equipment with the latest technology designed to combat the environmental pollution and by contractual agreements. Some countries use contractual agreements for temporary relocation of citizens requesting relief from construction noise.

5.6.9 GOVERNMENT EFFORTS

Every citizen has the freedom to choose the nature of their acoustical environment, and others should not force it, but the role of government should be to protect citizens from the adverse effects of noise. Legislators should take protective measures for noise pollution as seriously as other forms of pollution. National standards often promise the availability of quieter vehicles by limiting their maximum allowable noise emission through legislation. Local authorities are also taking the responsibility to set maximum operating noise levels and to guarantee the replacement of faulty silencer systems. For instance, aircraft noise emission is controlled by adjusting flight events to protect the maximum number of people from flyover noise, mainly during night hours.

5.6.10 SEVERE PUNISHMENT OR PENALTY

In China, instead of hanging men for dangerous crimes, noise was used as a source of torture. This implies that the annoyance level of noise is more painful. Therefore, severe punishments or penalties should be enforced on citizens so they refrain from loud activities that result in noise-induced damage to the community. Governments should take steps for constructing healthy housing, as the health of the entire family suffers while living in poorly designed physical environments. Doing so can only control individual noise exposures, but it often does not reduce noise source levels because the recipients of noise bear the load to reduce noise while the creators of noise continue, as they have no incentive to reduce emissions. Therefore, this requires cautious preparation and detailed analysis.

6 Solid Waste Management
Source and Treatment

Meenakshi Suhag

CONTENTS

6.1 INTRODUCTION

Discarded leftover materials that are generally supposed to be of no economic value are described as "waste." Likewise, solid waste is a non-liquid, unwanted material generated from the routine activities of people in society. According to the 2016 Solid Waste Management (SWM) rules,

> Solid waste is defined as solid or semi-solid domestic waste, sanitary waste, commercial waste, institutional waste, catering and market waste and other non-residential wastes, street sweepings, silt removed or collected from surface drains, horticulture waste, agriculture and dairy waste, treated bio-medical waste excluding industrial waste, bio-medical waste and e-waste, battery waste, radioactive waste generated in the area under the local authorities.

Rapid population expansion, escalating urbanization and prompt developments of infrastructure with varying lifestyle and economic situations have greatly accelerated the rate of per capita waste generation globally. Worldwide around two billion

DOI: 10.1201/9781003131427-6

tons of municipal solid waste (MSW) is annually generated, and this quantity may be expected to increase drastically by 2025. India is the second largest populated country in the world, with more than 42% of its urban population living in metropolitan cities like Mumbai, Delhi, Bangalore, Kolkata and Chennai and a high per capita waste generation rate due to high living standards and economic activities.

6.2   TYPES OF SOLID WASTE

Depending upon the sources and sectors to which they belong, solid waste is categorized into the following types:

1. *Residential*: This includes waste generated mainly from dwellings and apartments and includes leftover foodstuff, vegetable or peeled material, clothes, plastic and wood pieces.
2. *Municipal*: This refers to leaf litter, dust, building debris and sediment from treatment plants generated during different municipal activities like street sweeping, land scraping, construction and demolition work.
3. *Commercial*: This mainly includes grocery items, leftover food and metal parts produced from stores, restaurants, theatres, markets and medical amenities.
4. *Institutional*: This includes paper, plastic and glass generated from school, administration buildings, colleges and offices.
5. *Agricultural*: This consists of spoiled crop grains, crop residues, vegetables, grasses, litter from fields, farms and yards.
6. *Industrial*: This includes waste from different types of production processes, industrial operations, ashes, demolition waste, e-waste and hazardous wastes.

6.3   IMPACT OF SOLID WASTE

An elevated amount of generated waste and its proper disposal is a big challenge for a country, as is the protection of community health, aesthetics and the natural environment. Improper waste management activities may result in many problems, some of which are mentioned as follows:

1. *Contamination of ground and surface water resources*
   Waste discarded or dumped near any surface water resources (river, ponds etc.) result in the degradation of toxic chemical constituents that cause undesirable changes in water quality and pollute the same. Contamination of groundwater sources that are used for drinking and domestic purposes occurs by the leaching of harmful substances if the landfills are not capped with impermeable materials.
2. *Greenhouse gases and pollutant emissions*
   Uncontrolled burning of MSW creates thick smoke containing toxic gases and soot particles that can significantly contribute to urban air pollution and are hazardous to people living nearby. Burning of polyvinyl

chlorides (PVCs) produces highly toxic dioxins, which are carcinogenic in nature. Moreover, decomposition of organic wastes at various landfill sites emits a large amount of methane gas, a powerful greenhouse gas (GHG) that plays a vital role in global warming.

3. *Damage to aquatic ecosystems*

   The high concentrations of nutrients from organic wastes deplete dissolved oxygen in water bodies, which significantly damages native fishes and aquatic animals. Solid residues aggravate the sedimentation process, changing stream hydraulic parameters, which may destroy the underlying habitats and the services provided by them.

4. *Increased disease transmission and threat to community health*

   Untreated disposed waste and decomposing organic materials attract different types of insects, flies, rats and mosquitoes that can transmit various infectious diseases. Improperly managed dumpsites located within urban areas serve as propagation grounds for vector-borne diseases, creating public health problems for nearby residents. Due to poor collection and insufficient transportation arrangements, waste accumulating along the street corners or in slum areas not only cause unpleasant odors but may also clog drains, making living conditions upsetting. Associated workers like waste sweepers, handlers and pickers are more prone to direct health risks and hazards. Decomposition and open burning of mixed MSW on dumpsites cause various environmental and health hazards.

6.4 MUNICIPAL SOLID WASTE

The Ministry of Environment and Forests (MoEF) (Government of India, 2000) defined MSW as "waste generated from the residential and commercial area which included treated biomedical wastes may be solid or semi-solid without including any type of hazardous industrial waste". Municipal solid waste usually comprises materials that are (a) biodegradable (mainly paper, straws, backyard waste, food and textiles waste), (b) moderately biodegradable (like woodchips, sludge, disposable napkins etc.) and (c) non-biodegradable (like plastic, leather, metals, rubber, glass etc.).

The typical MSW stream includes waste generated from *household*, *commercial*, *industrial*, *institutional*, *construction*, *demolition* and *municipal and sanitation service* sources. Sharholy et al. (2008) reported the following substances in MSW: *recyclables* (paper, plastic, glass, metals etc.), *toxic substances* (pesticides, paints and used battery parts), *compostable organic matter* (food waste, vegetable and fruit peels) and *grubby waste* (stained cotton plugs, disposable syringes and sanitary napkins).

6.5 SOLID WASTE MANAGEMENT

During 1989, for effective implementation and sustainable waste management practices, the US Environmental Protection Agency organized some elements as reduction at source, recycling, combustion and landfilling. Tchobanoglous et al. (1993)

defined integrated solid waste management (ISWM) as the selection of appropriate processes, technologies, management programs and their application in order to achieve specific waste management objectives and goals. Solid waste management includes a diverse range of activities encompassing reduction, recycling, separation, modification, treatment and disposal at varying levels of sophistication. Municipal solid waste management (MSWM) involves activities consisting of six basic principles of waste generation: storage, collection, transfer, transport, processing and disposal.

6.6 POLICY REGULATIONS AND SCHEME IN INDIA

In India, the Ministry of Environment and Forests and Climate Change (MoEF & CC) collectively with Central and State Pollution Control Boards deal with the issues associated with solid waste management, and these basic services are provided by respective urban local bodies to the public.

After 1980, much attention has been paid by the government and local bodies towards waste generation issues and waste management activities, while enactment of the Environment Protection Act of 1986 creates a landmark by specifying Hazardous Waste Management Rules in 1989. Realizing the need for proper and scientific management of solid waste, Municipal Solid Waste (Management & Handling) Rules, 2000, was appraised by the Ministry under the Environment (Protection) Act of 1986. Important directions were made under these rules, and the main goal was to help municipal and local authorities become well organized and equipped in order to perform better in their areas with respect to collection, storage, transportation, processing and disposal of municipal solid wastes.

Successful waste management requires the participation of citizens, local governments and private entrepreneurs. Moreover, pre-planned strategies, assessment of composition and quantity of waste are beneficial for the management of waste in a more sustainable manner. However, it has been widely observed that the Municipal Corporations in India do not have adequate resources or the technical expertise necessary to deal with the problem. The growing rate of waste generation, inadequate budget allotted, lack of well-equipped infrastructure and skilled/trained workforce, and costly technologies for waste disposal are some major challenges.

Swachh Bharat Mission (SBM) is a national-level campaign being implemented in 2014 by the Ministry of Urban Development (MoUD) and by the Ministry of Drinking Water and Sanitation (MoDWS). The main objectives of this mission include ensuring cleanliness and sanitation and managing solid waste more scientifically for urban and rural areas in India.

Various important legislation regarding the management of waste in India are:

- Environment Protection Act—1986
- Hazardous Waste Management and Handling Rules—1989
- Manufacturing, Storage and Transportation of Hazardous Waste Rules—1989
- Bio-Medical Waste Management and Handling Rules—1998, 2016
- Municipal Solid Waste (Management and Handling) Rules—2000, 2016

- Plastic Waste (Management and Handling) Rules—2011, 2016
- E-Waste (Management and Handling) Rules—2011, 2016
- The Construction and Demolition Waste Management Rules, 2016

For strengthening solid waste management, the Municipal Solid Waste (Management and Handling) Rules, 2000, were replaced in 2016 with the more efficient MSW Rules, 2016. Some of the remarkable directions framed under MSW Management Rules, 2016, include source segregation is obligatory to ensure safe disposal of solid waste and waste material should be recycled/recovered or co-processed and organic waste must be composted for better management of waste, with a timeline of two years for its implementation. In addition, apparent regulations constituted for biological/thermal-based waste treatment technologies are encouraged with clauses to support the sale of compost and RDF (refuse-derived fuel) and the purchase of power from waste-to-energy plants.

For the implementation of MSW Rules, authorities must include integrated scientific approaches and infrastructure amenities for conducting collection, handling, processing and disposal activities. However, the Central Pollution Control Board report 2017–18 observed that (1) most of the states until now have not formulated any strategies and policies with compliance of rules, (2) their waste processing, disposal and treatment facilities are at inadequate levels, (3) most of their dumpsites are not scientifically designed and operated and (4) assessment and monitoring of groundwater quality, ambient air and treatment of leachate at waste processing sites have not been conducted.

6.7 WASTE GENERATION IN INDIA

According to the Planning Commission Report, 2014, 62 million tons of municipal solid waste per annum was generated by the urban population; by 2031 and 2050 this annual generation rate is projected to reach 165 million tons and 436 million tons, respectively. These projections are based on an average of 0.45 kg/capita/day waste generation for India's urban population. The composition and characterization of solid waste change considerably from place to place depending upon the various factors like way of living, seasonal conditions, economic status, government policies, regulations and industrial activities. Rise in community living standards has significantly accelerated MSW generation rate in the Indian cities. Mumbai generates about 7000 metric tons (MT) of waste per day (Municipal Corporation of Greater Mumbai, 2014); Bangalore generates about 5000 MT (Bruhat Bengaluru Mahanagara Palike, 2014); and other large cities such as Pune and Ahmadabad generate waste in the range of 1600–3500 MT per day (Pune Municipal Corporation, 2014). In developing countries, open dumpsites are common due to the low budget for waste disposal and the non-availability of a trained work force. Safe and cost-effective management of MSW is a significant environmental challenge for modern society. Inadequately managed waste disposal facilities have the potential to affect the health and environment. The composition of MSW at generation sources and collection points on a wet weight basis comprises a large organic fraction (40%–60%), ash and fine earth (30%–40%), paper (3%–6%) and plastic, glass and metals (each less than 1%). The composition of MSW of Pune is organic fraction 40%–60% (weight basis), paper 5% by weight, plastics 5% by weight, and ash and fine earth 15% by weight. The generation of solid waste ranges from 0.1 to 0.5 kg per capita per day with a predictable annual increment of 1.33%.

6.7.1 WASTE COLLECTION, SEGREGATION, STORAGE AND TRANSPORTATION

A consistent municipal solid waste collection and transportation system is a keystone for good quality waste management services and needs a separate collection of different wastes known as source-separated collection system. Haphazard littering of roads and drains is also common in most cities and towns. Scavenging of the wastes by rag pickers and stray animals leads to further scattering of solid waste. The problem is most sensitive in slum areas where the lower-income groups live. According to the Waste Management Rules, 2016,"waste generators shall segregate and store the waste generated by them in three separate streams namely bio-degradable, non-biodegradable and domestic hazardous waste in suitable bins" (MoEFCC 2016). Source-separated systems frequently occur in mainly high-income areas of the world like North America, Europe and Japan, where good facilities are available for transportation and separation of waste streams. MSW collection efficiency in developing countries is only 41%, whereas for the developed countries the waste collection efficiency is around 90% (World Bank, 2012). However, Saxena et al. (2010) reported that the average collection efficiency for solid wastes, particularly in Indian cities, is about 70%. In Indian towns, people deposit their waste in bins or enclosures usually located at street corners. The enclosures usually are made up of metal, concrete or brick. While source segregation of waste is a prerequisite for successful disposal methods, this activity has not been planned and operated in most households or at the community level. Moreover, waste sorting and recycling are typically accomplished by the unorganized sector of rag pickers. Transportation modes employed for solid waste management services in India are varying and include rickshaws, compactors, mini-trucks, tractors and dumpers and maintenance via municipalities or urban local bodies. Cucchiella et al. (2014) reported that running old vehicles without any scientific vehicle routing and planning decreases waste collection and transportation efficiency and adds pollutants such as high emissions of carbon dioxide to the air. Also, the practice of source separation of municipal waste is limited or absent in most cities and villages in India. This results in a mixed composition of waste, leading to several constraints in introducing and implementing technologies for treatment in a sustainable way.

6.7.2 GOVERNMENT/PRIVATE INITIATIVES

The Government of India in December 2005 launched the Jawaharlal Nehru National Urban Renewal Mission (JnNURM) with an objective to improve basic services under solid waste management. Under JnNURM through different solid waste management projects, opportunities are provided for expanding public–private partnership for enhancing primary collection, waste transportation and waste disposal. Local bodies implement activities such as door-to-door collection, source segregation, composting and recycling. Some examples are

1. SWaCH–PMC model of door-to-door collection in Pune city since 2008. This is a collaborative partnership between Pune Municipal Corporation (PMC) and solid waste collection and handling (SwaCH) Seva Sahakari Sanstha Maryadit, a cooperation of self-employed wastepickers for waste collection in Pune, where waste pickers of SWaCH provide door-to-door waste collection and resource recovery services at the local level.

2. The Municipal Corporation of Greater Mumbai (MCGM) is one of the strong urban local bodies in the country. The Parisar Vikas Program was initiated through the efforts of Stree Mukti Sanghatan (SWS). In this program, under the Advanced Locality Management (ALM) scheme of MCGM, mainly waste-picker women are educated, organized and skilled to provide the best waste management facilities to housing societies, apartments and campuses.

3. The door-to-door collection by Patna Municipal Corporation was introduced in 2008. Here, Nidan Swachhdhara Pvt. Ltd offers an integrated approach for SWM through door-to-door services to households, restaurants and shops by connecting the waste-picker workers' communities.

6.8   SOLID WASTE DISPOSAL

Regardless of the best efforts employed to reduce, reuse and recycle, the final functional component in the solid waste management system is waste disposal. All the waste from all sources is collected and transported to a disposal site. It may be a landfill site, an incinerator or some other mode of disposal. The solid waste disposal problem is a growing concern in many developed and developing countries. A developing economy like India is facing the waste disposal problem due to the rapid industrialization, population growth and changing lifestyle. Currently, a large number of cities do not have any processing facilities and dump their collected solid wastes, whether from residential or commercial uses, into dumpsite(s). SWM is a growing problem in Sri Lanka due to the absence of proper SWM systems within the country, and open dumping is also very common. The US Environmental Protection Agency (EPA) has collected and documented data on the generation and disposal of waste in the United States that helps most American communities to access and deal with SWM issues by adopting an integrated waste management system. The European Union meanwhile has been promoting policies with respect to sustainable waste management via high rate waste collection approaches, waste reduction, recycling, composting and energy conversion technologies.

6.8.1   METHODS AVAILABLE FOR TREATMENT, PROCESSING AND DISPOSAL OF SOLID WASTE

1. Open dump and landfills
2. Biological waste treatment (composting, vermicomposting, biomethanation)
3. Thermal waste treatment or waste-to-energy technologies (incineration, gasification, landfill gas recovery) (see Figure 6.1)

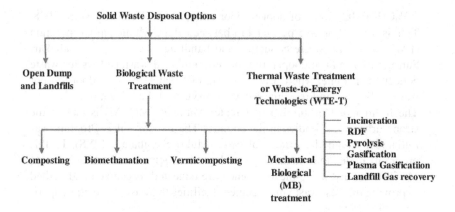

FIGURE 6.1   Solid waste disposal options.

6.8.2 Open Dumps and Operated Landfills

The most accepted method for disposal of solid waste is the landfill method, in which waste is usually cast off into the low-lying areas of any wasteland, as absolute reuse, recycle or incineration of the entire solid waste stream is not possible. Kumar et al. (2009) in their study found that uncontrolled open dumping is a common practice in nearly all cities. MSW is disposed of by means of depositing it in low-lying areas outside the city without following the principles of sanitary landfilling. Therefore, open dumpsites or open landfills are a common feature in all developing countries. The most significant components of the landfill are the bottom linings and top cover. Semi-controlled landfills are designed in such a way that the waste is first compacted, and then a topsoil cover is provided on a daily basis to prevent consequences. The aim is to avoid any contact between the waste and the surrounding environment, particularly the groundwater. Generally, this practice is considered as a non-engineered disposal method unable to manage the leachate discharge or emissions of landfill gases. The discharge of leachate into the soil is the key problem associated with landfills and landfill gases (LFGs), primarily methane and carbon dioxide (up to 90%), carbon monoxide, nitrogen, alcohols, hydrocarbons and heavy metals. Leaching of contaminants also causes surface and groundwater pollution, which may harm human and natural systems. The Central Pollution Control Board (CPCB) India (2017) reported that in 2014–2015, only 22% (32,871 tons per day [TPD]) out of the 80% (117,644 TPD) of the total generated waste collected was processed or treated (see Figure 6.2). In addition, most of the dumpsites are designed and operating unscientifically without following SWM Rules, 2016.

6.8.3 Biological Waste Treatment

Different treatment methods are practiced (Table 6.1) depending on the type and nature of the waste residue generation, cost, risk associated, safety and other environmental aspects. Organic solid waste, such as food waste, garden (yard) and park

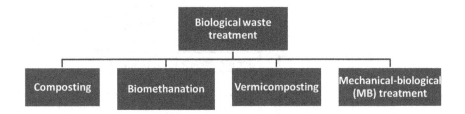

FIGURE 6.2   Biological treatment of MSW.

waste and sludge, is common, and their generation dramatically increases each year in both developed and developing countries like India. Biodegradable waste constitutes 58–76% and recyclables 14–23% of the total household waste generated. Urban MSW on average contains 51% of organics, 28% of inerts, 12% of recyclables and 9% of others. The larger proportion of organic matter in MSW comes from household activities, indicating the desirability of biological processing of waste. In case of readily biodegradable solid waste, the best-known *treatment techniques* include either composting (aerobic treatment) or biomethanation (anaerobic treatment) generally adopted in India. The objective of bio-treatment is to (1) improve or alter its physical and chemical characteristics for energy and resource recovery, (2) offer waste stabilization and nutrient retrieval and (3) reduce toxicity, final volume and reduce greenhouse gas emissions. Hence, almost 65–70% of the waste can be either processed by aerobic or anaerobic digestion (biodegradable waste) or recycled, and only 30–35% needs to be landfilled. *Composting* is a biological process of decomposition and stabilization of organic matter by microorganisms under controlled aerobic conditions under moist and warm environment, resulting in the production of humus or compost rich in nutrients. The solid compost produced can be used as fertilizer for crops, and the combustible gas is used to produce heat and electricity. Currently, a large amount of waste is treated by this method, which is the most simple and cost-effective technology for treating the organic fraction of MSW. It is further classified as *aerobic composting*, and anaerobic composting is also known as biomethanation. It can also be classified as open or window, mechanical or closed etc. depending upon the operating condition and design of the plant. During aerobic composting, aerobic microorganisms oxidize a large fraction of the degradable organic compounds into carbon dioxide, nitrite and nitrate. Being one of the traditional methods, there has been some problems associated with it mainly due to transportation, poor acceptance by farmers (maybe because of quality concerns), marketing, price etc. *Biomethanation* is a process in which organic waste matter is converted into methane and manure by microbial action in the absence of air (Figure 6.3). It is a two-stage process of acidification, followed by methanation. The production of biogas reduces the amount of waste and, therefore, the need for disposal in landfills. The organic fraction of MSW is a potential feedstock for the biomethanation process. It is estimated that by this method of processing, 1 ton of MSW produces two to four times as much methane in three weeks as 1 ton of waste in a landfill produces in six to seven years. Biomethanation requires less energy than

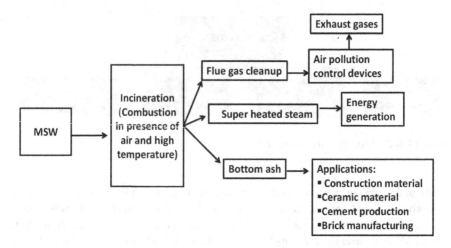

FIGURE 6.3   Anaerobic digestion (biomethanation) process of organic waste.

the aerobic process does and creates much lower amounts of biological heat. This biogas is usually used in two ways—first to generate electricity and second to produce heat in different required processes. The advantages of the biological treatment include reduced volume in the waste material, stabilization of the waste, destruction of pathogens in the waste material, and production of biogas for energy use and high-grade soil conditioner. Generation of less amounts of electrical and thermal energy per ton of treated MSW (kWh/ton MSW), long retention times (typically 20–40 days) and low removal efficiencies of organic compounds are some of the shortcoming of biogas to energy plants. Large-scale biomethanation has failed due to the absence of source separation, as in the case of Lucknow, while the same process has seen huge success on a small-scale level. Traditional bio-treatment processes for solid wastes are generally low rate, often because of either low or negligible rates of mixing and relatively large particle sizes result in low surface area to volume ratios. The former problem can be overcome by enhanced mixing, while the latter can be alleviated by greater feedstock diminution. Both measures involve increased capital investment and increased operating costs in the form of energy requirements. Several schemes have been planned for biomethanation of vegetable wastes and fruit waste in various cities.

In numerous industrialized countries, waste management practices have changed a lot over the last decade. In Europe, *mechanical-biological* (MB) treatment of waste is becoming popular, in which the waste substance undergoes a series of mechanical and biological operations that aim to diminish the waste volume but also stabilize it to reduce emissions from final disposal. Typically, the mechanical methods (may include separation, shredding and crushing of the material) separate the waste material into fractions that will be further operated biologically (by composting, anaerobic digestion, vermicomposting, recycling). The composting can be performed either through heaps or in composting facilities where optimized process conditions are provided along with filtration of the gas produced. The chances to decrease the amount of organic material to be disposed of at landfills are about 40%–60%. Because of the reduced waste mass, high organic content and biological activity, the MB-treated waste can reduce up to 95% methane emission as compared to other methods. Some methane is formed in anaerobic sections of the compost, but the same will be oxidized to a large extent in the aerobic sections of the compost. The estimated amount of methane emitted into the atmosphere ranges from < 1% to a few percent of the initial carbon content in the material. Vermicomposting is a process in which the biodegradable portion of solid waste is composted with the assistance of earthworms (mainly *Pheretima* sp., *Eisenia* sp. and *Perionyx excavatus*).

6.8.4 Waste-to-Energy Technologies (WTE-T)

This is an attractive and recognized approach that may help to solve the problems of waste management by creating sustainable waste treatment systems as well as possible renewable energy sources, particularly for developing countries. Most of the existing WTE systems operating in many countries, such as the USA, Germany, Japan, Poland, Malaysia, Italy and Canada, not just helped them to manage waste sustainably but also to have energy security. In India, waste is available in enormous quantities. It is now believed that waste can be a valuable resource to replace fossil resources and has tremendous potential to generate energy. As per the Ministry of New and Renewable Energy (2018), in order to produce cumulative 405 MW, presently 53 waste-to-energy plants are purposed for construction and operation.

Incineration is the most popular thermal conversion technology used globally, through which reduction of waste mass and volume can achieve 70% and 90%, respectively. It is a *thermal waste treatment process* usually helpful for the management of dry waste with a high percentage of non-degradable organic matter with high calorific value. Sharholy et al. (2008) describe it as a simple combustion process taking place in the presence of air at high temperature ranging between 980 to 2000°C (Figure 6.4). The final product of the process is CO_2, water vapors and bottom ash with a huge amount of heat. The final product of the incineration process is superheated steam, which if recovered properly can turn out to be a potential source of energy generation. The electric energy is produced by a turbine connected to a generator and the heat by a district heating system. The bottom ash can further be used as raw material in cement production, feedstock for producing ceramic material and in construction. Incineration can be positioned within city limits, has a low land area requirement and reduces the cost of waste transportation. However, incinerator is not as effective in the case of Indian municipal solid waste because of high organic composition, excessive moisture content, low calorific value (ranges 800–1100 Kcal/kg) and the requirement of auxiliary fuel to sustain combustion. Small incinerators are more efficient but used mainly for burning hospital and medical waste. The major environmental risk of incineration is the generation and release of carcinogenic, toxic emissions containing air pollutants like SO_x, CO_x, and NO_x, which may result in air pollution and health hazards. Thus, it is of prime importance to equip the incinerator with emission control accessories. An additional major factor for waste combustion is its low electrical efficiency, normally in the range of 18%–26%, which

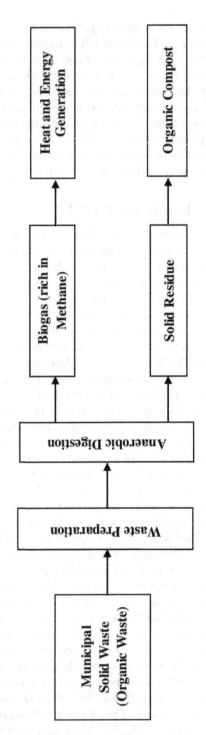

FIGURE 6.4 Incineration process of MSW.

can be potentially enhanced only when municipal waste is pretreated previous to combustion to produce *refuse-derived fuel* (RDF). Combustible waste such as paper, wood, plastic, leather and textile waste can be converted into RDF, and the process involves segregating, crushing and mixing high- and low-heat-value organic waste material and solidifying it to produce fuel pellets or briquettes, also referred to as RDF, which serve as an alternative fuel for energy plants and can be co-fired with coal. As of the year 2015–16, India has 26 operational RDF plants (out of 45 plants installed) that are producing around 156 MW of energy (CPCB 2016). Most of the RDF plants encountered operational problems such as lack of financial planning and skilled personnel, high operational and maintenance costs and overall low efficiency for small power stations. RDF technology appears as an eco-friendly way to dispose waste, and pellets are frequently used for paper, wood and sawmill industries. The Municipal Corporation, Chandigarh, transported solid waste to a waste processing plant producing RDF by treating waste and one compost plant of 300 TPD capacity; in Telangana State, two RDF plants have a capacity of 1200 TPD each and three compost plants each of 680 TPD; in Madhya Pradesh, one RDF plant is operated at Ujjain with a 25 TPD capacity; Municipal Corporation Bilaspur (Chhattisgarh) has constructed and will operate a waste to RDF plant; it is in progress as per the CPCB annual report (2017–18).

Pyrolysis is a thermo-chemical process carried out at high temperature (300–500°C) in absence of oxygen and may recover 80% of the stored energy of carbonaceous waste into liquid fuel, syngas and char. The syngas can be utilized for operating engines, turbines and heat pumps. Pyrolysis is a high-energy consuming process mostly used in developed countries specifically for recyclable plastic waste.

Gasification seems to be a viable alternative for combating the pollution impacts as well as reducing the burden of rising energy demands. Gasification is better than the conventional process because it is a comparatively faster process, requires less land area, reduces environmental contamination and is capable of replacing incinerators that emit toxic dioxin and furan. All types of solid wastes can be directly utilized as a low-cost feedstock for gasification in an efficient way to obtain synthesis gas or produce gas of adequate calorific value. Compared to other biological treatment processes like biogas generation, gasification is free from microbial cultures, and susceptible operating parameters can produce higher calorific value gas more rapidly and reduce about 70% mass and 90% volume of waste.

The process is carried out at high temperature (500–1000°C) with a limited supply of oxygen. The partial combustion of organic matter is accomplished to produce mainly three components in streams:

1. *Syngas or producer gas*: Mixture of carbon monoxide (CO), hydrogen (H_2), carbon dioxide (CO_2) and methane (CH_4). The gas is further cleaned up and used as fuel to generate electricity and to produce different chemical products.
2. *Liquid fuel*: Contains tar, light oil and organic chemicals like acetic acid, acetone, methanol.
3. *Char*: Made up of elemental carbon along with some inert materials in the waste.

Electrical conversion efficiencies can be reached up to 26% to 28% by the application of a gas engine or up to 30% with the help of a gas turbine. Moreover, gasification technology is more flexible in its application to vary from small and medium-scale WTE conversion projects by using agro-based biomaterial as feedstock with an operational capacity up to about 120,000 tons/year.

Plasma gasification is adopted for mixed nature waste like MSW, typically performed by means of some gas (e.g. air, oxygen, nitrogen, noble gases) to create the plasma in the jet or arc for the stipulation of heat in the absence of gasification oxidant. At very high temperatures (>5000°C), feedstock is broken down into its main component atoms of carbon, oxygen and hydrogen. They rapidly react to form hydrogen and carbon monoxide gases, thus generating a very high-quality syngas with no methane, hydrocarbons or tars. Treatment of solid waste by various other methods may involve certain difficulties for the complete disposal of residues. A distinctive feature of the plasma process is its ability to produce very high temperature, which is not achievable with conventional gasification and combustion. Due to its high temperature, a total disposal of residues is possible with all the organic material converted into syngas. In order to make it more accessible and affordable to all, the equipment related to the process may be made available at reasonable expenses by innovative design.

6.8.5 SANITARY LANDFILLS AND GAS RECOVERY

Around the world, gaseous emissions from municipal landfills are the main causes of major environmental impacts. As per the Intergovernmental Panel on Climate Change (IPCC) report 2001, methane gas generated from solid waste disposal sites contributes about 3% to 4% to the yearly global greenhouse gas emissions. Liquid expulsion generally contains extracts of decomposed organic waste and liquid run-off from percolated rainwater, collectively called leachate. If a landfill site is not lined properly, the leaching of toxic substances pollutes the surface and groundwater resources and the fertility and quality of the respective surface and subsoils. In developed countries, sanitary landfills are used very effectively, and they have facilities to manage/control the risks or hazards connected with waste disposal, to treat leachate and to recover landfill gas. Recovered gas can be utilized for energy generation or for direct heat applications for domestic purposes. It is a cost-effective technique that utilizes and converts any low covered waste area for useful work. Unlike open dumping sites, scientifically designed and operated landfills need to have pre-planned installations, systematic parameter monitoring and skilled personnel. Landfills must have amalgamated liners to control percolation of leachate to subsurface water level along with all the equipment required for proper collection and a ventilation system to restore gas production. As per annual report CPCB 2016–17, out of 2120 accessible dumpsites, 40 are reclaimed/capped, whereas 21 dumpsites have converted into sanitary landfills, including 17 in Tamil Nadu, one in Sikkim, one in Meghalaya, one in Madhya Pradesh and one in Chandigarh. The Government of India has made it obligatory to install landfill gas (LFG) control systems as per guidelines under MSW rules, 2016. The landfill gas, mainly methane (CH_4), is important as it can be utilized

TABLE 6.1
Different Treatment Technological Options for MSW

Sr. No.	Technology	Opportunities	Limitations
1.	Composting	Disintegration of organic MSW with the help of microorganisms under controlled aerobic environment. The final product is rich in nutrients known as compost. Being a traditional method, it is attractive for garden and large-scale facilities.	Large-scale composting plants are often unsuccessful due to contamination and need more source segregation. Some problems associated because of transportation, less recognition by farmers, quality concerns, marketing issues and cost factors. If not properly done, odor and insect problems.
2.	Anaerobic digestion	Organic waste material (chiefly vegetable and food waste) is decomposed by microbes in anaerobic conditions. Reduces waste amount and produces biogas, which is used to generate heat and electricity. Associated with production of high-grade soil compost. Can be done at small scale and free from bad odor. Control greenhouse gas emissions and require low power.	Problem of leakage of methane gas. Require waste segregation for enhanced efficiency. Least suitable for wastes containing a lesser amount of organic matter.
3.	Incineration	Most commonly used waste treatment technology. Widely used in developed countries. Reduces waste mass, waste volume and transportation cost significantly. Suitable for sorted waste with high calorific value. Small incinerators are more efficient. Produces energy used to generate electricity.	Emissions contain air pollutants. Little or no stack emissions monitoring. Lack of advanced air pollution control equipment. High capital, operation and technical costs. Inappropriate for low calorific content and organic waste with high moisture content.
4.	Refuse-derived fuel (RDF)	Frequently used for flammable waste like wood, plastic, paper, leather and textile waste. Offers safe and eco-friendly disposal. Used as alternate fuel for energy generation and co-fired with coal. Low handling and transportation cost.	Skilled personnel required. Sorting and preparation of waste is important. High operational and maintenance costs. Less efficient for small power stations.

TABLE 6.1

(Continued) Different Treatment Technological Options for MSW

Sr. No.	Technology	Opportunities	Limitations
5.	Pyrolysis	Solid waste is heated in the absence of air.	High-energy consumption and noise and air pollution.
		Thermo-chemical conversion process yields a gas mixture, bio-oil and solid char.	Product stream is more complex.
			Gas stream has a higher concentration of carbon monoxide. Product yields depend mainly on type of feedstock, temperature range and type of reactor used.
		Appropriate for carbonaceous waste and significantly decreases the waste volume sent to landfills.	
		Bio-oil could be used in diesel engines as fuel after some upgrading.	High viscosity of pyrolysis oil.
			High maintenance and capital cost.
6.	Gasification	Efficient technology independent of microbial culture.	Used mostly for agro waste.
		Partial combustion of waste to generate energy.	Optimization of efficient reactor design, heating parameters and operating temperature.
		Requires high temperature and limited air supply.	Occurrence of extremely high-molecular-weight tars in syngas and bio-oil may result in incomplete combustion and other problems.
		Final products include char, tar and syngas.	
		Syngas has much higher calorific value and can produce petrochemical products.	
		MSW acts as low-cost biomass for gasification.	
7.	Landfill gas recovery	It is a cost-effective technique common in developed countries.	Not very common and generally not successful for high-moisture-content waste, high capital, operation and high technical costs.
		Best alternative for open landfills.	
		The gas produced is utilized for thermal applications and power generation.	
		Natural resources are recycled and returned back to the soil.	Requires large area.
		Helps to convert waste/low-lying areas into useful land.	If not operated scientifically, untreated leachate may cause soil and groundwater pollution.
		It is mandatory to install an LFG control system as per MSW Rules, 2016.	Additional cost due to pretreatment and upgrading of gas quality.

for power generation. Hence, LFG must be used either for power production or for direct heat recovery or should be flared to upgrade air quality.

For example, a modernization project was undertaken at the Achan landfill site, Srinagar city in Kashmir. Previously, open dumping of solid waste was practiced on the site. After 2011, the site was maintained by J&K State Pollution Control Board and operated scientifically, satisfying all the arrangements required for environmental pollution control and quality monitoring. Similarly, in Tamil Nadu, assessment and monitoring of air, water and leachate are performed in Kadungaiyur at a dumping site in a Madurai Corporation solid waste disposal facility.

6.9 CONCLUSION

Sustainable waste management, safe disposal practices and waste-to-energy conversion technologies are the major areas taken up by the government/urban bodies in order to offer better solid waste management services to the public. An ideal waste management must include the principles of waste minimization, reducing, recycling and treatment/processing. Moreover, the door-to-door collection system should be encouraged, especially intended vehicles should be deputed for transportation of waste to a landfill site and, most importantly, the landfill sites should be scientifically designed and operated. The SWM Rules, 2016, mainly emphasize the promotion and installation of various waste-to-energy plants in India, as these promising technologies can be helpful to achieve sustainable waste management and mitigation of major environmental problems like pollution and global warming.

7 Bioremediation
Concepts and Application

Jing He, Xueyan Chen and Varenyam Achal

CONTENTS

7.1 INTRODUCTION

Bioremediation is defined as a process that uses microorganisms, plants and lower animals or their enzymes to remediate polluted sites for recovering their original conditions. This technology utilizes some natural characteristics of organisms, including metabolism activities and extracellular secretions, for environmental remediation.

Bioremediation was used for the first time in 1972 to treat gasoline leakage in Pennsylvania, USA. Following this, the concept of using metal-accumulating plants to clean up sites contaminated with heavy metal(loid)s was also presented in 1983. Since 1991, the United States started implementing bioremediation technology to control hazardous environmental pollutants in soil, groundwater and marine environments. Since then, bioremediation has been identified independently as an emerging technology for pollution remediation on a large scale. Furthermore, many types of remediation technologies, such as bacterial remediation, phytoremediation, fungal remediation, animal remediation and even combined remediation methods, have developed in the field of bioremediation.

DOI: 10.1201/9781003131427-7

7.2   IMPORTANCE OF BIOREMEDIATION

Bioremediation usually utilizes the functions of biological metabolism (of plants, microorganisms and protozoa), absorption and transformation to remove or degrade environmental pollutants and achieve the goals of environmental purification as well as ecological recovery. Therefore, it is significant for people to know how organisms remediate contaminated soil, water or marine areas and why many researchers widely apply this method.

As energy needs may drive microbial transformation, and many pollutants, especially organic contaminants, contain high-level nourishment like carbon, nitrogen and other essential elements, pollutants would be decomposed by microbial metabolism activities before achieving detoxification. Also, unlike decomposing organic pollutants, microbial extracellular secretions contribute to precipitate complex inorganic matter such as toxic metal(loid)s. A unique character of fungi is their ability to break down organic compounds using extracellular oxidoreductases. Their enzymes are probably related to supporting fungal growth on recalcitrant substrates such as lignocellulose that are not accessible for most of the bacteria via removal and depolymerization of lignin and its humic substance derivatives. Because of the universal nature of microorganisms, their large biomass relative to other living organisms in the earth, diversity and abilities in their metabolic mechanisms, including the capability to function even in the absence of oxygen and other extreme conditions, the research/application of exact pollutant-degrading microorganisms have become a priority research area. In addition to microorganisms, plants are also used to remove pollutants. For instance, many researchers have reported several hyperaccumulator plants for contaminant removal, which could adsorb numerous pollutants exceeding their limits. Those toxic contaminants will be held in the plants' trunks/branches by metabolic and transformation activities.

As an important part of the plant, roots also synthesize and secrete a wide variety of substances, which could act as chemical attractants for a vast number of diverse and active soil microbial communities in the rhizosphere, constituting a significant part of the plant for environmental remediation. Microorganism activities in the rhizosphere can promote root growth and plant development, by which they take up some microbe-oriented molecules, and microorganisms can metabolize a fraction of these plant-derived small organic molecules in the vicinity as carbon and nitrogen sources. Therefore, this coadjutant behaviour is beneficial for pollutant treatment by phytoremediation. Some animals, especially protozoa, also play an important role in ecological remediation. For example, earthworm biomass represents more than 60% of total animals in soil and plays an irreplaceable role in remediating the pollution of the soil ecosystem. Earthworms can increase the number of humic acids and organic acids as they degrade soil organic matter, which may promote immobilized metals, transforming them into free status for accumulating by organism adsorbents. Furthermore, some other protozoa also could degrade oil and organic pollutants in wastewater into carbon dioxide, energy and water, which are harmless.

As discussed earlier, bioremediation is a green technology for pollutant removal that rarely produces extra contaminants during the remediation process. When bioremediation technology is applied for treating pollutants, the final products are harmless and stable, such as carbon dioxide, water and nitrogen. This recovering method can immobilize contaminants directly and avoid multiple transformation processes of pollutants as well. Certainly, bioremediation still has many drawbacks. For example, some compounds such as those that are polychlorinated are hard to disintegrate, and the search for specific remediation organisms is complicated. Additionally, if the organisms are applied improperly, some intermediate products may be even more toxic and readily mobilized. However, these cases rarely occur, and compared to the disadvantages of physiochemical technologies, the weaknesses of bioremediation are easily accepted.

7.3   CLASSIFICATION OF BIOREMEDIATION

Generally, bioremediation could be divided into three types based on the kinds of organisms: microbial remediation, phytoremediation and animal remediation (Figure 7.1). However, animal remediation (especially protozoa) technology is not widely used compared with the other two technologies, due to limitations of living conditions and nutritional sources. Despite this, the other two technologies still have diverse removal efficiencies for different kinds of pollution.

7.3.1   MICROBIAL REMEDIATION TECHNOLOGY

Microbial remediation technology could utilize the metabolic function of indigenous microorganisms, or foreign microbial communities supplied artificially with

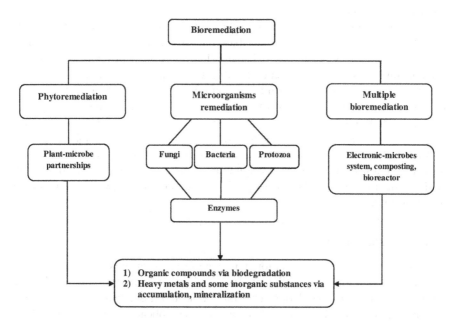

FIGURE 7.1   An overall review of bioremediation.

the ability to degrade and transform pollutants, or eliminate pollutants eventually by creating appropriate environmental conditions and strengthening the microbial function of metabolic activities. The essence of microbial remediation is biodegradation or biotransformation, which is the decomposition of organic pollutants or the inactivation of inorganic pollutants. Microbial remediation technology can also be grouped into many types. Based on the contaminants, microbial remediation technology can be divided into organic pollutants remediation and inorganic pollutants remediation. Also, microbial remediation can be classified as *in situ* remediation and *ex situ* remediation based on the remediated location. Bacterial and fungal remediation all belong to microbial remediation as well.

The usual microorganisms of used to degrade organic pollutants include bacteria (*Pseudomonas, Bacillus, Arthrobacter, Acinetobacter, Alcaligenes, Staphylococcus* and *Corynebacterium*), fungi (*Aspergillus, Penicillium, Trichoderma*, white-rot fungi) and *Actinomycetes* (*Streptomyces* etc.); of these, *Pseudomonas* is the most active and could decompose a variety of organic pollutants, such as pesticides and aromatic compounds. In some cases, due to the lack of dissolved oxygen and/or other electronic receptors, the degradation speed of microorganisms would be slow and need to be strengthened by various measures, like providing O_2 (especially aeration in the activated sludge method and biological membrane system) or other electronic receptors (e.g. NO_3^-), adding the nitrogen and phosphorus nutrient, and inoculating efficient cultural microbes. Additionally, some complicated organic pollutants, such as trichloroethylene and polychlorinated biphenyl, are still hard to degrade directly by microbe but can be decomposed by microorganisms first into other primarily nitrogen and carbon nutrients, which is called co-metabolism. This co-metabolism has been reported as a significant mechanism for degrading refractory pollutants.

Microbes remove heavy metals from soils via various mechanisms that generally include bioaccumulation, biosorption, bioleaching, biofiltering and biomineralization. Firstly, microbes can absorb toxic metal(loid)s to store in different parts of their cells or combine with the extracellular matrix to precipitate or chelate the metalloid into microbial polymers; they can also be enriched by binding a protein or a peptide. Microorganisms could then fix metal(loid) ions by electrostatic adsorption or complexation through the negative charge on the cell surface as well. Additionally, metal(loid)s are removed by bioleaching via the oxidoreduction reaction component of microbial metabolism. Biomineralization is also a method that transfers free cations into sediment through the precipitation of cells, metabolites and macromolecules, which utilize specific urease-producing bacteria to induce the formation of metal carbonate.

7.3.2 PHYTOREMEDIATION

Phytoremediation uses the roots (or stems and leaves) of plants to absorb, enrich, degrade and/or immobilize pollutants from soil and water. Generally, plants have different degrees of degradation, transformation and absorption for organic and inorganic pollutants, and sometimes for some complicated contaminants, these mechanisms would work together at the same time. Currently, phytoremediation technology is widely applied to remediate contamination of heavy metals and/or radioactive elements. Certainly, phytoremediation can be grouped into types based on different goals, such as organic and inorganic contaminant remediation and soil, water and wetlands phytoremediation.

Phytodegradation or phytotransformation is mostly suitable for organic pollutants since plants can break down or degrade the contaminants by metabolic processes. Additionally, plants produce many secondary plant metabolites (SPMEs), which include allelopathic chemicals, root exudates, phytohormones/phytoalexins, phyto-siderophores and phytoanticipins. Structural analogy between organic contaminants and SPMEs allows the possible uptake and subsequent degradation of pollutants by microorganisms living around plants roots.

Phytoremediation of inorganic pollutants is generally categorized into phy-tostabilization, phytoevaporation and phytoextraction based on different uptake mechanisms. In phytovolatilization, detrimental metal(loid)s are assimilated into volatile organic compounds and/or converted into less toxic vapors, which are then released into the atmosphere as biomolecules. Phytostabilization is the use of plants to decrease the mobility and bioavailability of heavy metal(loid)s. The mechanism is mainly through the accumulation, precipitation or surface absorption of metals with the role of rhizosphere to strengthen the immobilization of toxic metal(loid)s. Phytostabilization is the ability to stabilize heavy metal(loid)s by means of (i) restrict-ing heavy metal(loid) leaching, (ii) inhibiting the shift of toxic metal(loid)s to other areas and (iii) preventing direct contact with polluted area by heavy metal(loid)s. Phytoextraction involves the clean-up of heavy metal(loid)s from soil by means of plant uptake and accumulation. However, phytoextraction is suitable for those areas that are polluted at low to moderate levels, because there are rare plants that could sustain a high concentration of pollutants. Therefore, hyperaccumulator species are a kind of plants that could accumulate 100–500 times more heavy metal(loid)s in their shoots tissues (without visible toxicity symptoms) than can common nonaccumulator plants and with no effects on their growth or biomass.

7.3.2.1 Enhancement of Phytoremediation

Plants of phytoremediation would suffer the risk of toxication in any cases when they are used to remediate pollutants. The toxication of plants decreases the removal efficiency of phytoremediation directly. Adding natural pollutant-degrading bacteria and applying growth-promoting bacteria are good ways to enhance phytoremediation efficiency. Most plants associate with various microbes that live around the plant's root, and some endophytic bacteria live in the plant bodies. These microbes have a capacity to assist the remediation of contaminants in addition to providing other benefits to the plants such as nitrogen fixation, phosphate solubilization and stress tolerance. When contaminant degradation occurs in the rhizosphere, this method is called rhizodegradation. In addition, rhizofiltration is a remediation technique for water pollution, which involves the uptake of pollutants by plant roots either by pre-cipitation onto roots or by absorption into them.

Although microbial remediation and phytoremediation are considered environ-mentally friendly and multi-functional methods of remediation, there are still some

limitations. For instance, the efficiency of phytoremediation and microbial remediation depends on numerous adequate organisms and influence factors such as the
physico-chemical properties of the polluted soil, bioavailability of pollutants, microbial and plant biomass, and the capacity of living organisms to uptake, accumulate,
sequester, translocate and detoxify metals.

7.4 TYPE OF BIOREMEDIATION

According to the spatial location of pollutant distribution, bioremediation methods
can be generally divided into soil, water and sediment bioremediation. The following
mainly generalizes the origin and types of contaminants, the hazards and the current
treatment mechanisms.

7.4.1 SOIL BIOREMEDIATION

Soil is an extremely complex system with a dynamic and living medium, which
mainly includes mineral particles, organic matter, water, air and living organisms. Further, soil is also a micro-ecosystem that has its own self-purification
capacity when subjected to slight pollution. Anthropogenic activity and excessive
utilization—the long-term application of pesticides, fertilizer, plastic film and sewage irrigation—tend to bring a series of problems, such as biomass reduction, land
degradation and the rising number of toxic/carcinogenic crops in the market, which
are a health threat.

Up to now, synthetic pesticides have been recognized as the main source of soil
organic pollutants, and most of them are persistent organic pollutants (POPs), which
are characterized by high toxicity, persistence and bioaccumulation in the environment. They can directly influence soil properties and restrain microbial enzyme
activities. Another organic contaminant from petroleum leakage accidents is generated by undergoing a perennial biochemical process under the subsoil, which is
composed of mixture of aliphatic, aromatic and heterocyclic hydrocarbons. Most
oil contaminants (60%–90%) are classified as biodegradable, while a small number
(10%–40%) are crude and can cause cancer in living beings, especially because of
their recalcitrant properties. The negative effects also include the damage of altered
biotic and abiotic conditions, sharply reducing habitat and biodiversity because of
abiotic stresses like heat, hypoxia and oxidative and osmotic stresses.

The development of updating and obsoleting electronic products and the discharging of printing and dyeing wastes are also responsible for soil with excessive
heavy metals. Heavy metals are non-biodegradable, have long-term persistence and
are bioaccumulative and migratory, which determine the complexity of governing
polluted soil. Although traditional chemical and physical treatments are quick and
feasible, the negative effects on soil cannot be overlooked—high cost and energy
consumption with low efficiency.

So far, there are two prevailing ways to restore the properties of contaminated
soil: *ex situ* and *in situ* bioremediation.

7.4.1.1 *Ex Situ* Bioremediation of Soil

Environmental factors play a principal role in the bioremediation of contaminated soil. Therefore, the conditions can be artificially controlled through *ex situ* methods in which biological organisms typically choose optimum conditions where they could operate favourably. In addition, in the presence of severe contamination of a highly hazardous nature, the method of ectopic repair can segregate and prevent the further spreading of pollutants.

Biopiles, landfarming and bioreactor are the three common *ex situ* soil bioremediations approaches. Composting is currently one of the most cost-effective bioremediation technologies that provokes the enlargement of microbes to clean up or stabilize pollutants in soil. In general, it consists of putting contaminated soil into a pile with aeration and nutrients amendment. Currently, organic wastes as biostimulated reagents are usually applied into the initial composting materials to provide essential nutrition for organisms, while simultaneously offering the benefits of improving soil fertility and quality. Bacteria and fungi are widely considered the most crucial factors for composting contaminated soil. Extra microorganisms (bioaugmentation) added to the soil matrix strengthen biopiles' efficiency when indigenous microbes cannot work well. Thus, they deposit and degrade the pollutants efficaciously through adsorption by organic matter or degradation by microorganisms. Moreover, the addition of organic amendments can directly increase the number of indigenous microbes by providing more available carbon resources, which lead the pollutants in soil be biotransformed or broken down. Researchers have pointed out that micro-organic communities are dynamic during the pollutants' biodegradation process of composting. The composting approach, however, comes with some limitations as well. In long-term bioremediation with spiked contaminants, bioavailability is the crucial factor for bioremediation efficiency compared to the density of microbes. Thus, the aged pollutants pose a great challenge for the remediation of soil. Also, the pollutants' stability could be reversed with a change of soil pH and as organic matter mineralizes. Compost quality should also be considered when evaluating the capability of microorganisms to remove organic compounds and metals. A low proportion of organic chemicals to spiked heavy metals can cause low-quality compost. Landfarming, which consists of degrading pollutants though excavation and placement of contaminated soils into lined beds, is commonly applied on petroleum hydrocarbon contaminated soil because of its low technological footprint and, consequently, low cost. Nutrient amendments along with periodic aeration will enhance the degrading efficiency. The advantages of modest equipment, low energy consumption and a relative low requirement of cover area had become the main reason for many large-scale soil remediation applications. However, the process emits a great amount of volatile gas and increases the stress of air pollution. As for off-site bioreactors, as a large biological process the system offers better control of temperature and pressure to enhance the degradation process of polycyclic aromatic hydrocarbons (PAHs) in soil. Contaminated materials break down better with the assistance of suitable organisms under a slurry-phase and two semisolid-phase (blade-agitated and rotary vessel) bioreactors.

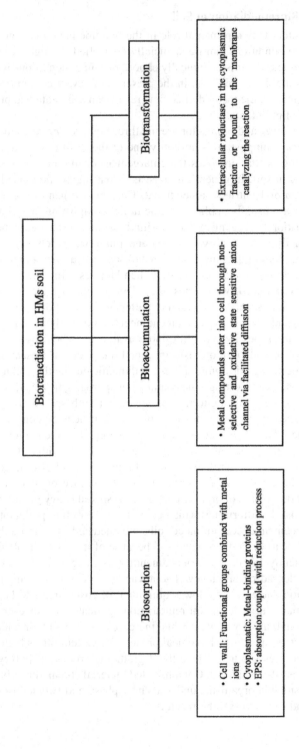

FIGURE 7.2 The mechanisms of microbial remediation of heavy metals.

7.4.1.2 *In Situ* Bioremediation of Soil

Bioremediation in *in situ* restoring methods has been shown to be feasible, cost-effective and environmentally friendly and is currently a promising way, through applying microorganisms or plants, to enhance and rehabilitate the properties of mono- or multi-polluted soil. Initially, most focused on the cell cultures of bacteria in to remediate soil because they are easier to cultivate, grow rapidly and are more convenient to manipulate genetically. In general, bacteria in heavy metals polluted soil rely on their functional groups and excretive extracellular polymeric substance (EPS) distributing in cell walls to bind metal ions. Some proteins spread over cytoplasm can also be associated with metals. Metals can be accumulated in internal organisms by entering into the cell through non-selective and oxidative state-sensitive anion channels via facilitated diffusion. Extracellular reductase in the cytoplasmic fraction or bound to the membrane catalyzes highly toxic metals into low toxicity, as shown in Figure 7.2. In addition, some bacteria can secrete a special enzyme that alters the surrounding pH, while the bacterial cell with a negative charge with existing calcium ions can provide a nucleation site for precipitation, which can be considered as cohesive material to bind heavy metals. However, a high concentration of metals might inhibit the growth and multiplication of microorganisms. Fungal bioremediation was also found to be an effective method for generating less-toxic, inert or fully degraded compounds. However, in the case of multiple contaminants, mixing or combining with multifarious bio-methods is the better option. One advantage is that plants or plant-associated microflora (especially in rhizospheric microorganisms, including mycorrhizal fungi, symbiotic bacteria and free-living rhizobacteria) can convert hydrocarbons (HCs) to nontoxic forms or stabilize and precipitate some heavy metals or highly toxic metals into low-toxic or harmless form. In the process, plants act indirectly in phytoremediation, and their rhizosphere provides favourable conditions for the growth of microorganisms. Plant roots can secrete organic and inorganic substances that spread over their exterior during normal metabolism. These types of exudates can be beneficial for soil microbes, thereby improving the decomposition of toxic organic compounds and immobilizing heavy metals. Limited oxygen levels and the low activity of the indigenous microbial community in soil restrain the application of phyto-rhizo-remediation. Thus, the establishment of a microbial electrochemical system (MES) can bring about the depth development of *in situ* bioremediation. In principle, soil microbes degrade organic matter and release an electron, which then flows through an electric circuit to the cathode and finally is accepted by oxygen or oxidized metals.

7.4.2 Water Bioremediation

Water is the carrier of all livings. The increasing concentrations of contaminations in current groundwater and surface water like various organic matter, heavy metals, nitrates and other compounds range from ng/L to mg/L levels. High-salt and alkaline marine environments are especially subjected to crude oil leakage and spills pollution; in cases like the armed conflict between Iraq and Kuwait, millions of crude oil was released into the sea. Also, frequent navigation close to the coastal areas leads to the increase of effluent discharge into the sea, which creates a deterio-rating marine environment, with chronic pollution and decreasing biota. Some ocean organisms directly affected by the toxic matter die of pneumatorexis or poisoning. As some water bodies contain soluble heavy metals and organic substances, all or most of which are highly toxic, mutagenic and carcinogenic, they could enter into organisms for accumulation through direct and indirect means. Thus, these events have disastrous consequences for ecosystems, and the remediation of water bodies has been the subject of intense study and constitutes perhaps the most significant grand challenge for the future of human society.

The treatment of such hydrocarbon pollutants has been performed using physi-cal approaches and the mechanical removal of contaminants. The application of chemical dispersants has largely been employed on the Gulf of Mexico. Accordingly, besides unsatisfactory efficiency, these chemical dispersants can constitute another source of contamination in the form of the accumulation of other toxic compounds in the environment. Application of microorganisms results in the decomposition of organic pollutants in fluid, and then take them as a carbon source of nutrition for their growth. Also, in a process called intrinsic or natural attenuation, some micro-creatures secrete enzymes or extracellular polymers during the metabolic process, changing the surrounding environment that leads to stabilizing of the pollution. Microbial-induced calcium carbonate precipitation (MICP) is an emerging bio-mediated technology which has the potential to repair water-heavy metals pollution. This process is dominantly controlled by urease activity. Researchers have screened some urease-producing bacteria and fungi resistant to high concentrations of heavy metals, showing that they can reduce the stress posed by metals through transform-ing highly toxic metal ions into less toxic ones. For those types of bioindicators for monitoring marine aquatic environments, sponges such as *Chondrilla nucula* have exhibited a remarkable ability to retain high quantities of bacteria (7×10^7 bacterial cells/h). One square meter patch of this sponge can filter up to 14 L/h of seawater and reveals a great ability to bioaccumulate heavy metals and other pollutants. Various algal strains can also be used to treat wastewater, not only as a based treatment for acid mine drainage (ADM), but also the eutrophication of the aquatic environment. Initially, algae species were also used as alkalinity boosters to monitor the levels of heavy metals in an environment. However, their efficiency and ease of maintenance for accumulated heavy metals make them much more attractive. A biomolecular study has examined the formation of metallothione in peptides in the algae and indi-cates that metals can be absorbed and can bond with algae's inner cell. Along with the removal of superfluous nutrients, algae species can take up dissolved nutrients for its growth. Those species can be extracted for further application, such as land fertilizer, fish feed or biogas production.

In one context, it was manifested that using microorganisms exhibited the capa-bility of mineralizing PAH pollutants to water-soluble products. Among them, marine-derived filamentous fungi showed the excellent ability of oxidizing PAHs, which can produce a non-specific enzymatic extracellular complex including lignin-peroxidase, manganese peroxidase and laccases, and then leading to lignin depo-lymerization. While the addition of surfactants will increase the solubility of oil, it also facilitates the use of biological degradation. Studies on biosurfactants have demonstrated that microbial surfactants not only exhibit a great surface activity and low poisonousness, but also strengthen biodegradation. Moreover, microbial fuel cell (MFC) technology is typically regarded as one of the most effective and sustainable strategies in bioremediation to clean up seawater and sediment contaminated with PHCs (petroleum hydrocarbons). Exoelectrogens are used as an inexpensive anode catalyst to influence the electro-generation and energy recovery performance of MFCs, while the organic matter acts as an electron donor under anaerobic conditions coupled with nitrate reduction. When faced with irresistible factors of high salinity and pH value in the marine environment, exploring and screening new exoelectrogen strains are obviously important. Theoretically, isolating these kinds of strains from the autochthonous environment is the optimum choice.

7.4.3 Sediment Bioremediation

The contaminants of urban river sediment and marine sediment are generated by rapid industrialization, poor sewage treatment facilities and oil spills because of exploitive anthropogenic petroleum activities. Such contamination severely affects freshwater resources and aquatic organisms and indirectly poses a serious threat to human health. Further, it might cause the destruction of an area of aquatic biological system. Studies revealed that encapsulated sediment contaminants in water can be released into aquatic surroundings again through ion exchange and surface dissolution. The release of nutrients such as nitrogen and phosphorus from sediment to the water column has an obvious effect on these nutrients from the external input sources. Therefore, it is highly recommended to take measures to deal with the series of ecological and environmental effects brought by sediment pollution. Indeed, compared to the upper layers of the water column, the bottom ones are characterized by a higher concentration of pollutants because of the resuspension of sediment and of associated pollutants. Generally, dredging is an *ex situ* common method for cleaning river sediment, but the disposal of contaminated sediment is always expensive and may impact ecological stabilization and further utilization. However, organic pollutants left in sediment can be used as a source of microbial growth nutrients, so that they are able to be degraded and consumed. Furthermore, sufficient nitrates reduced by receiving electrons can assist in regulating the microbial community, which in turn stimulates the degradation of organic pollutants. However, the addition of extra nitrates has a negative effect due to conditions created for eutrophication. The degradation of pollutants in most sediment is carried out in ananoxic condition. Therefore, researchers, from another perspective, stimulate the transformation of organic pollutants in sediment by adding methanol. Previous studies have also showed that methanol, as an external carbon source, has a good effect on nitrogen removal in sewage. At this time, under ananoxic environment, organic matter (such as PAHs, pesticides, POPs and so on) are converted into biogas in the presence of methanogens, and a small part of them are used by bacteria to grow and reproduce, emitting carbon dioxide. But an *in situ* biological covering technique with the growth of free bacteria has some shortcomings; for instance, a low concentration of bacteria can be satisfied with the degradation of pollutants but are competitively outed by the native bacteria. The immobilized microorganism technology provided a brand new and alternative way, which has the advantage of a high concentration of microbes, low cell loss and high resistance to impact water quality and quantity. By means of making a special material, solid microbial activated beads, which are spread on the surface of the sediment, provide the appendiculate point as carriers for the growth of more microorganisms and enable the pollutants to be utilized and degraded through the strength of the microorganisms' metabolism. Furthermore, aquatic sediment can store a portion of biomass and organic contaminants which can be developed as a potential energy library through microbial fuel cells, a promising tool for harvesting energy from various environments. A long-term scale-up of a microbial fuel cells system was verified as an efficient tool for simultaneous bioremediation and power supply. However, there are still some gaps in stimulating biodegradation efficiency and regulating the microorganism network in sediment in terms of low current. The electrokinetic (EK) process will be coupled with microbes to enhance the efficiency of biodegradation of organic compounds. Positive effects with EK include four aspects: (i) assist the release of organic compounds bound to mud particles in the sediment; (ii) improve transports of nutrients and terminal electron acceptors; (iii) increase the bioavailability of contaminants; and (iv) generate oxidizing and reducing zones close to the electrodes favourable for the biodegradation of pollutants. Coastal sediment, as an important carrier of heavy metals, plays a significant role in the storage of potentially deleterious materials. The researchers relied on previous work developing a more optimum bio-stimulant ball coated with polysulfone (BSB, mixing polymer and N,N-dimethylacetamide), then provided imperative nutrients for enhancing the microbes' community and increasing metal stabilization. This novel method is expected to further scale up the metal-remediation application. A huge pilot bioremediative project is widely used to treat sediment pollution, known as ecological wetlands. Ecological wetlands combine the advantages of plants to improve water bodies and bottom sediment through the use of super pollution-tolerant and pollution-rich aquatic plants. The presence of nitrogen and phosphorus in sediment is one of the key factors affecting water eutrophication. Even if the concentration of P and N in water has been lowered, their presence will lead to catastrophic eutrophication once the water bodies become active. Thus, aquatic plants can utilize the developed plant root-microorganism system to degrade, absorb and accumulate the pollution. Furthermore, while enhancing the overall quality of water environment, they are also a good biomass energy raw material, which can provide additional advantages.

7.5 BIOREMEDIATION APPLICATIONS

From the aforementioned description, bioremediation can be categorized into different remediation methods on the basis of organisms (microbial remediation, phytoremediation and animal remediation), polluted area (soil bioremediation, water bioremediation and atmosphere bioremediation) and implement sites (*in situ* bioremediation and *ex situ* bioremediation). The applications of bioremediation are wide, and for each pollutant, the bioremediation mechanisms and application of organisms are different. There are numerous examples of bioremediation applications for pollutant removal. Table 7.1 shows the already applied bioremediation methods for contaminants.

TABLE 7.1
Bioremediation Technology of Various Environmental Pollutants

Pollutant Types	Pollutants	Organisms	Mechanisms	Polluted Sites	Ref.
Oil/Petroleum	Crude oil	Indigenous inoculums	Biodegradation	Sand beach (shoreline)	Venosa et al., 1996
		Alcanivorax-Thalassolituus, *A. borkumensis SK2T*	Biostimulation, bioaugmentation	Seawater	Hassanshahian et al., 2014
	Petroleum	Indigenous microbes, elephant grass	Biostimulation, plant growth promoting bacteria (PGPB)	Soil	Ayotamuno et al., 2009
	Oil	*Actinobacteria*, *Gammaproetobacteria*, *Marinobacter*, *Alcanivorax*, *Geobacillus*	Oil mineralization, biostimulation, biodegradation	Desert soil	Abed et al., 2015
	Total Hydrocarbon Content (THC)	Indigenous microbes	Biostimulation	Soil	Kogbara, 2008
	Total Petroleum Hydrocarbons (TPH)	Indigenous microbes	Biostimulation, bioaugmentation	Soil	Suja et al., 2014
Organic pollutants and pesticides	Chlorpyrifos	*Enterobacter* strain B-14	Microbial degradation	Soil	Singh et al., 2004
		Poplar (*Populus* sp.) and willow (*Salix* sp.)	Phytoaccumulation and metabolized	Nutrient solution	Lee et al., 2012
	Hexahydro-1,3,5-trinitro-1,3,5-triazine (RDX)	*Acetobacterium paludosum*, *Clostridium acetobutylicum*	Microbial degradation	Soil and underground water	Sherburne et al., 2005
	PAHs	*Pseudomonas* sp., *Pycnoporus sanguineus*, *Coriolus versicolor*, *Pleurotus ostreatus*, *Fomitopsis palustris*, *Daedalea elegans*, fungi	Microbial degradation	Water, soil	Arun et al., 2008
	Hydrocarbon	*Actinobacteria, Ascomycota*	Biostimulation	Soil	Ros et al., 2010

TABLE 7.1

(Continued) Bioremediation Technology of Various Environmental Pollutants

Pollutants Types	Pollutants	Organisms	Mechanisms	Polluted Sites	Ref.
	Phenanthrene	*Agrobacterium, Bacillus, Burkholderia, Pseudomonas, Sphingomonas*	Microbial degradation	Soil	Aitken et al., 1998
	$^{13}C_4$-4,5,9,10-pyrene	Local bacteria community α-, β-, γ-Proteobacteria, Actinobacteria	Biostimulation Microbial degradation	Soil Compost	Puglisi et al., 2007 Peng et al., 2013
	Atrazine	Maize (*Zea mays*), switchgrass	Phytoextraction	Soil	Ibrahim et al., 2013, Albright III et al., 2013
	Dichlorodiphenyl-trichloroethane (DDT)	*Cucurbita pepo*	Plants uptake	Soil	Lunney et al., 2004
	Polychlorinated biphenyl (PCB)	Alfalfa, flatpea, *Sericea lespedeza*, deertongue, reed canarygrass, switchgrass, tall fescue	Phytodegradation	Soil	Chekol et al., 2004
Heavy metal (loid)	Cu	*Arthrobacter creatinolyticus, Arthrobacter oxydans, Nocardioides rotundus, N. dokdonensis, Pseudomonas azotoformans, P. aeruginosa, Streptomyces luteogriseus, S. acidiscabies*	Biostimulation, microbial-induced carbonate precipitation (MICP)	Soil	Chen and Achal, 2019
	Ni	*Bacillus cereus* NS4 *Sporosarcina globispora* (UR53)	MICP Precipitation (SRB)	Soil	Zhu et al., 2016 Li et al., 2013

Metal(s)	Plant/microorganism	Process	Medium	References
	Bacillus thuringiensis GDB-1, *Pseudomonas putida*, *Bacillus subtilis* strain SJ-101, *Microbacterium oxydans* AY509223	PGPRs		Babu et al., 2013, Kamran et al., 2016, Zaidi et al., 2006, Abou-Shanab et al., 2006
	S. macrogoltabidus, M. liquefaciens, M. arabinogalactanolyticum, Arthrobacter nicotinovorans SA40	Rhizobacterial effects		Abou-Shanab et al., 2003, Cabello-Conejo et al., 2014
Pb, Zn	*Brassica juncea*	PGPB	Pb-Zn tailing	Wu et al., 2006
Pb	Creeping zinnia, *Alternanthera phyloxeroides, Samvitalia procumbens, Portulaca grandiflora*	Phytoextraction	Soil	Cho-Ruk et al., 2006
As	*Brassica juncea* var. Varuna, Pusa Bold	Rhizosphere complexation	Solution	Gupta et al., 2009
Cd	*Salix fragilis* "Belgisch Rood", *Salix viminalis* "Aage"	Phytoaccumulation	Soil	Vandecasteele et al., 2005
	Amaranthus sp., *Rahnella* sp. JN27	Phytoextraction enhanced by endophytic bacteria	Soil	Yuan et al., 2014
Se	*Brassica* genus	Phytovolatilization	Soil	Terry et al., 1992
As, Hg	*Arabidopsis thaliana*	Phytovolatilization	Soil	Rugh et al., 1996
Cd, Mn, Pb, Zn	*Alternanthera philoxeroides, Artemisia princeps, Bidens frondosa, Bidens pilosa, Cynodon dactylon, Digitaria sanguinalis, Erigeron canadensis,* and *Setaria plicata*	Phytostabilization	Mn mine tailing	Yang et al., 2013
Cu, Cd, Cr, Zn, Fe, Ni, Mn, Pb	*Triticum aestivum* L., *Brassica campestris* L.	Phytoaccumulation	Soil	Chandra et al., 2009
Cd, Pb, Ni, Zn	*Berkheya coddii* Roessler (Asteraceae)	Phytofiltration	Water and waste stream	Mesjasz-Przybylowicz et al., 2004

7.6   CONCLUSION

Bioremediation is a green technique based on sustainable development and eco-friendly recovery which is widely accepted to treat contaminated environments. Moreover, under natural conditions, it has corresponded well to economic effectiveness with the superiority of autochthonous microorganisms and plant-animals when faced with the need to remediate polluted areas. From the current development trend of ecological restoration, bioremediation can roughly be divided into bioaugmentation, biostimulation, cell immobilization, production of biosurfactants, design of defined mixed cultures and the use of plant-microbe systems. To address the question of whether the technology is mature, researchers have provided some enlightenment from natural phenomena, where the secretion of microbial enzymes can change the surrounding environment, and then precipitate with ions that exist in the environment to form a stable sediment. Based on such principal to stabilize heavy metals, microbial-induced calcium carbonate precipitation (MICP) is currently a hot research topic. At the same time, the secretion of extracellular polymers is beneficial to form the tendentious combination of metals group and achieve stable function of heavy metals. Moreover, using the cell surface of microbes, some highly toxic metals are transformed into low-toxic metals through the metabolic activities of microorganisms or by generating bioligand complexes. In addition, fungi's filamentous hyphae can be applicable in the fixation and passivation of heavy metals and the complete degradation of chemicals. Another technique is research on plant growth-promoting rhizobacteria (PGPR), especially when the developed plant root system gathers a large number of fungi and bacteria. They have natural sensitivity to the pollutants as they could rely on pollutants' elements, for example, carbon, nitrate, phosphate, after the degradation of pollutants. Furthermore, some bacteria and fungi can release substances, and in turn promote the plant biomass for better plant growth.

In spite of aforementioned favourable methods, ecological restoration technology still has the following limitations. These are mainly summarized as technical defects and the complexity of environmental pollution: (1) some extreme environmental pollution is difficult to remediate, such as the extensive petroleum pollution of ocean with high salt and low temperature and the heavy metals pollution of high temperature and dry desert; (2) new bioremediation technologies leave many gaps for researchers to fill, for example, the application of immobilization cell technology is still in its infancy, recombinant DNA technology can enhance the natural environment but under the situation of the low efficiency of microbial enzyme production, and microbial fuel cell (MFC) uses highly expensive equipment, but results in low power output; (3) multi-environmental pollution is currently a worldwide problem, and heavy metals will inhibit the growth of microorganisms, which affects the efficiency of microbial degradation of organic pollutants. Thus, a single means of ecological restoration is unable to solve this kind of pollution problems. It's urgent to design a series of ecological pollution repair strategies for the application of hybrid practical environment pollution repair; (4) bioremediation schemes create uncertainty, in that the product formed after the repair needs to be treated with caution, and the toxicity test should be carried out to verify its impact on the environment, because they may show higher toxicity than its precursor; and (5) genetic engineering approaches bring certain challenges for environmental remediation due to issues associated with environmental release of genetically modified organisms. Thus, how to effectively overcome the limitations and achieve the maximum benefits of ecological restoration remind us not only to focus on the current economic benefits, but also the long-term benefits of the sustainability and large-scale application of ecological restoration technology in environmental improvement in the future.

8 Biopesticides and Biofertilizers

Trends, Scope and Relevance

Abhinay Thakur, Ajay Sharma
and Anil Kumar Sharma

CONTENTS

DOI: 10.1201/9781003131427-8

8.1   INTRODUCTION

In the past 50 years, world food production has doubled due to the green revolution. It is assumed that the population will increase by 30% to about 10 billion in 2050. The increasing global population has raised the demand for food production in developing countries by 70%. There is a potential loss of crop of about 35% due to pre-harvest pests as well as the high loss of food chain in the environment. There is an increased intensity of crop protection due to a 15–20 times increase in the level of pesticides used. Around USD 40 billion has been spent on an annual production of three million tons of chemical pesticides in the global market. According to Bosch (1978), this increasing dependence on chemical pesticides has been called a "pesticide treadmill".

In India, the production of pesticides started with the establishment of a plant near Calcutta for benzene-hexa chloride production in 1952. After China, which is known as the second largest pesticides manufacturer in Asia is India, which ranks 12th globally. The pesticides have primary benefits involving direct gains from their use as well as secondary benefits, which are less immediate or less obvious than primary benefits.

8.2   LIMITATIONS OVER USE OF CONVENTIONAL PESTICIDES

Due to the persistent nature of pesticides, it is estimated that 200,000 people are killed every year worldwide due to pesticide poisoning, according to the World Health Organization (WHO). Therefore, a reduction in the usage of pesticides has been observed because of their carcinogenicity, high toxicity, cause of hormonal imbalance, residual problem, spermatotoxicity and longer degradation. Indiscriminate use of pesticides leads to the killing of beneficial organisms, for example, natural enemies and pollinators, disruption of the food chain, contamination of groundwater and secondary outbreaks of pests. Environmental pollution is also caused by the use of synthetic pesticides due to their non-biodegradable nature. The use of some pesticides has been banned from agriculture due to their adverse effects on the environment, for example, methyl bromide, which is also responsible for ozone layer deterioration resulting in climate change. In another case, dichloro-diphenyl-trichloroethane (DDT), having carcinogenic properties, has also been banned due to its high toxicity that further led to the poisoning of birds and marine as well as human life. Therefore, keeping safety measures and health issues in mind, there is an increasing demand for eco-friendly biopesticides in agriculture.

8.3   BIOPESTICIDES AND INTEGRATED PEST MANAGEMENT

Biopesticides are derived from natural materials or their products, for example, insects, microorganisms, nematodes, plants as well as semiochemicals. There are different types of biopesticides based on nature and origin, for example, botanicals, predators and pheromones. Since a large number of bioactive compounds are present in plants and microorganisms, they fall under major sources of biopesticides. These mainly include active compounds such as quinones, steroids, phenols, alcohols, alkaloids, terpenes and saponins. The microbial biopesticides mainly include some of the bacterial species viz. *Rahnella, Bacillus, Pseudomona, Xanthomona* and *Serratia* or fungi such as *Metarhizium anisopliae, Verticillium, Trichoderma* and *Beauveria*. Various types of biological control agents like pathogens, predators, parasitoids and insects have also been utilized as biopesticides.

8.3.1 TYPES, ORIGIN AND MODES OF ACTION OF BIOPESTICIDES AND THEIR EFFICACY

8.3.1.1 Some Traditional Insecticides

Pyrethrum: Pyrethrum is a commonly used insecticide which is also known for its fast knockdown aerosolic potential. It is obtained from flowers of *Tanacetum cinerariaefolium* (Asteraceae). In technical grade form, it is found to be less toxic, while in pure form it is moderately toxic to mammals.

Pyrethroids: Although they are prepared to mimic pyrethrum, these are toxic to the environment as compared to pyrethrin and are harmful for natural enemies, honey bees and fish, with a long-lasting impact.

Nicotine: These compounds and their related alkaloids, for example, nornicotine and anabasine, are extracted from tobacco and *A. aphylla*. These are synaptic poisons and mimic acetylcholine (a neurotransmitter) and show highly insecticidal effects with poisoning symptoms similar to organophosphate and carbamate.

Rotenone: This is extracted mainly from the roots of legumes of *Tephrosia*, *Derris* and *Lonchocarpus*. It is mainly sold for use in gardens and houses. In Southeast Asia, it is mainly used to paralyze fish. It is sold in a dust form for home use and contains 1%–5% active ingredients, while its liquid form (containing 15% total rotenoids and 8% rotenone) is used in gardens.

8.3.1.2 New Plant Insecticides

Neem: *Azadirachta indica* are mainly found in India though they are widely grown in America, Africa and Australia. Neem's efficacy has been mainly linked to azadirachtin activity, which causes mortality in different larval insects by insect growth regulator (IGR) activity. This happens because it disrupts the endocrine system, which involves a decrease in the level of ecdysteroid in haemolymph by blocking the release of prothoracicotropic hormone (PTTH). Neem has been found to be effective against various medically and veterinary important arthropods, insect pests as well as some plant diseases, for example, sweet potato whiteflies, western flower thrips, lice, bugs, cockroaches, mite, tick, fleas, flies gypsy moths, leaf miners and mealybugs.

8.3.1.3 Essential Oils

A variety of secondary metabolites, for example, alcohols, terpenes and aromatic compounds, have been produced by plants or pathogens to protect themselves from herbivore attack. The complex form of these volatile compounds mainly forms essential oil. Many plant families have been known for their insecticidal activities, for example, *Poaceae, Rutaceae, Lamiaceae, Zingiberaceae, Asteraceae, Apiaceae, Cupressaceae, Myrtaceae, Lauraceae* and *Piperaceae.* The volatile compounds from plants may also involve interplant communication, attracting various pollinators and protecting them from some diseases. These compounds may also interfere with the various metabolic, physiological and biochemical activities of insects that may alter the behavior of an insect. They may be neurotoxic in action because of their interference with the neuromodulator chloride channels, for example, GABA-gated channel. They have the potential to cause larval mortality as well as the ability to delay the development and suppress the emergence of the adult stage.

8.3.2   Microorganisms as Biopesticides

Microbial pesticides either occur naturally or are genetically engineered and are highly selective in nature as compared to chemical pesticides. The microbial biopesticides in wide use include bioherbicides (Phytophthora), biofungicides (*Bacillus*, *Pseudomonas*) and biopesticides (Bt). They mainly produce toxic metabolites against pests while simultaneously preventing the establishment of other microorganisms.

8.3.2.1   Bacteria

These biocontrol agents include pathogenic bacteria belonging to the *Bacillaceae, Enterobacteriaceae, Streptococcaceae, Pseudomonadaceae* and *Micrococcaceae* families. Though members of *Bacillaceae*, which include *Bacillus* sp., have received maximum attention from the scientific community, approximately 90% of the biopesticide market in the United States use *B. thuringiensis* (Bt) and its various strains as microbial pesticides. The synthesis of Cry proteins or δ-endotoxins having insecticidal properties is the principal characteristic of these bacteria. Due to their specificity as well as safety in the environment, they are more efficiently used over chemical pesticides in controlling insect pests. Some of the species and subspecies of *Bacillus* and their targeted pests are listed in Table 8.1.

TABLE 8.1
Species and Subspecies of *Bacillus* and Their Targeted Pests

Bacterial Strain	Targeted Pests
Bacillus thuringiensis israelensis	Blackfly and mosquito
B. moritai	Dipteran
B. thuringiensis kurstaki	Larva of Lepidoptera
B. thuringiensis tenebrionis	*Leptinotarsa decemlineata*
B. thuringiensis galleriae	Larva of Lepidoptera
B. thuringiensis aizawai	Larva of moth
B. popilliae	Japanese beetle grub
B. sphaericus	Larvae of mosquito

Sources: Kunimi, 2007; Kabaluk et al., 2010.

8.3.2.2   Viruses

Various entomogenous viruses have been isolated from various insect species of different insect orders. They are mainly of two types—inclusion viruses producing inclusion bodies in the host cells and non-inclusion viruses that do not produce them. Inclusion viruses are further divided into granulosis virus (mainly producing granular bodies) and polyhedron viruses (PV) that produce polyhedral bodies. Based on their presence in the nucleus or cytoplasm, polyhedroses are of two types—nuclear polyhedrosis viruses (NPV) and cytoplasmic polyhedrosis virus (CPV). Among all insect viruses, the most commercial ones belong to the baculovirus family. A number of baculoviruses (more than 600) have been successfully isolated from Hymenoptera, Lepidoptera and Diptera. Some of the viruses along with their hosts are listed in Table 8.2.

8.3.2.3   Fungi

Entomopathogenic fungi play a very important role as biological control agents in controlling the insect population. Globally, many commercial products are available that utilize less than 10 species of fungi. It has been found that the number of insects infecting fungi have grown to more than 90 genera and 700 species. Entomopathogenic fungi generally kill insects by attacking and infecting either through the integument, by ingestion, wounds or the trachea. The list of some entomopathogenic fungi and their target host insects are mentioned in Table 8.3.

8.3.2.4   Nematodes

The nematodes that occur naturally in soil and kill insects are known as entomopathogenic nematodes. They locate their host with the help of chemical signals or other means, for example, CO_2 response, vibration and so on. *Heterorhabditidae* and *Steinernematidae* families have been found to be more effective in the management of pests. These nematodes are very specific to their pests but are known to be nontoxic to humans. They mainly penetrate into the host via mouth, spiracles, anal region and intersegmental membranes of the cuticle and finally enter into the hemocoel. Some entomopathogenic nematodes which kill their host insects are listed in Table 8.4.

TABLE 8.2
List of Viruses and Their Respective Hosts

Nature of Virus	Respective Host
Cytoplasmic polyhedrosis virus	Cotton bollyworm
Pox virus	AmEPV
Granulosis virus	Croton caterpillar, tobacco splitworm, sugarcane shoot borer, rice leafroller
Nuclear polyhedrosis virus	Cotton bollyworm, tobacco cutworm, lawn armyworm, rice moth, tomato looper, hairy tussock moth, snake gourd semilooper

Source: Ramakrishnan and Kumar, 1976.

TABLE 8.3
List of Fungi and Their Respective Targeted Hosts

Fungi	Targeted Host Insect
Beauveria bassiana	Brown plant hopper, pod borer, Colorado potato beetle, citrus root weevil, cotton bollworms, mango mealy bug, codling moth, Japanese beetle, boll weevil, cotton leaf hopper, yellow stem borer, rice leaf folder, chinch bug etc.
Hirsutella thompsonii	Phytophagous mite
Metarhizium anisopliae	Grasshopper, termite, sugarcane hopper, locust, rhinoceros beetle, spittle bug etc.
Nomuraea rileyi	Castor semi-looper, cotton bollworm, tobacco caterpillar, cabbage looper
Verticillium lecanii	Whitefly, aphid, soft-scale insects
Pandora delphacis	Brown plant hopper and green leaf hopper of rice
Aschersonia aleyroides	Whitefly

Sources: Pawar and Singh, 1993; Zimmermann, 1993.

TABLE 8.4
List of Entomopathogenic Nematodes and Their Targeted Hosts

Name of Nematode	Host
Steinernema glaseri	Banana root borer, white grub
S. carpocapsae	Armyworm, cutworm, sod webworm, crane fly, chinch bug, banana moth, codling moth, peach tree borer, cranberry girdler, billbug, black vine weevil, shore fly
S. kraussei	Black vine weevil
S. scapterisci	Mole cricket (*Scapteriscus* spp.)
S. riobrave	Mole cricket, citrus root weevil
S. feltiae	Western flower thrip, fungus gnat, shore fly
Heterorhabditis zealandica	Scarab grub
H. bacteriophora	Scarab grub, flea beetle, corn root worm, cutworm, black vine weevil, citrus root weevil
H. megidis	Weevil
H. indica	Root mealybug, fungus gnat, grub
H. marelatus	Black vine weevil, white grub, cutworm

Source: Tofangsazi et al., 2015.

8.3.2.5   Protozoa

There is another diverse group of entomopathogenic protozoans referred to as microsporidians having around 1000 species that attack invertebrates, including insect species. They are slow acting and host-specific in nature. The infectious stage is spore forming; the spores are ingested by the host, further leading to pathogenicity. The main symptoms involve reduced feeding, fecundity and the longevity of the insect, for example, *Nosema pyrausta* infection causes larval mortality in the European corn borer and also reduces adult fecundity and longevity.

8.4 PRODUCTION, FORMULATION AND COMMERCIALIZATION OF BIOPESTICIDES

The commercialization of biopesticides has several barriers, for example, the production of plant resources first requires the standardization of extracts, the late response of biopesticides of plant origin, availability of competing products and regulatory approval compared with other products.

Some specific physical properties of essential oils, for example, increased molecular weight, low vapor pressure and high boiling point, become the biggest obstacles in their large production.

The method of extraction of botanical extracts is also related to its quality. Under field conditions, there are no standard guidelines or methods for testing efficacy. In the laboratory, due to the presence of a suitable substrate and material, the microorganisms multiply easily, but the conditions vary in the field. The products or active compounds could be identified with the help of techniques like gas chromatography–mass spectroscopy (GC–MS), thin-layer chromatography (TLC), high-performance liquid chromatography (HPLC) and gas chromatography (GC). The lack of such standardized techniques may lead to a substandard product. Moreover, the registration of the active products requires toxicity data, packaging date and type of formulation, which is not always easily available. It is very hard to invest in biopesticides due to the lack of a readily available market. The biggest challenge is their competition with synthetic pesticides because of their high costs. Moreover, lack of sufficient awareness, policymakers and stakeholders also pose hurdles in this field.

8.5 BIOFERTILIZER AND SUSTAINABLE AGRICULTURE IN INDIA

8.5.1 INTRODUCTION

A considerable level of pressure on agricultural land for food and other resources is increasing daily with the growing population in India and throughout the world. Decadal growth of 17.64% has been observed during 2011 (15th census) in the Indian population, that is, 121 million, including 68.84% rural population. In Asia, it is estimated that there is a 0.48% reduction in the number of poor people because of a 1% increase in the production of crops. It is further estimated that a 0.4% and 1.9% decline in poverty in India—short and long term, respectively—has been observed because of a 1% rise in agricultural value per hectare. Generally, fertilizers and pesticides are the major inputs necessary for the production of crops. Pesticides work as medicine, while fertilizers act as food for plants. The green revolution has increased crop productivity and satisfied the population's need in India post-independence. This revolution started with the production of nitrogen fertilizers produced by means of the Haber–Bosch process, which further increases the use of fertilizers in the agricultural field. The indiscriminate use of fertilizers of chemical origin has adversely affected microflora and fauna along with soil texture and productivity. It has been found that there is an increase in nitrogen, phosphorus and potassium application to 61.4:18.7:1 and 61.7:19.2:1 in Haryana as well as Punjab, respectively, from ideal ratios of 4:2:1. Ultimately, this distortion of ratio leads to impaired health of human beings and animals.

Not all of the nutrients supplemented in chemical fertilizers are taken up by plants; some of them are mixed with water bodies. Therefore, there is a need to implement cost-effective, selective and eco-friendly nutrients for sustainable agriculture where biofertilizer could be the best option.

8.5.2 BIOFERTILIZERS

"Biofertilizers", called as "microinoculants", derived from the term "biological fertilizer" where biological depicts the use of a living organism that colonizes in the plant's rhizosphere and speeds up the growth by increasing the accessibility and uptake of minerals and various nutrients to the host. Recently, zinc and sulfur solubilizers like *Thiobacillus* sp., potash mobilizers like *Frateuri aaurentia*, and Mn solubilizer fungi like *Pencillium citrinum* have been identified for use at the commercial level.

8.5.3 HISTORY OF BIOFERTILIZER USE AND PRODUCTION SCENARIO IN INDIA

In 1920, N.V. Joshi started a study on biofertilizers. *Rhizobium* was first isolated from many cultivated legumes and has been extensively studied for improving the production of crops. Some milestones in this area of research in India are mentioned in Table 8.5. Some biofertilizers can be considered established, for example, *Rhizobium* and blue-green algae (BGA); while others are in an intermediate stage of development, for example, *Azolla, Azotobacter, Azospirillum*. Some marketed biofertilizers in India are available to farmers viz. nitrogen fixers, for example, *Azotobacter, Rhizobium, Azospirillum, Acetobacter, Bradyrhizobium, Azolla* and BGA; phosphate mobilizers, for example, VA-mycorrhiza (VAM) like *Glomus*; phosphorus solubilizers, for example, *Bacillus, Aspergillus* and *Pseudomonas*; K-solubilizers, for example, *Frateuria aurantia*; plant growth-promoting biofertilizers, for example, *Pseudomonas* sp.; and silicate solubilizers, for example, *Thiobacillus thiooxidans*. There is an approximate requirement of 550,000 metric tons of biofertilizers in India, which was increased by 50,000–60,000 tons by 2020 based on crop area. Nowadays, biofertilizer industries have been boosted by providing subsidies. Also, the biofertilizer production units achieving efficient productivity have been recognized with a national productivity award by the government. The milestones in the history of biofertilizers in India and annual biofertilizer production (in tons) in India are included in Tables 8.5 and 8.6.

8.5.4 BIOFERTILIZERS TYPES

Biofertilizers include substances that contain living microorganisms or latent cells that increase the nutrients of the host and increase their growth via converting phosphorus, nitrogen etc. through various biological processes, for example, nitrogen

TABLE 8.5
Milestones in the History of Biofertilizers in India

Year	Events
1920	First study on symbiosis of legumes and *Rhizobium* (N.V. Joshi)
1934	Production of *Rhizobium* inoculant documented by M.R. Madhok
1939	Nitrogen fixation by blue-green algae discovered by P.K. Dey
1956	Production of first commercial biofertilizer
1957	Phosphate solubilization by microorganisms by Sen and Pal
1958	First attempt to standardize quality of legume inoculant by Sankaran
1960	First-time isolation of new non-symbiotic nitrogen-fixing organism *Derxiagummosa*
1964	Spurt in demand of biofertilizer in M.P. for soyabean
1968	Establishment of All India pulse improvement project and soyabean project by Indian Council of Agricultural Research (ICAR)
1969	Use of Indian peat as carrier by V. Iswaran
1975	J.N. Dube reported coal as alternate carrier to peat
1976	Indian standard specification for *Rhizobium*
1977	ISI marks for *Rhizobium* in use
1979	Initiation of All India coordinated project on biological nitrogen fixation
1983	National project on development and use of biofertilizer setup by Ministry of Agriculture, India
1985	First award of National Productivity on Biofertilizer
1988	Setup of national facility center for blue-green algae at Indian Agricultural Research Institute

Source: Panda, 2011.

TABLE 8.6
Biofertilizer Production in India (in tons)

Total Production (Zone)	Year						
	2008–09	2009–10	2010–11	2011–12	2012–13	2013–14	2014–15
East Zone	661.270	561.370	887.310	1276.700	1604.800	2846.920	3210.270
West Zone	3601.683	5563.770	12,960.720	13,566.220	10,138.100	19,547.565	23,580.414
North Zone	2114.237	2332.420	2485.480	12,183.010	12,212.170	11,914.860	13,512.696
South Zone	18,525.804	11,162.610	20,660.600	11,674.100	22,261.800	30,764.030	39,982.630
Grand Total Production	**24,902.994**	**19,620.170**	**36,994.110**	**38,700.030**	**46,216.870**	**65,073.375**	**80,286.010**

Source: National Centre of Organic Farming, Department of Agriculture & Cooperation DAC, 2013, 2014 and 2015.

TABLE 8.7
Different Types of Biofertilizers Based on Their Functions and Nature

Sr. No.	Groups	Examples
1	**Plant growth-promoting rhizobacteria**	
	Pseudomonas	*P. fluorescens*
2	**P-mobilizing biofertilizers**	
	Ectomycorrhiza	*Boletus* sp., *Laccaria* sp., *Amanitia* sp., *Pisolithus* sp.
	Orchid mycorrhiza	*Rhizoctonia solani*
	Arbuscular mycorrhiza	*Acaulospora* sp., *Glomus* sp., *Sclerocystis* sp., *Gigaspora* sp., *Scutellospora* sp.
	Ericoid mycorrhiza	*Pezizella*
3	**Nitrogen-fixing biofertilizers**	
	Free-living	*Clostridium, Azotobacter, Klebsiella, Anabaena*
	Symbiotic	*Anabaena-Azolla, Rhizobium, Frankia*
	Associative symbiotic	*Azospirillum*
4	**Biofertilizers for micronutrients**	
	Ciliates and Zn solubilizers	*Bacillus* sp.
5	**Phosphate-solubilizing biofertilizers**	
	Bacteria	*Bacillus subtilus, P. striata, B. circulans*
	Fungi	*Aspergillus awamori, Penicillium* sp.

Source: Singh et al., 2014a,b.

fixation. Various types of biofertilizers based on their different kinds of nature and functions are mentioned in Table 8.7.

Some important biofertilizers are explained next.

8.5.4.1 Nitrogen-Fixing Bacteria

a. *Rhizobium*

This belongs to the Rhizobiaceae family and is mainly symbiotic in nature. It has the capability to fix nitrogen 50 to 100 kg/ha approximately. The association of *Rhizobium* and leguminous plants is very crucial in the agriculture system.

Rhizobium is a gram-negative soil bacterium and has a potential role in the fixation of atmospheric nitrogen into ammonia in symbiotic association with leguminous plants and some non-leguminous ones as well, for example, Parasponia.

b. *Azospirillum*

This is associative and heterotrophic in nature and belongs to the family Spirilaceae. Its nitrogen-fixation capacity ranges from 20 to 40 kg/ha.

There are a number of species under this genus, but the important ones are *A. brasilense* and *A. lipoferum*. It mainly shows associative symbiosis with those plants showing the C4 pathway of photosynthesis, since they fix nitrogen using malic and aspartic acid.

c. *Azotobacter*

The nitrogen-fixation process mainly occurs in free-living *Azotobacter* in soil. It is mainly present in neutral or alkaline soil. The most common species is *A. chroococcum*, which is found in arable soils. It acts as a beneficial tool in the area of sustainable agriculture as it enhances soil properties and microbial diversity. It can fix atmospheric nitrogen in the soil up to 20 to 25 kg/ha.

8.5.4.2 Phosphate Solubilizers

Concerning applications of fertilizers in inorganic form during the production of the crop, around 80% of an applied fertilizer is converted into a non-available form with the help of calcium, aluminum, iron (Ca-P), (Al-P), (Fe-P) and others. *Bacillus*, Pseudomonas, and fungi dominate in soil among P-solubilizing microorganisms in the soil. In India, phosphate-solubilizing bacteria has the potential to utilize low-grade rock phosphate into an enriched phospho-compost which has ultimately will reduce the P-fertilizer import in the country.

8.5.4.3 Potassium Solubilizers

The potassium solubilization from the main potassium ore has been done with the help of various soil microorganisms that produce organic substances while solubilizing the potassium (K). Crop growth and production is also increased by different types of growth promoters like indole acetic acid, gibberellic acid and amino acids produced by microorganisms.

8.5.4.4 Blue-Green Algae

Cyanobacteria, commonly known as blue-green algae (BGA), are widely distributed in the world and play a very important role in agriculture by increasing soil fertility. They are free-living organisms that are symbiotically associated with *Azolla* and increase soil fertility. In the paddy field, BGA has been utilized as a major component of the fixation of nitrogen. After water, nitrogen is the second most important factor for the growth of plants whose deficiency is met by using biofertilizer. In comparison to the temperate regions, BGA is mainly found in tropical and subtropical regions.

8.6 MARKETING AND REGULATORY PROBLEMS

Biofertilizers contain living organisms which have restricted shelf life, which may lead to problems in their marketing. Some of the biofertilizers that mainly occur in powder form have a limited shelf life and are difficult to store and transport. Moreover, there is no appropriate standardization in their labeling and packing. The packets become useless when they are transported to villages in spoiled or outdated form, as microorganisms die very quickly. The activity of microorganisms is generally very specific to either the crop or a particular location. Therefore, it is very important to select the strains for the specific locations to survive in adverse conditions. The lack of a good microbiologist or skilled person in biofertilizer production units may lead to substandard biofertilizers. In addition, alternative carriers to good-quality peat, like lignite and charcoal, use either unsterilized or contaminated forms and may cause problems for the working of good-quality biofertilizers.

Generally, biofertilizers available in markets have a low microorganism count and are in contaminated form; therefore, they cannot express their actual potential.

Recently, specifications for *Rhizobium* and *Azotobacter* are available in the Indian Standard Institute (ISI), while formulations have also been prepared for *Azospirillum* and phosphobacteria. However, there is currently no regulatory act available for biofertilizer production in India.

8.7 CONCLUSION

Although the adoption of biopesticides has been facing many challenges, they are still found to be the best alternative to conventional pesticides. The indiscriminate use of chemical pesticides has raised many problems because of their adverse effects on beneficial organisms as well as on the environment. Due to biopesticides' persistence in the environment, less toxicity and biodegradability, they are considered the best alternative to chemical insecticides. The overexploitation of natural resources due to the growing population may create a food crisis in the future, so there is an urgent need to conserve agricultural crop production without deteriorating the natural resources as well as the environment. The use of low-cost and eco-friendly biofertilizers over chemical fertilizers must be the target for future research, keeping in mind the environmental aspects. The investment of more funds in developing new strains of biofertilizers, maintaining quality and increasing shelf life is necessary for their propagation and use. Wide publicity is needed through media with the help of government and non-government organizations to cope with the present situation.

9 Biosensors for Environmental Monitoring

Alok Jha

CONTENTS

9.1 INTRODUCTION

Biosensors are analytical tools used for the analysis of biomaterials from a mixture of analytes to get information about their composition, structure and function by converting a biological response into an electrical signal. These tools consist of a biological recognition element spread over a signal transducer and measure the

DOI: 10.1201/9781003131427-9

concentration of a specific analyte in the sample mixture. Biosensors allow detection of a broad spectrum of analytes in complex samples and show great promise in different areas of investigation, for example, clinical diagnostics, food analysis, bioprocess and environmental monitoring as given below. Depending on the method of signal transduction, the biosensors may be classified into six groups, including electrochemical, mass, magnetic, optical, micromechanical and thermal.

Types of Biological Recognition Element	Biosensors
Enzymes	Enzyme electrode
Proteins	
Antibodies	Immunosensors
DNA	DNA sensors
Organelles	
Microbial cells	Microbial sensors
Plant and animal tissues	
	Potentiometric
Electrochemical	Amperometric
	Voltametric
Electrical	Surface conductivity
	Electrolyte conductivity
Optical	Fluorescence
	Adsorption
Mass sensitive	Reflection
	Resonance frequency of piezo crystals
Thermal	Heat of reaction
	Heat of adsorption

A typical biosensor consists of the following components:

- *Analyte*: This is the substance of interest to be detected. For example, glucose is an "analyte" in a biosensor designed to detect glucose.
- *Bioreceptor*: This specifically recognizes the analyte. Some examples of bioreceptors include enzymes, cells, aptamers, deoxyribonucleic acid (DNA) and antibodies. The interaction of bioreceptor with the analyte is associated with the process of signal generation (in the form of light, heat, pH, charge or mass change, etc.), which is also known as biorecognition.
- *Transducer*: The transducer converts one form of energy into another. The role of a transducer in a biosensor is to convert the effect of biorecognition into a quantifiable signal. The process of energy conversion is also known as a signal generation. A transducer produces optical or electrical signals proportional to the amount of analyte–bioreceptor interactions.

- *Electronics*: A biosensor has an inbuilt system that processes the transduced signal and setup for display. It comprises complex electronic circuitry that executes conditioning of signals such as amplification and conversion of signals into the digital form. The display unit of the biosensor then quantifies the processed signals.
- *Display*: The display unit consists of a user interpretation system such as the liquid crystal display of a computer or a directly connected printer that generates numerical values or curves. This part of a biosensor often consists of a combination of hardware and software that generates results in a user-friendly mode. The output signal on display can be numeric, graphic, tabular or an image, depending on the demand of the end user.

9.2   CHARACTERISTICS OF A BIOSENSOR

The performance of the biosensor largely depends on the optimization of certain characteristics:

a. *Selectivity*

Selectivity is the most important feature of a biosensor, which is the ability of a bioreceptor to detect a specific analyte in a sample containing other admixtures and contaminants. The best example of selectivity is the interaction of an antigen with the antibody in which antibodies act as bioreceptors and are immobilized on the surface of the transducer. A solution (commonly a buffer) containing the antigen is then exposed to the transducer where antibodies interact only with the antigens.

b. *Reproducibility*

Reproducibility is the ability of the biosensor to generate identical responses for a repeat or similar experimental set-up. The reproducibility is defined by the precision and accuracy of the transducer and electronic circuitry used in a biosensor. Precision is the quality of the sensor to provide similar results every time a sample is measured, and accuracy represents the capacity of the sensor to provide a mean value approximate to the real value when a sample is measured repeatedly. Reproducible signals provide high dependability and robustness to the interpretation of the biosensor response.

c. *Stability*

Stability is the most crucial feature for a biosensor application because it requires steps involving long incubation or continuous monitoring. The stability of a biosensor can be affected by the temperature-sensitive response of transducers and electronics. Therefore, to ensure a stable response of the sensor, appropriate regulation of electronics is required. Another factor, that is, the affinity of the bioreceptor, can also influence the stability of a biosensor, which is the degree to which the analyte binds to the bioreceptor. Another factor that might affect the stability of measurement is the susceptibility of the biosensor to degradation over a period.

d. *Sensitivity*

The minimum amount of analyte that can be detected by a biosensor defines its limit of detection (LOD) or sensitivity. In several medical and environmental monitoring applications, a biosensor is required to detect analyte concentration of as low as ng/ml to fg/ml to confirm the presence of traces of analytes in a sample. For instance, a prostate-specific antigen (PSA) concentration of 4 ng/ml in the blood is associated with prostate cancer, for which doctors suggest biopsy tests. Hence, sensitivity is an important property of a biosensor.

e. *Linearity*

Linearity is the characteristic that shows the accuracy of the measured response (for a set of measurements with different concentrations of the analyte) to a straight line, mathematically depicted as $y = mc$, where c is the concentration of the analyte, y is the output signal, and m is the sensitivity of the biosensor. The linearity of the biosensor depends on the resolution of the biosensor and concentration range of analytes. The resolution of the biosensor is defined as the smallest change in the concentration of an analyte that is required to bring a change in the response of the biosensor. Depending on the application, a good resolution is required, as most biosensor applications involve not only detection of the analyte but also the measurement of the concentrations of the analyte. Another term associated with linearity is a linear range, which is defined as the range of analyte concentrations for which the response of biosensor changes linearly with the concentration.

9.3   TYPES OF BIOSENSORS

The key part of a biosensor is the transducer, which uses the physical change associated with the reaction for conversion into a quantifiable signal. It can be:

1. The heat absorbed or liberated in the reaction (calorimetric biosensors)
2. Changes in the distribution of charges causing an electrical potential to be produced (potentiometric biosensors)
3. The transfer of electrons produced in a redox reaction (amperometric biosensors)
4. The emission of light during the reaction or a difference in the absorbance of light between the reactants and products (optical biosensors)
5. Effects due to the mass of the reactants or products (piezo-electric biosensors)

There are three so-called generations of biosensors: First-generation biosensors in which the normal product of the reaction diffuses to the transducer and causes the electrical response; second-generation biosensors which involve specific "mediators" between the reaction and the transducer in order to generate improved response; and third-generation biosensors where the reaction itself causes the response and no product or mediator diffusion is directly involved.

9.3.1 ELECTROCHEMICAL BIOSENSORS

Electrochemical biosensors are simple devices which measure electronic current by changes in conductance carried by bio-electrodes (Figure 9.1). The transducer often produces a low electrical signal superimposed on a relatively high and noisy baseline. Here, noise means a random high-frequency signal component that is generated due to electrical interference or within the transducer itself. The signal processing normally involves subtracting a "reference" baseline signal, derived from a transducer without any biocatalytic membrane, from the sample signal, amplifying the resultant signal difference and electronically filtering (smoothing) out the unwanted signal noise. The relatively slow biosensor response considerably eases the problem of electrical noise filtration. The analogue signal produced at this level may be released directly but is usually converted to a digital signal and passed to a microprocessor stage where the data is processed, converted to units of concentration and output to a display device or data storage.

9.3.2 CALORIMETRIC BIOSENSORS

Many enzyme-catalyzed reactions are exothermic, that is, generating heat (Table 9.1), which may be used as a basis for measuring the rate of reaction and also the concentration of the analyte. The temperature changes are usually dictated by employing thermistors located at the entrance and exit of small packed bed columns containing immobilized enzymes within an environment of constant temperature (Figure 9.2). Under such closely controlled conditions, up to 80% of the heat generated in the reaction may be recorded as a temperature change in the sample flow. This may be simply calculated from the amount of reactant that is involved in the reaction and the corresponding change in enthalpy.

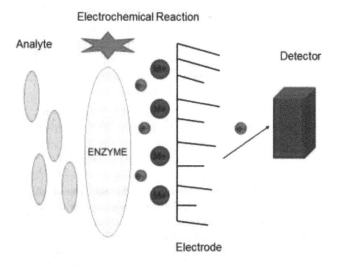

FIGURE 9.1 Electrochemical biosensor. (From: www.watelectrical.com/biosensors-types-its-working-and-applications/.)

TABLE 9.1

Heat Output (Molar Enthalpies) of Enzyme-Catalyzed Reactions

Reactant	Enzyme	Heat Output $-\Delta H$ (kJ^x mol^{-1})
Cholesterol	Cholesterol oxidase	53
Esters	Chymotrypsin	4–16
Glucose	Glucose oxidase	80
Hydrogen peroxide	Catalase	100
Penicillin G	Penicillinase	67
Peptides	Trypsin	10–30
Starch	Amylase	8
Sucrose	Invertase	20
Urea	Urease	61
Uric acid	Uricase	49

Source: Adapted from www1.lsbu.ac.uk/water/enztech/calorimetric.html.

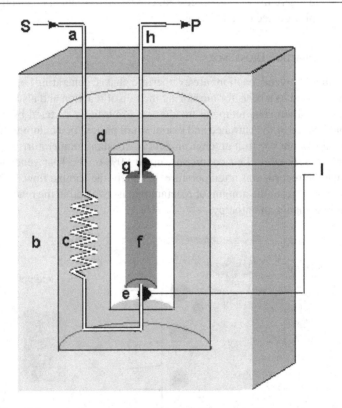

FIGURE 9.2 Schematic diagram of a calorimetric biosensor. The sample stream S passes through (a) and the outer insulated box (b) to the heat exchanger (c) within an aluminium block (d). From there, it flows through the reference thermistor (e) and into the packed bed bioreactor (f, with 1 ml volume), containing the biocatalyst, where the reaction occurs. The change in temperature is determined by the thermistor (g) and the solution through (h) moves to waste. External electronics (l) determine the difference in the resistance, and hence temperature, between the thermistors. (From: www1.lsbu.ac.uk/water/enztech/calorimetric.html.)

For most applications, the sensitivity (10^{-4} M) and range (10^{-4} to 10^{-2} M) of thermistor biosensors are quite low, although sensitivity can be increased using more exothermic reactions (e.g. catalase). The low sensitivity of the system can be increased substantially by increasing the heat output in the simplest case by linking together several reactions in a reaction pathway, all of which contribute to the heat output.

9.3.3 POTENTIOMETRIC BIOSENSORS

Potentiometric biosensors use ion-selective electrodes to transduce the biological reaction into an electrical signal. It consists of an immobilized enzyme membrane surrounding the probe of a pH meter where hydrogen ions are generated or absorbed due to catalyzed reactions. The reactions occurring in the vicinity of thin sensing glass membrane causes a change in pH which can be read directly from the pH meter. The typical use of such electrodes is that the electric potential is determined at very high impedance, allowing effectively zero current flow without any interference with the reaction.

There are three types of ion-selective electrodes which are used in biosensors:

1. Glass electrodes for cations (e.g. normal pH electrodes) in which the sensing element is a very thin hydrated glass membrane which generates a transverse electrical potential due to the concentration-dependent competition between the cations for specific binding sites. The composition of the glass determines the selectivity of this membrane. The sensitivity to H^+ is greater than that for NH_4^+.
2. Glass pH electrodes coated with a gas-permeable membrane selective for CO_2, NH_3 or H_2S. The diffusion of the gas through this membrane causes a change in pH of a sensing solution between the membrane and the electrode, which is then determined.
3. Solid-state electrodes in which the glass membrane is replaced by a thin membrane of a specific ion conductor made from a mixture of a silver sulphide and a silver halide. The iodide electrode is useful for the determination of I^- in the peroxidase reaction and also responds to cyanide ions.

9.3.4 AMPEROMETRIC BIOSENSORS

The function of amperometric biosensors largely depends on the application of a potential between two electrodes and subsequent generation of the current. They usually have response times, dynamic ranges and sensitivities similar to the potentiometric biosensors. The simplest amperometric biosensors in common usage include the Clark oxygen electrode (Figure 9.3) that consists of a platinum cathode at which oxygen is reduced and a silver/silver chloride as reference electrode. When a potential of -0.6 V, relative to the Ag/AgCl electrode, is applied to the platinum cathode, a current proportional to the oxygen concentration is produced. Commonly both electrodes are bathed in a solution of saturated potassium chloride and separated from

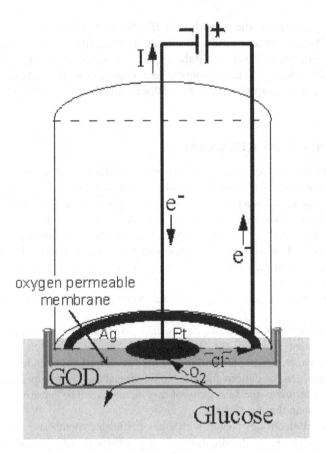

FIGURE 9.3   Schematic diagram of a simple amperometric biosensor. A potential is applied between the central platinum cathode and the annular silver anode. This generates a current (I) which is carried between the electrodes by means of a saturated solution of KCl. This electrode compartment is separated from the biocatalyst (glucose oxidase, GOD) by a thin plastic membrane, permeable only to oxygen. The analyte solution is separated from the bio-catalyst by another membrane, permeable to the substrate(s) and product(s). This biosensor is normally about 1 cm in diameter but has been scaled down to 0.25 mm in diameter using a Pt wire cathode within a silver-plated steel needle anode and utilizing dip-coated membranes. (From: www1.lsbu.ac.uk/water/enztech/amperometric.html.)

the bulk solution by an oxygen-permeable plastic membrane (e.g. Teflon, polytetra-fluoroethylene). The following reactions occur:

$$\text{Ag anode } 4Ag^0 + 4Cl^- \rightarrow 4AgCl + 4e^-$$
$$\text{Pt cathode } O_2 + 4H^+ + 4e^- \rightarrow 2H_2O$$

A typical application for this simple type of biosensor is the determination of glucose concentrations using an immobilized glucose oxidase membrane. The reaction results in a decrease in the oxygen concentration as it diffuses through the biocatalytic membrane to the cathode, which can be detected by a reduction in the current between the electrodes. Other oxidases may be used for the analysis of their substrates (e.g. alcohol oxidase, D- and L-amino acid oxidases, cholesterol oxidase, galactose oxidase, and urate oxidase).

The major problem with these biosensors is their dependence on the concentration of dissolved oxygen. This may be overcome by the use of "mediators" which transfer the electrons directly to the electrode, bypassing the reduction of the oxygen co-substrate.

In order to be generally applicable, these mediators must possess a number of useful properties:

1. They must react quickly with the reduced form of the enzyme, whereas the reduced form of the mediator should not readily react with oxygen.
2. They must be sufficiently soluble, in both the oxidized and reduced forms, to be able to rapidly diffuse between the active site of the enzyme and the electrode surface. Essentially, the solubility of the mediator should not be much to cause significant loss of the mediator from the biosensor's micro-environment to the bulk of the solution.
3. The overpotential for the regeneration of the oxidized mediator at the electrode should be low and independent of pH.

9.3.5 OPTICAL BIOSENSORS

In the optical biosensors, determining changes in absorption of light between the reactants and products of a reaction or measuring the light output by a luminescent process is possible. The former usually involves the use of widely established colorimetric test strips that are disposable single-use cellulose pads infused with enzyme and reagents. The most common use of this technology is for whole-blood monitoring in diabetes control, in which the strips include glucose oxidase, horseradish peroxidase (EC 1.11.1.7) and a chromogen (e.g. o-toluidine or 3,3',5,5'-tetramethylbenzidine). The hydrogen peroxide, produced by the aerobic oxidation of glucose, oxidizes the weakly colored chromogen to a highly colored dye.

$$\text{chromogen (2H)} + H_2O_2 \xrightarrow{\text{peroxidase}} \text{dye} + 2H_2O$$

The evaluation of the dyed strips is best achieved by the use of portable reflectance meters, although direct visual comparison with a colored chart is often used. A wide variety of test strips involving other enzymes is commercially available. A most promising biosensor involving luminescence uses firefly luciferase (*Photinus-luciferin* 4-monooxygenase [ATP-hydrolysing], EC 1.13.12.7) to detect the presence of bacteria in food or clinical samples. Bacteria are specifically lyzed and the ATP released (roughly proportional to the number of bacteria present) reacts with

D-luciferin and oxygen in a reaction which produces yellow light in high quantum yield.

$$ATP + D\text{-luciferin} + O_2 \xrightarrow{\text{luciferase}} \text{oxyluciferin} + AMP + \text{pyrophosphate} + CO_2 + \text{light} \left(562 \text{ nm}\right)$$

The light produced may be detected photometrically by use of high-voltage, and expensive, photomultiplier tubes or low-voltage in expensive photodiode systems. The sensitivity of the photomultiplier-containing systems is somewhat greater ($<10^4$ cells ml^{-1}, $<10^{-12}$ M ATP) than the simpler photon detectors which use photodiodes. Firefly luciferase is a very expensive enzyme, only obtainable from the tails of wild fireflies. Use of immobilized luciferase greatly reduces the cost of these analyses.

9.3.6   OPTICAL BIOSENSORS FOR BLOOD GLUCOSE

For diabetes patients, monitoring blood glucose level is very important. In this type of biosensor, a simple technique is used—paper strips saturated with the reagents, for example, glucose oxide, horseradish peroxidase and a chromogen. Using the portable reflectance meter, the intensity of the color of the dye can be measured. The glucose strips, in which the calorimetric test strips of cellulose covered with the suitable enzymes and reagents can be used at home for the measurement of blood and the urine parameters, are very popular worldwide. Other optical fibre biosensors, which measure pCO_2, are used in critical care and in surgical monitoring.

9.3.7   IMMUNOSENSORS

Biosensors may be used in conjunction with enzyme-linked immunosorbent assays (*ELISA*). The principles behind the ELISA technique to detect and amplify an antigen-antibody reaction; the amount of enzyme-linked antigen bound to the immobilized antibody is determined by the relative concentration of the free and conjugated antigen and quantified by the rate of an enzymatic reaction. Enzymes with high turnover numbers are used in order to achieve rapid response. The sensitivity of such assays may be further enhanced by utilizing enzyme-catalyzed reactions which give an intrinsically greater response; for example, those which produce highly colored, fluorescent or bioluminescent products. Assay kits using this technique are now available for various types of analyses.

Recently, ELISA techniques have been combined with biosensors to form *immunosensors*, in order to increase their range, speed and sensitivity. In a simple immunosensor configuration, the biosensor merely replaces the traditional colorimetric detection system. However, more advanced immunosensors are being developed, which rely on the direct detection of antigen bound to the antibody-coated surface of the biosensor. Piezoelectric and field-effect transistor (FET)-based biosensors are particularly suited to such applications.

9.3.8   Aptasensors

Aptamers (derived from the Latin *aptus*, meaning to fit) are artificial nucleic acid (DNA or RNA) ligands which can be selected from combinatorial libraries of synthetic nucleic acids, possessing specific binding characteristics to their targets. Using the systematic evolution of ligands (SELEX) process, aptamers can be isolated from randomly synthesized pools of RNA or DNA. Numerous high-affinity and highly specific aptamers have been generated against a miscellany of target molecules, including small organics, peptides, proteins and even complex targets such as the whole cell. In addition, aptamers can be selected against non-immunogenic and toxic targets, which make the technique superior to antibodies, enabling the design and synthesis of probes or capture molecules for such targets.

Biosensors that employ aptamers as a recognition element are called aptasensors. DNA and RNA aptamers can be modified chemically to undergo analyte-dependent conformational changes. For instance, aptamers can be modified for immobilization purposes and also incorporate particular reporters, without influencing their affinity, which has aided a number of design methods. Aptamers can also be engineered to withstand repeated cycles of denaturation and renaturation; this opens up a possibility to regenerate the immobilized biocomponent function for reuse. In addition, they can be easily labelled for their use in diagnostics.

9.3.8.1   Signal Transduction

Development of aptasensors has been carried out with various detection strategies, including label-free methods, such as surface plasmon resonance (SPR) and quartz crystal microbalance (QCM) measurements, and other methods often requiring labels, such as electrochemistry, fluorescence, chemiluminescence and field effect transistors, which have been reported on and have enhanced the progress of this field. Alternatively, sensors often utilize labelled target molecules or secondary capture agents (in the case of a sandwich-type assay) which themselves contain some forms of label or reporter molecule. There are mainly two types of aptasensors: electrochemical and optical.

9.3.8.2   Electrochemical Aptasensors

A typical electrochemical aptasensor consists of an electrode surface as the platform to immobilize a biological sensing aptamer, for which the analyte-binding event is monitored based on electrochemical current variations. Electrochemical transduction has the advantage of high sensitivity, which can be enhanced by attaching biocatalytic labels to the aptamer-target complexes to amplify the detection signal, can be readily miniaturized and has a low cost of production since they do not require expensive optical instruments. Additionally, it is possible to use label-free and reusable detection systems.

9.3.8.3   Optical Aptasensors

Optical aptasensors include aptamers labelled with fluorescence, luminophore, enzyme, nanoparticles or aptamer with label-free detection systems (e.g. SPR, optical resonance), but most of the methods are based on a labelling approach. Some targets can be optically detected using not only colorimetry, chemiluminescence or the most developed fluorescence mode but also more recent non-conventional optical methods such as surface plasmon-coupled directional emission (SPCDE).

Different label-free techniques have recently been shown to be appropriate for developing aptasensors or aptamer-based microarrays, such as SPR, diffraction grating, evanescent-field-coupled (EFC) waveguide-mode, optical resonance and Brewster angle straddle interferometry (BASI). SPR biosensors equipped with optical fibres can be used as a cost-effective and relatively simple-to-implement alternative to well-established biosensor platforms for monitoring biomolecular interactions *in situ* or possibly *in vivo*.

9.3.9   BIOSENSORS FOR ENVIRONMENTAL MONITORING

Biosensors such as immunosensors, aptasensors, genosensors and enzymatic biosensors have been used for the detection and monitoring of various environmental pollutants using recognition elements, for example, antibodies, aptamers, nucleic acids and enzymes, respectively.

9.3.9.1   Pesticides

Due to their evidential presence in the environment, pesticides are considered as the most critical environmental pollutants. For example, the organophosphorus insecticides are extensively used in agriculture and also represent a class of pesticides with major environmental concern due to their high toxicity.

9.3.9.1.1   Organophosphorus Pesticides

Disposable amperometric enzymatic (acetylcholinesterase) biosensors are used for the detection of organophosphorus insecticides, for example paraoxon, in which cysteamine is assembled on gold screen-printed electrodes as a monolayer. This type of biosensor shows high sensitivity and a high linear range with a low limit of detection; this kind of performance is attributed to the highly oriented enzyme immobilization in the form of a self-assembled monolayer. Recently, different types of biosensors based on gold nanorods and colorimetric biosensor such as amperometric acetylcholinesterase biosensors that employ iodine and starch color reaction and multi-enzymes (acetylcholinesterase and choline oxidase) have been used for the detection of paraoxon in real water samples. Another organophosphorus pesticide, methyl parathion, has been detected by using a sensitive and selective enzymatic biosensor that uses hydrolase and a uniform nanocomposite based on magnetic Fe_3O_4 and gold nanoparticles of diameters 120 nm and 13 nm, respectively. The cysteamine can also be used as a linker. This type of biosensor can be used repeatedly, and the advantage of using hydrolase is that it remains non-poisonous in the presence of organophosphorus pesticide. The potential of a biosensor to detect methyl parathion in natural water samples has been compared with other analytical techniques, for example, solid-phase microextraction-gas chromatography/mass spectrometry (SPME-GC/MS). A disposable enzymatic biosensor comprised of iridium oxide nanoparticles with tyrosinase enzyme and screen-printed carbon electrodes has been used for the detection of chlorpyrifos in river water samples. Recently, an electrochemical aptasensor based on a composite film (carbon black and graphene oxide–Fe_3O_4) has been proposed for the detection of chlorpyrifos in vegetable samples, which can obtain a detection limit in picomole range. This low limit of detection can be achieved by amplifying the current signal of the aptasensor because functionalized carbon black has a high specific surface area; also its dispersibility, electrical conductivity as well as the adeptness to capture more graphene oxide–Fe_3O_4 enhances its performance. The analytical performance of the aptasensor for chlorpyrifos detection is better in comparison to the enzymatic biosensors because of the lower limit of detection.

9.3.9.1.2   Other Pesticides

Acetamiprid has been detected in real environmental samples, for example, fresh surface soil samples by colorimetric aptasensors and water samples by impedimetric aptasensors. This type of biosensor uses gold nanoparticles, multi-walled carbon nanotubes (MWCNT) and reduced graphene oxide nanoribbons as composite for supporting the acetamiprid aptamer at the electrode surface, which can be responsible for higher electron transfer and better analytical performance. A similar limit of detection (33 fM) is obtained by an aptasensor based on silver nanoparticles anchored on nitrogen-doped graphene oxide nanocomposite fabricated for the detection of acetamiprid in wastewater samples. The low limit of detection may be observed due to excellent electrical properties and the large surface area of silver nanoparticles and nitrogen-doped graphene, which display more effective electron transfer and high loading capacity than the nanomaterials alone. In addition, the nanocomposite constitutes ideal support for the immobilization of aptamer, which promotes the amplification of response signal and further detection of acetamiprid with high sensitivity. A label-free electrochemical (voltammetric) immunosensor based on gold nanoparticles and a disposable and label-free electrochemical immunosensor based on field-effect transistor with single-walled carbon nanotubes (SWCNT) have been used for the detection of atrazine in environmental samples such as crop samples and seawater/river water samples, respectively. The FET immunosensor displays a lower limit of detection (0.001 ng mL^{-1}) in comparison to voltammetric immunosensor (0.016 ng mL^{-1}). Recently, a highly sensitive impedimetric aptasensor based on interdigitated electrodes and microwires moulded by platinum nanoparticles has been proposed for the exclusive detection of acetamiprid and atrazine in real water samples. Another atrazine biosensor (electrochemical immunosensor) has been suggested that uses a recognition element ingrained by a recombinant M13 phage/antibody complex and magnetic beads functionalized with protein G. The biosensor provides an enhanced limit of detection (0.2 pg mL^{-1}), which may be attributed to the high sensitivity of phage/antibody complex and reaction kinetics of the covalent assembly of magnetic beads with protein G.

9.3.9.2   Pathogens

The presence of pathogens in environmental matrices, mainly in water compartments, can be a potential hazard for human health, and some biosensors have been recently proposed for their environmental monitoring. For example, rapid and specific optical biosensors based on surface plasmon resonance have been proposed for the detection of metabolically active *Legionella pneumophila* in various types of environmental water samples. In another example, the RNA detector probe immobilized on the biochip gold surface has been used for the detection of bacterial RNA. Streptavidin-conjugated quantum dots are used for signal amplification, and the detection of bacteria in a range of 10^4–10^8 CFU mL^{-1} is possible in much less time. In another type of biosensor, the gold substrate is functionalized with a self-assembled monolayer of protein A and an antibody solution against *L. pneumophila*. In another example, an electrochemical immunosensor has been used for the detection of *E. coli*, which uses a polydopamine surface-imprinted polymer-sensing surface with nitrogen-doped graphene oxide quantum dots. The lower limit of detection (8 CFU mL^{-1}) is essentially due to the enhanced photo-electrochemical electron transfer and photoluminescence of graphene quantum dots and specific binding of polyclonal antibodies to the bacteria. Also, an electrochemical immunosensor based on SWCNT-gold electrodes has been used for the detection of pathogenic bacteria (*B. subtilis*) in airborne dust, explicitly during Asian dust events.

9.3.9.3 Potentially Toxic Elements

The pollution of the natural water environment caused by heavy metals and various ions poses severe hazards to human health. To solve this problem, there is a need of portable, low-cost tools for fast analyses of heavy metals. A DNA optical biosensor can be used for the detection of heavy metal ions, for example, mercury ions (Hg^{2+}), which are ubiquitous and highly toxic pollutants in the environment. This type of biosensor is portable, fast and low-cost with *in situ* screening of Hg^{2+}. The principle of detection is based on the ability of some metal ions to selectively bind to some nucleotide bases to form stable metal-mediated DNA duplexes; in the case of Hg^{2+}, they are capable of selectively binding to thymine bases to form stable thymine-Hg^{2+}-thymine complexes. Recently, two optical biosensors based on fluorescence have been proposed for the detection of Pb^{2+} in water samples (pond and lake water samples) using DNAzymes/carboxylated magnetic beads and DNA aptamers. The better limit of detection provided by the DNAzymes-based biosensor can be attributed to the use of label-free specific dye (SYBER Green I), which is intercalated into the double stranded DNA, displaying strong fluorescence intensities. Recently, an electrochemical biosensor is proposed for the detection of Zn^{2+} using paper-based microfluidic channels with reduced graphene oxide and chitosan. The biosensor is considered as a lab-on-chip system since immobilization, separation and rinse, as well as high-throughput analysis, can be performed on the same sensing platform.

9.3.9.4 Toxins

The algal blooms of cyanobacteria produce harmful toxins such as brevetoxins and microcystins due to the eutrophication of aquatic systems, and therefore reliable and cost-effective systems are needed for the early detection of such toxins. An electrochemical aptasensor has been applied to the sensitive detection of brevetoxin-2, a marine neurotoxin, using gold electrodes functionalized with cysteamine self-assembled monolayer. Biosensors have also been reported for the detection of the toxin okadaic acid in water samples mixed with seawater and contaminated with alga and in shellfish. For okadaic analysis in algal and seawater samples, and also for the detection of saxitoxin and domoic acid, a multiplex surface plasmon resonance biosensor has been proposed. For the detection of okadaic acid in real seawater samples, disposable electrochemical immunosensors based on field effect transistors with graphene have been proposed. Another toxin, domoic acid, has also been detected in seawater samples with disposable carbon nanotube field-effect transistor immunosensors.

9.3.9.5 Endocrine-Disrupting Chemicals

As an endocrine-disrupting chemical, bisphenol A, has been detected in water samples by aptasensors based on the fluorescence principle with functionalized aptamers (by fluorescein amidite) and gold nanoparticles and based on evanescent-wave optical fibre. The evanescent-wave optical fibre aptasensor is portable, fast, cost-effective, sensitive and selective for bisphenol A detection in water samples, and it has the advantage that there is no requirement of pre-concentration and treatment of samples. Recently, another endocrine-disrupting chemical, 4-nonylphenol, has been analyzed in seawater samples by a disposable and label-free electrochemical immunosensor based on a field-effect transistor with SWCNT. 17β-estradiol has been detected in environmental water samples (lake water) with a photo-electrochemical aptasensor based on CdSe nanoparticles modified with vertically aligned ordered TiO_2 nanotube arrays; a femtomolar level (limit of detection of 33 fM) and highly selective detection of 17β-estradiol is performed using an anti-17β-estradiol aptamer as biorecognition element. The specific recognition reaction between the 17β-estradiol and aptamer leads to the *in situ* formation of complexes on the surface of the photoelectrochemical sensing interface, which increases the steric hindrances for the diffusion of the electron donors, thus leading to the decrease of the photocurrent. There is a variation in the photocurrent which is correlated with various concentrations of 17β-estradiol, and a gradual decrease in the photocurrent is observed with higher concentration—15 pM. The outstanding selectivity of the biosensor to 17β-estradiol has been reported after testing the biosensor to other endocrine-disrupting compounds that have similar structure coexisting with 17β-estradiol, including estriol, ethinylestradiol, atrazine, m-dihydroxybenzene, bisphenol A, 4-nonylphenol, and diethyl phthalate.

9.3.9.6 Other Environmental Compounds

The high frequency of harmful algal blooms has led to the development of rapid and reliable analytical methodologies for their early detection and monitoring. Biosensors have been developed for the detection of algal RNA using nucleic acid probes with excellent sensitivity and specificity to their complementary binding partners. Recently, an electrochemical genosensor based on screen-printed gold electrode has been reported for enhanced selective and sensitive detection of RNA from different harmful algal species; the genosensor is able to discriminate RNA targets from environmental samples (spiked seawater samples) containing 10^5 cells, which is also considered as the limit of detection. Halogenated compounds have also been detected by environmental biosensors, for example, by a fluorescence-based enzymatic biosensor proposed for the monitoring of 1,2-dichloroethane, 1,2,3-trichloropropane, and γ-hexachlorocyclohexane in water samples with pH ranging from 4 to 10 and temperature from 5 to 60°C. In real conditions, the biosensor is used for rapid quantification of 1,2-dichloroethane contamination in water reservoir with the quality of mapping the distribution of contamination using GPS.

9.4   CONCLUSION AND FUTURE PERSPECTIVES

Although environmental biosensors can be constructed using the improved characteristics of nanomaterials and novel nanocomposites, there is an increased focus on *in situ* and real-time monitoring of pollutants by different technologies. Recently, environmental monitoring has been a field of interest for the application of drones mainly in the monitoring of water and air quality, the surveillance of agriculture and volcano gas measurements, as reported in some daily news, and a few scientific articles have also been published. The main limitation for the recent environmental biosensors is based on the lack of application in real environmental samples since the majority of identified "environmental biosensors" has been applied to tap water samples or synthetic samples. There are still few commercial biosensors for environmental monitoring, contrary to clinical applications, which could be due to the interdisciplinary context of fabrication as well as limitations on the *in situ* operation and on the analytical performance, primarily in reproducibility. Thus, future sensing systems based on the conjugation of biosensors and drones are required for environmental monitoring in remote locations due to their ease of distribution, low cost and low power requirements. Also, the study of biosensors for environmental monitoring is an important research area with a better perspective.

10 Bioleaching and Biomining
Concept, Applications and Limitations

Indu Sharma

CONTENTS

DOI: 10.1201/9781003131427-10

10.1   INTRODUCTION

Metal contamination of the environment and the valuation of its possible hazard to the global environment and human health are some of the most challenging responsibilities. Use of microbial approaches for the recovery of base and valuable metals is known as "bioleaching" or "bio-oxidation". Bioleaching is the extraction of minerals by microorganisms. Bacteria are effective agents for ore extraction and are used in various biological processes. The biomining process is the use of microbes for the commercial extraction of excessive metals from ores and mines with the least effect on the environment. In bio-oxidation, microorganisms prepare metal for removal. Biomining is the extraction of particular minerals from their ores through biological means, usually microbes. Biomining is done in two steps, often called bioleaching and bio-oxidation. Bioleaching commonly refers to biomining technology applied to metals, whereas bio-oxidation is usually applied to refractory gold ores and concentrates. Even though it's a new method used by the mining enterprise to extract minerals equivalent to copper, uranium and gold from their ores, currently biomining occupies first place among the available mining applied sciences. Biomining methods are inexpensive, safe, effective and also environmentally agreeable. With the use of biotechnology, the effectiveness of biomining can also be extended with the help of genetically modified microorganisms.

10.2   HISTORY OF BIOLEACHING

Soaking colored minerals and soils in water and transferring the colored liquid for clothing/rug fibre dying is likely the oldest preparation of leaching process used by humans.

In earlier times, around 1000 BC, miners in the Mediterranean basin used microbial activity for copper extraction from mine drainage waters, but they were unaware of their operation. The previous records of leaching as a mining technique can be found in V. Biringuccio's book *Pirotechnica* published in 1540 and Georgius Agricola's book *De Re Metallica* published in 1557 illustrating a heap leach for saltpetre (sodium/potassium nitrate) and alum (aluminium sulphate) recovery, respectively. In the 16th century, copper extraction by dump/heap leaching was known to be practiced in the Harz Mountains area in Germany and Rio Tinto mines in Spain. Using microorganisms is an old procedure which was first introduced in Roman time and then used by the Welsh in the 16th century and the Spanish in the 18th century for recovering metals from ores. However, the contribution of bacteria to metal discharge was not recognized until the 20th century.

The use of cyanide for the leaching of gold and silver ores was reported in England in 1887. The Bayer method is still used for bauxite ore beneficiation. Pressure leaching has been in use since 1890 for recovery of many metallic ores. The contributions of certain unknown bacteria in the leaching of zinc and iron sulphides were reported in 1920. The universal application of the cyanidation method with heap and vat leaching and gold recovery procedures increased significantly during the 1900–1920 period. The essential role of bacteria in the leaching of mineral ores was ignored until 1947. Colmer and Hinkle from West Virginia University designated a bacterium (*Thiobacillus ferrooxidans*) mostly responsible for the leaching of metal sulphide ores. Different microbes such as bacteria and fungi are mostly used for the extraction of metals like copper, gold, and uranium from low-grade mines and soluble metal deposits. Some *thermophilic* bacteria which can tolerate a temperature range from 15°C to 80°C are also used to extract metals. The most common genera of bacteria reported for this process were *Thermobacillus* and *Leptosprillium*.

10.3   MICROORGANISMS USED IN BIOLEACHING

Thermophilic bacteria are widely used for metal recovery by bioleaching processes from solution. A consortium of mostly pure bacteria is used for superior recovery of metals. *Acidophilic, chemolithotrophic* and *heterotrophic* bacteria and fungi are commonly used in bioleaching. The chemical conditions under which natural bacteria helps sulphide mineral leaching takes place under normal acidic conditions, and further sulphuric acid is used as a solvent for metal mining. Mesophilic, moderate and extreme thermophilic microorganisms are also reported to be used for bioleaching. They have chemosynthetic metabolism and the capacity for the oxidation of inorganic sulphur and its compounds to produce energy for bacterial growth. The bacteria *Thiobacillus ferrooxidans, Thiobacillus thiooxidans, Thiobacillus acidophilus, Thiobacillus organoparus, Leptospirillum ferrooxidans, Rhodococcus, Ferrimicrobium, Sulphobacillus, Thermophilic thiobacilli,*

TABLE 10.1
Groups of Organisms Used in the Bioleaching of Metals

Microbes	Mode of Nutrition	Leaching Agent	Metals	pH	Temp. °C	Reference
Acidianus ambivalens	Facultative heterotrophic	Sulfuric acid	Fe	1.3–2.2	45–75	Johnson, 1998
Acidithiobacillus ferrooxidans, Acidithiobacillus thiooxidans	—	—	Fe, Mn, NiMo, V, Al, Pb, As, Cu, Mo, Zn, As	2.0–3.0	45–75	Rastegar et al., 2014; Natarajan, 2016; Adekola et al., 2011
Acidithiobacillus caldus	—	—	Cu			Watkin et al., 2009
Bacillus coagulans,Bacillus licheniformis, Bacillus megaterium	Heterotrophic	Citrate	Cu, Pb, Sn	5.4–6.0	22–37	Hahn et al., 1993
Leptospirillum ferrooxidans	Chemolitho-autotrophic	Ferric iron	Ni, Au, Ag, Cu, Cr	2.5–3.0	30	Patel et al., 2012
Aspergillus niger	Heterotrophic	—	Nickel (Ni)		30	Chaerun et al., 2017

Ferroplasma, Sulfolobus, Metallosphaera, Acidianus, Pseudomonas aeruginosa, Rhizopusarrhizus, Aspergillus niger, Aspergillus oryzae and other thermophilic species of *Sulfobacillus acidianus* are commonly used to leach metals such as copper, zinc, uranium, nickel and cobalt from a sulphide mineral.

10.4   BIOLEACHING

The use of microorganisms has empowered more effective extraction and purification of ore from its compounds. Bioleaching is one of the applications of bio-hydrometallurgy, which include bioremediation, biosorption, phytoremediation and wastewater treatment.

In bioleaching, living organisms are involved in the process of extraction of metal from high- and low-grade ores.

1. *High-grade ore:* Higher amount of Cu
2. *Lower-grade ore:* Small amount of Cu

10.4.1   LEACHING IN NATURE

Leaching is a natural physico-chemical process where minerals in rock masses are dissolved under saturating water and anion/cation exchange reactions to produce metal salts in solute/colloid phase that migrate and accumulate in the environment under hydrological forces. Lateritic ore deposits, the significant resources of aluminium, nickel, platinum, cobalt and even gold are a clear indication of current natural leaching process through geological times.

10.4.2   TYPE OF BIOLEACHING

The chemical transformation (bioleaching) of metals by microorganisms may occur by direct or indirect bioleaching.

1. *Direct leaching (contact leaching)*

   In direct leaching, bacteria make direct physical contact with ores and involve enzymes for the recovery of metal from the liquid. A biochemical reaction is known as oxidation, where bacteria changes the metal sulphide crystals into soluble sulphates, thus dissolving the metals (Figure 10.1). For example, pyrite is oxidized to ferric sulphate as below:

$$2FeS_2 + 7O_2 + H_2O \xrightarrow{\ T.\,ferrooxidans\ } 2FeSO_4 + 2H_2SO_4$$

2. *Indirect leaching (non-contact leaching)*

   In indirect leaching, no physical contact between bacteria and ore is required. Bacteria produce oxidizing agents called metal leachates that act on ores to produce a solubilized extract that is further purified to get the metals. A bacterium reoxidizes the ferrous iron back to the ferric form and oxidizes the elemental sulphur formed in some cases. The ferric ion then chemically oxidizes the sulphide minerals producing ferrous iron. The bacteria only have a catalytic function because they accelerate the reoxidation of ferrous iron, which takes place very slowly in the absence of bacteria.

$$FeS_2 + Fe_2(SO_4) \rightarrow 3FeSO_4 + 2S^\circ$$
$$2S^\circ + 3O_2 + 2H_2O \rightarrow H_2SO_4$$

Oxidation of ferrous (Fe^{2+}) to ferric (Fe^{3+}) by *T. ferrooxidans* at low pH is given below:

$$4FeSO_4 + 2H_2SO_4 + O_2 \xrightarrow{\ T.\ ferrooxidans\ } 2Fe_2(SO_4)_3 + 2H_2O$$

10.4.3 METHODS FOR BIOLEACHING

Traditional extraction methods include complex steps like roasting, smelting, bessemerization and electrolytic release with adverse effects on the environment. Biomining reduces waste in the environment and produces approximately 90% concentration of metallic compounds from ores.

10.4.3.1 Leaching Techniques

Leaching techniques employed in modern leaching technologies mimic the naturally occurring leaching processes under optimized operational conditions for improved productivity.

In situ leaching is a technique used to recover salt/trona and uranium ores in suitable hydrogeological settings. The *in situ* technique includes the pumping of air, which is forced under pressure into mines and ore deposits, making them porous through explosive charging. An earth surface drill is connected to the pipe and adds liquid containing metal leaching microbes, in which bio-oxidation take place. In this process, metal-supplemented solutions are injected into the wells, which are drilled into the earth's surface and recovered. The three different kinds of ore bodies considered for *in situ* leaching include (1) deep deposits below water tables, (2) surface deposits above water tables and (3) surface deposits below water tables.

Dump leaching is a technique used in ancient and early modern times, where generally run-of-mine sulfidic copper ore dumps are wetted with water and sulphuric acid as a lixiviant to leach copper salts. In gold and silver, the dump is sprayed with a dilute cyanide solution that penetrates through the ore to dissolve gold and silver. The solution containing gold and silver exits the base of the dump where it is collected and the precious metals are extracted.

Heap leaching is an industrial mining process used to extract precious metals like copper, uranium and other compounds from metal ores using a series of biochemical reactions that absorb specific minerals and re-separate them after their division from another earth material. In this technique crushed (>5 mm) and agglomerated ores are stacked over an engineered impermeable pad, wetted with lixiviant (solvent)

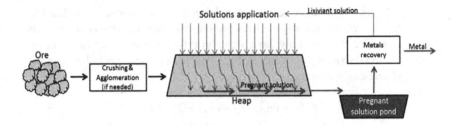

FIGURE 10.1 Heap leaching technique. (From: www.mining.com/heap-leach-minings-breakthrough-technology/.)

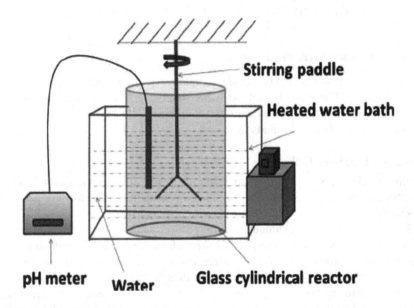

FIGURE 10.2 Tank leaching technique.

chemicals under atmospheric conditions and leachates (metal-loaded solutions) are collected for the metal recovery processes.

Tank leaching is a technique in which ground ores are chemically treated in open tanks under atmospheric pressure conditions to extract metal salts from the ore in an accelerated rate. In tank leaching, the ground, classified ores are previously mixed with water to form a slurry or pulp, and this is pumped into the tanks. Leaching reagents are added to the tanks for the leaching reaction. In a continuous system, the slurry will then be pumped to the next tank. Ultimately, the "expecting" solution is separated from the slurry using some form of liquid/solid separation process, and the solution passes on to the next phase of recovery (Figure 10.2).

Pressure leaching is a technique in which ground ores are chemically treated in reactors (autoclaves) under high pressure and temperature conditions to extract metal from the ore in an enhanced rate. Pressure leaching is the chemical suspension of soluble minerals within a solid ore or concentrates carried out at rising pressures and giving rise to a solution containing metals to be recovered. The process is carried out in a closed autoclave at higher temperatures (>220°C) and pressures (>20 atm) than are possible with open tanks. The increased demand improves the solubility rate of solids and increases the speed of dissolution into the leaching solution. Because of the rapid kinetics, the duration of the leaching is short (30 minutes to 24 hours) depending on the concentrate leached and conditions applied.

10.4.4 TECHNIQUES USED IN BIOMINING PROCESSES

Heaps and tank reactors are used for a large-scale leaching process. Agitation and percolation are two main techniques applicable in the mining process. The advanced technology includes the purification of soluble and insoluble particles via static bed hole agitation. The commercially used methods of *in situ* leaching are vat leaching, heap leaching and dump leaching.

10.4.4.1 Vat Leaching

Vat leaching units are rectangular containers such as drums, barrels, tanks and vats, which are usually massive and made of wood or concrete and lined with a material resistant to the leaching media. The treated ore is often coarse; the leaching reagents are added to the leaching object using different methods. The vats are usually run to maximize the contact time between the ore and the reagent. Then leachates collected from one container are added to another tank with fresher ore. The extraction increases the yield; subsequently, the fresh reagents meet the most leached ore in one end.

10.4.4.2 Dump Leaching

Dump leaching contains the piles of uncrushed waste ores or rocks. In these dumps, metals like copper are too low to be recovered conventionally and profitably. Heap leaching involves the preparation of the ore; first, size is reduced to increase the mineral lixiviate interactions and completely laying on the impermeable base to prevent the loss of lixiviant and water pollution. Lixiviants are those chemical solutions which are used to accelerate the dissolution of ores in mining processes. The more frequently used lixiviants are cyanide solution and sulphuric acid solution. Solvent extraction and electrowinning methods are employed for metal recovery.

10.4.4.3 Heap Leaching

Heap operation is simple and suitable for handling large volumes of minerals, but their efficiency and yields are restricted because of the severe complications in exerting an adequate process control. On the other hand, reactors can economically handle sufficient volumes of material. Heap leaching is carried out by piling the ore upon the impermeable lining. The acidic solution, especially cyanide solution, is sprinkled on to the top of the pile, which penetrates down into the heap and leaches the metals. Both procedures include using lixiviant from the top of the heap or dump surface, and minerals are recovered in the loaded solution. Lixiviants are chemical containing solutions used in leached mining to increase the dissolution of metals in ores. Sulphuric acid and cyanide salts are the most common lixiviants used in heap or vat (tank) leaching processes which are functional under atmospheric conditions. This solution leaches to the bottom due to the gravity flow. From the top, diluted sulphuric acid is spread which enters through the dump. The pH of the solution drops, which enhances the growth of the acidophilus bacterial species. The acid is collected from the bottom and shifted to the recovery station for solvent extraction. The wet leaching includes the crushing of ore material and its dilution in a confined tank bioreactor, which increases the recovery of metal. This is a high-cost method, but currently is still applied to the ore oxides. However, precious metals are actively extracted by this process.

10.4.5 MECHANISM OF BIOLEACHING

The mechanisms of bioleaching by using microbes is not known precisely, but both chemical and biological processes together are involved in the oxidation of metals in acidic conditions.

The classical understanding of the bio-oxidation reactions of *Acidithiobacillus ferrooxidans* involve two possible mechanisms: direct and indirect.

In the direct mechanism, bacteria directly oxidize the minerals and solubilize the metal.

$$MS + H_2SO_4 + \frac{1}{2}O_2 = MSO_4 + S + H_2O$$
$$S + 1\frac{1}{2}O_2 + H_2O = H_2SO_4$$

Bacterial attachment to the mineral surfaces is a prerequisite for the direct attack.

In indirect attack, the bacteria oxidize ferrous materials to ferric form, which is the oxidizing agent for the minerals.

$$MS + 2Fe^{+++} = M^{++} + 2Fe^{++} + S$$
$$2Fe^{++} + O_2 + 2H^+ = 2Fe^{+++} + H_2O$$

Both direct and indirect mechanisms may occur simultaneously along with other physicochemical reactions in real bioleaching systems. There are several reports on direct cell attachment on several sulphides and the role of ferric ions bound to the cell surface.

10.5 BIOMINING

Microbial processes are gaining interest in the mining industry. The term biomining has been designed for the use of microorganisms in mining processes. Biomining includes two interrelated microbial processes that are useful in the extractive metallurgy of heavy metals: bioleaching and bio-oxidation. Leaching is the solubilization of one or more components of a composite solid by exchange with a liquid phase. In bacterial leaching, the solubilization is facilitated by bacteria. So, bacterial leaching is a procedure by which the metal of interest is extracted from the ore by bacterial activity.

On the other hand, bio-oxidation indicates the bacterial oxidation of reduced sulphur species associated with the metal of interest. For many years, bioleaching technology was used for the recovery of metals from low-grade ores, flotation tailings or waste material. Today, bioleaching is functional as the primary process in large-scale operations in copper mining and as a significant pretreatment stage in the processing of gold ores.

The biomining process is the use of microorganisms for the commercial extraction of excessive metals from ores and mines with minimum effect on the environment. The bacterial cells play a dynamic role in detoxifying/replacing waste cyanide, marginal biomass and activated carbon in commercial mining. These procedures are chosen over conventional techniques because they are energy efficient, low cost, environmentally friendly and produce useful by-products. At an industrial scale, diverse microbial strains (acidophilic, *Sulphobacillus*, *Rhodococcus*, *Ferrimicrobium* and chemolithotrophic) are used to increase the processes of copper and uranium bioleaching.

These extraction procedures comprise oxidation of insoluble metal sulphides to soluble sulphates. Conventional metallurgical techniques are harmful to the environment because the depletion of high-grade ores releases toxic gases. Microbe-assisted gold mining is estimated to double the yield of gold, and there is a need to explore different array of microbes having such an ability. Bioleaching is a low-cost and straightforward method for developing countries with large ore deposits. About 30 isolates of bacteria have been discovered with advances in molecular genetics, physiology and microbial genomics. Further efforts are going to explore a diverse culture range of archaea and improve its genetic potential for commercial bioleaching.

Biomining reduces waste in the environment and produces a 90% concentration of metal from ores. Consequently, industries select biomining over other mechanical processes of ore extraction. The bacteria and fungi work as biocatalysts in the leaching processes by providing secondary raw material as metabolites, which help to extract metal from its ore and convert them into more or less water-soluble compounds. These procedures are used to recover metals like gold, copper, iron and uranium. Generally, sulphide ores are oxidized to soluble sulphates and sulphuric acid during the metabolic activity of bacteria.

Different bacterial strains are used to extract metals from a leaching solution. Additionally, acidophilic bacteria are used for the extraction of toxic metals from mine wastes by bio-hydrometallurgy. *Thiobacillus* bacteria can accelerate the solubility of heavy metals from minerals and increase the leaching process by a factor of about 10^4 under aerobic conditions.

10.5.1 Use of Fungi and Plants in Biomining

Phytomining (plants) and mycomining (fungi) are used for metal recovery, but these are not as popular as bacterial compression mining is. The bacteria used in biomining occur naturally but are also created by genetic engineering and conjugation procedures, which contain desired characteristic genes to increase the rate of bioleaching. Mainly bacterial (*Chemolethotrophes*, *Bacillus* sp., *Pseudomonas putida*, *Thiobacillus* and *Thiooxidans*) and fungal strains (*Aspergillus niger*, *Alternaria* sp., *Fusarium* sp., *Mucor racemosus* and *Penicillium simplicissimum*) are used in the biomining process (Table 10.1). Many biomining experiments have been conducted by using two fungal strains: *Aspergillus niger* and *Penicillium simplicissimum*. These fungi could collect copper up to 65% and aluminium, lead and zinc approximately up to 90%–95%.

Similarly, in phytomining, some plant species are used for metal extraction. *Alyssum murale*, which grows in ultramafic soils found in Albania, has the affinity to accumulate nickel at about 100 kg/ha. It is considered a hyperaccumulator plant, especially for nickel. Ultramafic soil is extremely augmented with cobalt, magnesium, nickel and zinc. The selection of such plants is based upon their ability to accumulate metal, which they do by storing huge amounts of metals in their cells. Such plant species are called hyperaccumulators. These plants are grown in regions that are highly augmented with heavy metal in high concentration. These plants are first grown in such areas and then allowed to burn. This leads to the subsequent production of metal-containing ash, which is then separated by electrolysis. This process is generally not used because it is too time-consuming. Similarly, plants help to remove the toxicity of the soil by taking up toxic metals from the soil. *Helianthus annuus* and *Nicotiana tobacum* are also used as metal accumulators.

Plants with mycorrhizal association and an antioxidant defence system provide resistance to the toxic metals. They secrete antioxidant enzymes, which decontaminate the effects of heavy metals. Moreover, mycorrhizal fungi increase phytostabilization and solubilization, which takes place by enzymatic action resulting in the solubilization of heavy metals like cadmium, iron and zinc.

10.5.2 Factors Affecting Biomining

Certain environmental factors are necessary for the biomining procedure, mostly pH, temperature and the growth of the organism. Microbial activity is the critical property for the biomining of metals, which is influenced by an optimum growth condition. Environmental conditions in the biomining process favour the growth and development of mesophilic microorganisms, while moderately thermophilic and extremely thermophilic microbial groups have also been observed.

10.5.2.1 Effect of pH and Temperature

The ranges of temperature and pH (1 to 3) differ among microorganisms, although the commonly used bacteria are *Thiobacillus ferrooxidans*, *Thiobacillus thiooxidans*, and *Leptosporillium ferrooxidans*. The primary purpose of their usage in biomining industries is due to their maximum activity (oxidation potential), that is, Fe^{3+}/Fe^{2+} ratio. The pH value range of 1.8–2.5 is considered as optimum pH for the growth and microbial activity of *T. ferrooxidans*, while *L. ferooxidans* has more acid-resistant capability with a pH range of 1–2. Mesophilic, moderately thermophilic and extremely thermophilic microbial groups have also been observed for the biomining procedure. Temperature range could improve bioleaching efficacy and the potential for more commercial application. The optimal temperature considered for the *T. ferooxidans* growth is 30–35°C, but it can be activated at 10°C. *Leptosporillium ferrooxidans* show their maximum activity at 45°C and minimum at 20°C. Most of the bio-oxidation of gold recovery procedures are carried out at 40°C and pH value of 1–6.

10.5.2.2 Biomining of Copper

Biomining is a progressively applied biotechnological practice for the processing of ores. Copper is present in the form of copper ores such as chalcopyrite ($CuFeS_2$), chalcocite (Cu_2S) and covellite (CuS). Currently, the production of copper from low-grade ores is the most important industrial application. Biomining of copper includes the conversion of water-insoluble copper sulphides to water-soluble copper sulphates. The copper minerals and ores of covellite and chalcopyrite are crushed and then acidified with sulphuric acid. For homogenization, the entire material is converted to a rotating drum before piling in heaps and dumps. The heaps are then allowed to make contact with the iron-containing solution, which penetrates the heap and provides growth media for the bacteria and additional microbes which catalyze the ore to release the copper. The produced ferric ions react with the copper to give copper sulphate. The charged solution contains 0.5 g/L to 2.0 g/L copper and 20 g/L iron which is collected and transferred to the recovery plant. The precipitation method is commonly used for copper extraction.

$$CuFeS_2 + 4H^+ + O_2 \rightarrow Cu^{2+} + Fe^{2+} + 2S + 2H_2O$$
$$4Fe^{2+} + 4H^+ + O_2 \rightarrow 4Fe^{3+} + 2H_2O$$
$$2S + 3O_2 + 2H_2O \rightarrow 2SO_4^{2-} + 4H^+$$
$$CuFeS_2 + 4Fe^{3+} \rightarrow Cu^{2+} + 2S + 5Fe^{2+}$$

10.5.2.3 Biomining of Gold

Metalophilic bacteria *Cupriavidusmetalli durans* has the capability of converting the gold dissolved into pure gold through the mechanism of biomineralization. Biomining of gold can be applied using manufacturing methods such as stirred tank bioleaching/bio-oxidation of minerals by using a bioreactor. Different gold ores like calcaverite ($AuTe_2$), sylvanite (Ag, Au)Te_2 and petzite (Ag_3AuTe_2) are available in nature. Solubilization of ore with cyanide solution recovers gold in the solution. The insoluble sulphides cover the small gold particles. In the first stage, the ore arsenopyrite is oxidized by bacterial breakdown to the sulphur; the second stage is then oxidized it to high oxidizing states, reducing dioxygen by H_2 and Fe^{3+}. All the soluble components are dissolved in these stages. The entire process takes place at the bacterial cell membrane where electrons pass into the cell to provide energy to the cell, which is used for the reduction of oxygen to water. Then the metal is oxidized to the high oxidizing state by bacterial activity. The electrons obtained from Fe^{3+} is oxidized to Fe^{2+}. The gold is separated from the ore particles of the solution; the bio-oxidation process increases the gold extraction from 15%–30% to 85%–95%.

In the leaching process, sodium cyanide converts gold to a soluble cyanide complex:

$$4Au + 8NaCN + 2H_2O + O_2 \rightarrow 4Na[Au(CN)_2] + 4NaOH$$
$$2Na\,[Au(CN)_2] + Zn \rightarrow Na_2[Zn(CN)_4] + 2Au$$

Bioleaching occurs in bioreactor or heap leaching using *Thiobacillus ferrooxidans*.

10.5.2.4 Uranium Leaching

Uranium leaching is more significant than copper, but less amount of uranium is obtained than copper. To obtain one ton of uranium, a thousand tons of uranium ore must be handled. Uranium can be extracted from uranium ores, for example, uraninite or pitchblende (UO_2) and brannerite (UTi_2O_6) by *Acidithiobacillus ferroxidans* or *T. ferrooxidans*. Uranium ores occur in low-grade ores and are insoluble. They can be converted to a leachable form by oxidation with ferric ion. However, uranium leaching from ore on a large scale is widely practiced using *in situ* technique. Insoluble tetravalent uranium is oxidized with a hot $H_2SO_4/$ Fe^{3+} solution to create soluble hexavalent uranium sulphate at pH 1.5–3.5 and temperature 35°C.

The recovery of uranium from ore is carried out using the same steps as copper extraction. Firstly, the insoluble uranium oxides are converted to soluble uranium sulphates by the activity of ferric iron and sulphuric acids produced by the bacteria. Biomining is applicable effectively for uranium extraction from waste gold ores. Mostly, other metals present in the form of insoluble sulphur are converted to soluble sulphate and sulphite form. The metals are then recovered from these soluble solutions by using different bacterial strains.

There are two processes. Direct leaching by *T. ferrooxidans* has been proposed in the following equation.

$$2UO_2 + O_2 + 2H_2SO_4 \rightarrow 2UO_2SO_4 + 2H_2O$$

However, where oxygen is limited, this process cannot operate, and the indirect bio-leaching process operates using pyrite.

The ferric (II) ion reoxidized by *Acidithiobacillus ferroxidans*, which reacts with the uranium ore and sulphuric acid,

$$UO_2 + Fe_2(SO_4)_3 \rightarrow UO_2SO_4 + 2FeSO_4$$
$$UO_3 + H_2SO_4 \rightarrow UO_2SO_4 + H_2O$$

10.6   ADVANTAGES AND DISADVANTAGES OF BIOLEACHING

Bioleaching and biomining is the operative and straightforward technology for metal extraction from ores by the use of microorganisms. Some of the advantages and disadvantages associated with the process are discussed next.

10.6.1   ADVANTAGES

1. It is flexible and can be used for single or mixed minerals.
2. Mining of metals from low-grade ores is possible and is an environmentally friendly process.
3. It is employed for collecting metals from waste or drainages.
4. Bioleaching is simpler and more cost-effective to operate and maintain than traditional methods.
5. It is used to extract refined and expensive metal which is not possible by other chemical processes.
6. Bioleaching can be applied from simple to complex bioreactors.
7. Liquid effluent generated in this process can be neutralized, and thus the process is more environmentally friendly than traditional extraction approaches.
8. Less landscape destruction occurs since the bacteria grows naturally. As such, the mines and surroundings grow naturally. Bioleaching is more environmentally friendly than conventional extraction methods, as it uses less energy and does not produce harmful emissions.

10.6.2   DISADVANTAGES

Bioleaching technology for metal extraction has some disadvantages:

1. Bioleaching is very slow compared to the smelting process.
2. Any change in atmospheric conditions may lead to a decrease in the efficiency of the process.
3. Toxic chemicals like sulphuric acid and H^+ ions are sometimes produced in this process and can leak into the ground and surface water, making it more acidic, and thus causing environmental damage.
4. Heavy metals such as iron, zinc and arsenic are extracted during acid mine drainage. When the pH of this solution increases, these ions precipitate after dilution by freshwater, causing "yellow boy" pollution. Therefore, a bioleaching procedure must be carefully planned since the process can also lead to bio-safety issues.

11 The Degradation of Organic and Inorganic Pollutants

Mahdieh Houshani, Sarieh Tarigholizadeh,
Vishnu D. Rajput and Hanuman Singh Jatav

CONTENTS

DOI: 10.1201/9781003131427-11

11.1 INTRODUCTION

The industrial and urban wastes produced by anthropogenic activities are one of the critical environmental concerns. Different organic and inorganic pollutants are severely polluting the world's soil and water resources, causing deterioration of both soil and food quality. These include polycyclic aromatic hydrocarbons (PAHs), pesticides, polychlorinated biphenyls (PCBs), explosives, metals, metalloids, and radionuclides. These pollutants are mutagens and carcinogens, which are present almost everywhere in the environment. Great attention is owed to them due to their presence in high concentrations, persistence in the environment, and long-term toxicity. Human and wildlife exposure to such persistent and toxic compounds occurs because of bioaccumulation in plants, which subsequently transfers to the natural and agricultural food chain.

Various physical and chemical approaches have been employed for the elimination of these pollutants from the environment. However, these techniques are mainly expensive, are energy consuming, and may create toxic products which contribute to secondary pollution in the environment. The use of biological approaches for removing toxic pollutants has received increasing attention because of their economic viability. Accordingly, developing new remediation technologies can be environmentally compatible and inexpensive compared to dangerous and expensive engineering-based remediation technologies. Bioremediation uses living organisms to improve and treat contaminated sites. The bioremediation agents include bacteria, fungi, and higher plants with the capability of decamping pollutants with the help of their natural biological activity.

However, as an alternative approach, phytoremediation has emerged as the most desirable technology for the elimination of environmental pollutants. This approach employs plants to degrade and eliminate organic and inorganic contaminants from polluted soil, water, and sediment. Plant-based methods include multiple processes such as metabolism, volatilization, and sequestration, followed by harvest. The compounds/enzymes produced by plants facilitate the breakdown of pollutants and activate microbial biodegradation in the rhizosphere. Phytoremediation uses resistant and hyperaccumulator plants to degrade, transform, translocate, extract, and detoxify organic and inorganic pollutants present in the soil. This approach has been successfully implemented in the bioremediation of various pollutants including PAHs, radionuclides, metals/metalloids, pesticides, and PCBs.

However, there is a profound gap of knowledge about the degradation, detoxification, and elimination of pollutants by higher plants. Accordingly, this study aims to discuss their uptake mechanisms and recent progress in the understanding of degradation, detoxification, and removal of organic and inorganic pollutants by plants. It is essential to understand the primary sources of contaminants and their possible concentration in soils (Table 11.1).

11.2 SOURCES OF SOIL POLLUTANTS

There are two main sources of soil pollutants: natural and anthropogenic. Natural events leading to the accumulation of pollutants into the environment can be volcanic eruptions, geographical alterations, forest fires, earthquakes, soil degradation, air pollution, and changes in rainfall patterns. Over centuries, anthropogenic activities (intentional or accidental) have been causing widespread pollution in soils along with natural processes. The primary intentional anthropogenic sources of soil pollution are fossil fuel combustion, industries, fertilizer and pesticide application, land application of sewage sludge, disposal of wastes, irrigation with untreated wastewater, smelting, mining, sport shooting, application of agrochemicals and sewage, and military training. Leaching or spills from landfills, nuclear explosions, and flooding by rivers or seas are examples of accidental emission of pollutants into the ecosystem. Soil pollution can also contribute to air pollution via emission of volatile compounds into the atmosphere and water contamination via leaching chemicals into groundwater or reaching to lakes, seas, and oceans. Some of the main human-caused reasons for soil pollution, including industrial, agricultural, and municipal are discussed next.

11.2.1 Industrial Activities

Industrial activities release pollutants into the environment in forms of gas, liquid, or solid. The most important gaseous pollutants, such as carbon dioxide (CO_2), carbon monoxide (CO), sulphur dioxide (SO_2), and nitrogen dioxide (NO_2), released into the atmosphere can readily enter into soils via wet or dry deposition of particles or washing out by acid rain. Direct discharge and incorrect chemical storage of some chemicals (such as fly ash, sawdust, plastics, and sludge) and garbage (e.g. wood, glass, and metals) can contaminate or cause physicochemical changes in soils, rivers, lakes, and oceans. Industrial wastes can be categorized into several groups, including mining wastes, smelting, chemical wastes, and solid wastes. These activities cause diffusion of organic pollutants (such as PAHs) and toxic heavy metals (HMs) into the soil.

11.2.1.1 Mining and Smelting

Mining operations (including metal smelting) is one of the major sources of toxic elements. In every stage of mining activity, vast amounts of chemicals, including HMs (e.g. Cu, Pb, Zn, etc.), can be released into the environment. Many documented examples can be found for heavily contaminated soils associated with mining activities around the world. Though mining operations are limited to smaller areas, the consequences of its related activities (smelting operations, tailings, etc.) have major impacts on the environment. Extraction, the processing of mineral ores, and the blowing of tailing particles (a waste product and a large part of ore known as tailings) as well as acid mine drainage are the most harmful activities. These can cover the topsoil layer of lands in and around the area and spoil natural resources. Tailings usually contain different undesired substances. They can contain a mixture of water, any residual chemicals applied in processing stages such as cyanide, xanthates, and other flotation agents, surfactants and sulfuric acid, and finely ground ore. Tailings may also contain HMs (As, Hg, Cu, Pb, Cd, Se, Zn, and Ni) that can create environmental concerns as well as sulphides due to the natural composition of the ore. The mining of gold creates severe environmental pollution due to the emission of very lethal substances such as Hg, As, and cyanide.

11.2.2 WASTE AND SEWAGE DISPOSAL

Together with mining activities, waste disposal is one of the oldest sources of contamination. The problem associated with sewage disposal has become a major concern, and the overall waste generation will rise drastically to 3.40 billion metric tons per year by 2050. The most common ways to manage waste are incineration and landfilling/open dumping. Both processes release many toxic chemical substances, such as HMs (Hg, Pb, Cd, Cr, etc.), PAHs, polybrominated biphenyls (PBBs), polybrominated diphenyl ethers (PBDEs), and PCBs. The landfill leachates contaminate soil and underground water. The ash fallout from incineration can change the quality of groundwater and subsequently the food chain. Nowadays, the fastest-growing category of solid waste is known as electronic waste (e-waste), which contains valuable elements, such as Cu and Au. However, inappropriately recycled e-wastes can become a source of the aforementioned hazardous substances. Other approaches for waste and sludge disposal include establishments for recycling of lead batteries, application of untreated sewage sludge in soil amendments, dumping and incineration of biodegradable and non-biodegradable domestic and urban wastes.

11.2.3 MILITARY ACTIVITIES

Conflicts and related military activities are ever-present forces which can overwhelmingly impact and alter the nature of adjacent soils as well as water resources. This type of pollution source differs remarkably from other types, such as waste depositories, industrial activities, and so on owing to the intensity of activities. These activities include direct armed conflict, military training and related production of chemical and metals contamination, rocket launching, blasting and armament destruction, projectile and mortar fire impact, and nuclear warfare, which commonly use hazardous substances, for example, trace or radioactive elements for weapons. Previously, conflicts were limited to local areas, resulting in pollution of small magnitude. However, since the 20th century, modern conflicts utilize non-degradable weapons of mass destruction, and the chemicals remain in the area for centuries after the end of the war. For instance, the devastating world wars not only caused destruction, death, and significant detrimental effects at the time of their incidence but also threatened the environment around the world with such leftovers as mines and unexploded ordnance (UXO). There are few studies on these type of pollutants, most likely due to restrictions imposed by governments related to security reasons.

11.2.4 INTENSIVE AGRICULTURAL ACTIVITIES

Intensive agriculture is a major cause of local and diffuse contamination, which releases massive amounts of pesticides and excessive chemical fertilizers. Despite their potential benefit for agriculture, it has been demonstrated that using excessive agrochemicals can alter the physical-chemical character of agricultural soils and make them infertile. Compounds such as PAHs and Pb, As, and Hg accumulate permanently in soils and subsequently influence the metabolism of plants. They also influence the productivity of soil as well as the native microflora and hence human health too. Another source for agricultural pollutants can be accidental spills of hydrocarbons used as fuels for machines and improper management of livestock production, which can result in the deposition of urine and faeces, leading to soil pollution.

11.2.5 ROAD AND TRANSPORT INFRASTRUCTURES

The major sources of soil pollution associated with transport include combustion processes, the wearing of vehicles (engine, tires, brakes), road surface degradation, and the leaking of oil and coolants. The chemicals involved in road maintenance (de-icing salts) disseminate pollutants, as do splashes generated by traffic during rainfall events, which also spreads specific polluting substances such as HMs, PAHs, and mineral oil into the environment. Moreover, all other sorts of organic and inorganic materials which fall from vehicles or originate from car accidents can also contaminate soils. The soil contamination that originated from these activities is higher in urban than in rural areas. Mainly, Pb and PAHs originate from combustion processes. It seems that most of the heavy metals have mixed origins. However, Zn and Cu are reported to have mainly come from tire dust and the corrosion of radiators and brakes, respectively. Cu and Mn are used in automotive engines and piping. Also, Cr and Ni are derived from the combustion of lubricating oils and also are used in chrome plating. Surprisingly, the soil pollution caused by transportation activities has received little attention in comparison with other types of activities.

11.3 PHYTOREMEDIATION PROCESSES

In recent years, the increase of pollutants in the environment has attracted a great deal of attention. The elimination of pollutants from a contaminated environment through biological resources seems to be an outstanding alternative. Phytoremediation or plant-assisted bioremediation means the use of plants to remove toxic inorganic and organic environmental pollutants presented in a solid and liquid substrate. Generally, phytoremediation of pollutants by a plant includes uptake, translocation, transformation, compartmentalization, and sometimes mineralization. Various mechanisms such as rhizoremediation, phytoextraction, phytostabilization, phytovolatilization, and phytodegradation are employed by plants to remove or degrade soil pollutants (Figure 11.1). Such mechanisms are used for the effective elimination of wastes such as phenolic compounds, metals, azo dyes, colorants, and various other organic and inorganic pollutants.

Organic pollutants are chiefly degraded by two mechanisms: phytodegradation or phytotransformation and rhizoremediation, whereas the degradation of inorganic pollutants is based on phytoextraction and phytostabilization. The approach known as rhizodegradatio/phytostimulation defines enhanced microbial degradation of contaminants in the root zone due to the exudates secreted by plant roots. The pollutant, in some cases, is immobilized in the root zone, which is known as phytostabilization.

TABLE 11.1
Concentrations of Organic and Inorganic Pollutants in Soil from Cities around the World

Pollutants		Source(s)	Location(s)	Concentration (µg/kg)	References
Organic	PAHs	Agriculture; diffuse pollution; industrial and traffic/vehicle emissions	China (Hong Kong)	30–170	Aichner et al., 2007; Banger et al., 2010; Baumard et al., 1998; Kumar & Kothiyal, 2011; Morillo et al., 2008; Trapido, 1999; Vane et al., 2014; Wang et al., 2013; Zhang et al., 2006
			China (Shanghai)	83–7220	
			USA (Miami, Florida)	251–2364	
			Spain (Seville)	89.5–4004	
			France (Arcachon Bay)	32–4120	
			UK (Greater London)	67,000	
			India (Jalandhar, Punjab)	4040–16,380	
			Nepal (Kathmandu)	184–10,279	
	PCBs	Burning of old appliances; electrical products; urban water runoff; industrial/commercial activities, spills, and disposal	France (Seine River basin)	0.09–150	Aichner et al., 2007; Martinez et al., 2012; Motelay-Massei et al., 2004; Rose et al., 2013; Wu et al., 2011; Vane et al., 2014
			Nigeria (Lagos)	3.6–23.6	
			China (Beijing)	11	
			Nepal (Kathmandu)	0.4–447	
			UK (Greater London)	123	
			USA (Cedar Rapids, Iowa)	56–160	
	Pesticides (e.g. DDT, HCN, AMPA)	Agriculture	China (Tianjin) (DDT)	628–2841	Gong et al., 2002; Kannan et al., 2003; Babu et al., 2003
			China (Tianjin) (HCN)	387–4689	
			USA (Southern California) (DDT)	0.11–45	
			USA (Southern California) (HCN)	0.1–0.54	
			India (DDT)	122–638	
			India (HCN)	13–238	
	Dioxin-like PCBs	Industrial activity	Brazil (Belo Horizonte-MG)	0.04–0.93	Pussente et al., 2017

	Element	Source	Location	Value	References
Inorganic (e.g. some HMs)	Zn	Wastewater	China (Beijing)	157	Abdu et al., 2011; Khan et al., 2008
			West Africa (Kano, Bobo-Dioulasso and Sikasso)	167	
	Pb	Digested sludge; industrial effluents; fertilizer and sewage sludge; wastewater	China (Tianjin & Beijing)	41.5–49	Khan et al., 2008; Meng et al., 2016; Qishlaqi et al., 2008; Sterrett et al., 1996; Wong, 1985
			USA	140–5210	
			Iran (Shiraz suburban)	412	
			India (Gujarat)	46	
			China (Hong Kong)	33	
	Cu	Digested sludge; pig manure compost; wastewater	China (Beijing)	33	Akoto et al., 2015; Khan et al., 2008; Wong, 1985
			China (Hong Kong)	52–105	
			Ghana (Kumasi)	7.4	
	Cd	Fertilizer and sewage sludge, wastewater	USA	3–10	Abedi-Koupai et al., 2015; Baker et al., 1979; Sterrett et al., 1996; Tiwari et al., 2011
			Iran (Isfahan)	6	
			India (Gujarat)	19.3	

Phytoextraction means to accumulate pollutants in harvestable plant tissues, especially shoot tissues, and this approach is notably utilized for inorganic pollutants.

In some cases, plants can degrade pollutants inside their tissues, and this method is called phytodegradation and is mostly suitable for organic pollutants since inorganic pollutants can only be moved and not degraded. Plants can also volatilize some pollutants, and the process is known as phytovolatilization. This method is used to volatilize organic compounds and specific metal such as Hg, As, and Se.

The mechanisms of phytoremediation are dependent on the type of pollutants, bioavailability, soil structures, plant species, and their tolerance to pollutants. The degradation of organic pollutants depends on many factors, such as their distribution and transformation. These factors, in turn, are governed by physical and chemical characteristics of the compound, such as water solubility, molecular weight, and octanol-water partition coefficient (K_{ow}). The degradation process is also governed by environmental circumstances such as temperature, pH, organic matter, and soil moisture.

11.4   ORGANIC POLLUTANTS

Most of the organic pollutants are xenobiotic and hence are not present naturally in the environment. Most of them are toxic even at low concentrations and may also bear a carcinogenic character. They are released into the environment through accidental discharge (using fuels, solvents), industrial activities (e.g. chemical, petrochemical), agriculture (e.g. pesticides), and military operations (e.g. explosives, chemical weapons). Moreover, polluted sites often contain a mixture of both organic and inorganic pollutants. Further, there are various types of organic pollutants: solvents (e.g. trichloroethylene), explosives (e.g. trinitrotoluene (TNT) and cyclotrimethylene trinitramine), PAHs, petroleum products including benzene, toluene, ethylbenzene, and xylene (BTEX), PCBs, and pesticides (e.g. atrazine, chlorpyrifos, 2,4-D).

11.4.1   Uptake, Translocation, and Accumulation of Organic Pollutants

Some of these pollutants, like chlorinated solvents and polycyclic aromatic hydrocarbons (PAHs), can be absorbed from the soil through plant roots and transpired from the shoot. The plants can take up organic pollutants from the environment through four mechanisms. These mechanisms include passive or active uptake through the root system, gaseous and particulate deposition to aboveground shoots, and direct contact between soil and plants' aerial parts. The mechanisms are influenced by the chemical and physical properties of pollutants such as their lipophilicity, water solubility, and vapor pressure; environmental conditions such as temperature and content of organic compounds; and the plant species and structure. Additionally, some plants have more capacity for uptake of some unique pollutants because of particular root exudates which actively mobilize them from the soil and make these compounds available for absorption and translocation.

The organic pollutants are taken up by plants simultaneously via different pathways. A pollutant may enter the plants from polluted soil to the roots and may be translocated by the xylem. Organic pollutants may also enter the vegetation from the atmosphere as gas or by particle deposition onto the waxy cuticle of the leaves or by uptake through the stomata and may be translocated by phloem. The low-molecular-weight pollutants enter the leaves through stomata, while high-molecular-weight pollutants enter leaves more often through the cuticle. Thus, the mechanism of uptake of high-molecular-weight compounds is based on adsorption on the cuticle leaf surface. Accordingly, leaves adsorb substances more selectively than roots do because of the easier mechanism of the entrance. The uptake via roots is determined by surface and the ratio of lipid components in roots. The lipid components in roots enable easier absorption of lipophilic pollutants. The plants with a high content of lipid components have high accumulation capacity for organic compounds such as PAHs. Higher accumulation of organic pollutants by plants was proved by comparing species with a large surface area of leaves to plants with a lower surface area. For example, vegetables like spinach and lettuce have higher potential for absorption and accumulation of organic pollutants because of the broad surface area of their leaves. It seems that organic pollutants can accumulate on the trichomes and then gradually diffuse to the base of the cell and spread into the adjacent cells of the trichome basement; finally, the trichome collapses. This fact that trichomes act as entering points for the organic pollutants or that they import organic pollutants coming from other cell types and tissues is still unclear. Trichomes of *Brassica juncea* accumulated cadmium and *Arabidopsis thaliana* trichomes contained a high glutathione concentration suggesting that trichomes may function as an efficient site of xenobiotic conjugation. Hence, trichome-specific engineering with organic pollutants degrading enzymes may be one strategy for future phytoremediation strategies.

11.4.2 Degradation of Organic Pollutants

11.4.2.1 Microbial Degradation

Biodegradation has been recognized as an effective approach for the remediation of organic pollutants in which microorganisms and their diverse metabolites break down toxic pollutants in different environments. In most cases, microbes transform organic pollutants to other soluble organic compounds or mineralize them into inorganic end products to use these pollutants as carbon and energy sources. The degradation of organic contaminants in microorganisms like fungi and bacteria have been vastly studied. The presence of various catabolic genes coding enzymes contributes to different potential activities in microorganisms. Besides, microorganisms have other adaptation strategies; for example, they can apply the efflux pumps to reduce the concentration of toxic pollutants within the cell and to generate biosurfactants. All these strategies and metabolic potentials make microorganisms a purification tool for the bioremediation of contaminants in the environment.

The physicochemical interactions of microbes with pollutants lead to the structural modification or degradation of the target pollutants. During an oxidation-reduction (redox) reaction, which involves the transfer of electrons between two compounds, the organic pollutants are oxidized by loss of electrons (electron donor), while the lost electron(s) is gained by other compounds (electron acceptor) that is reduced in this process. Microbes can also rebuild different strategies that enable them to clean up the environment and are generally classified into two types: aerobic and anaerobic processes. Aerobic processes occur in the presence of oxygen, where O_2 acts as the terminal electron acceptor. In the aerobic process, microorganisms use oxygen to oxidize the pollutants to CO_2 and biomass. Also, in this process, O_2 availability promotes the rate and efficiency of replication of aerobic microbes. The anaerobic process occurs in the absence of O_2, sub-categorizing into anaerobic respiration, fermentation, and methane fermentation. In this process, other oxidized inorganic or organic molecules replace O_2 as the terminal electron acceptor.

The microbial degradation of organic pollutants under aerobic and anaerobic conditions occurs through denitrification, methanogenesis, and sulphidogenesis. The initial enzymatic reaction in aerobic biodegradation is oxidation catalyzed by oxygenases (monooxygenases and dioxygenases) and peroxidases. A number of bacteria such as *Pseudomonas*, *Bacillus*, *Rhodococcus*, and *Mycobacterium* are recognized to break-down PAHs via metabolic enzymes such as methane and ammonia monooxygenases.

11.4.2.1.1 Polycyclic Aromatic Hydrocarbons (PAHs)

Biodegradation of PAHs as organic pollutants has been frequently studied. It was determined that a phthalic acid and its derivatives are produced through the PAHs' degradation by white-rot fungi and bacteria, which are reduced to CO_2 and polar metabolites. In bacteria, initiation reaction of PAHs degradation is via attack on the aromatic rings via dioxygenase leading to the formation of cis-dihydrodiole, which consequently is dehydrogenated to form pyrocatechol. Pyrocatechol is the primary intermediate of this cleavage. Such intermediates were also observed in fungal degradation processes. For example, wood-rot fungi are capable of releasing extracellular enzymes which degrade lignin in wood. But despite that, these enzymes are not only specific for lignin degradation but can also transform organic pollutants such as PAHs to quinone intermediates. In this process, some wood-rot fungi may then split the rings and finally degrade the compounds to CO_2 and water; however, for other fungi, it appears that the quinones are dead-end products. Moreover, fungi can also oxidize PAHs by the cytochrome P-450 enzyme system to form phenols and trans-dihydrodiols, which can conjugate and be eliminated from the microorganisms.

11.4.2.1.2 Aliphatic Hydrocarbons

Aliphatic compounds are also called linear alkanes, and the decomposition time of linear hydrocarbon compounds in the environment is shorter than that of aromatic compounds. Aliphatic hydrocarbons, as prevalent soil pollutants, define as open-chain methane derivatives which are both non-aromatic and non-cyclic organic compounds. These pollutants are carcinogenic at high concentrations and bring notable hazard to biological receptors due to the formation of toxic and carcinogenic metabolites during their biodegradation. These compounds can be subdivided into three structurally different groups, including alkanes (saturated hydrocarbons with single bonds), alkenes (unsaturated hydrocarbons with double bonds), and alkynes (unsaturated hydrocarbons with triple bonds). Alkanes as aliphatic hydrocarbons are one of the most significant percentages of petroleum-related contaminants being easily destroyed, especially those that have shorter chains and lower hydrophobicity. However, their degradation potential is decreased with an increase in the number of carbon chain length due to their decreasing solubility in an aqueous medium.

The fastest and the most complete breakdown of most organic pollutants occurs in aerobic conditions because of the metabolic benefits of having the O_2 availability as an electron acceptor. Aerobic bacteria like *Pseudomonas, Alcaligenes, Sphingomonas, Rhodococcus,* and *Mycobacterium* degrade both alkanes and polyaromatic compounds. The main process of aerobic degradation of aliphatic hydrocarbons is via the addition of one or two oxygen atoms to the hydrocarbon molecule. This process is catalyzed either by oxygenases or by cytochrome P450. The activated molecule is then changed into an alkanol, which is then oxidized to the related aldehyde and finally converted into fatty acid. The fatty acid is bonded to CoA, forming an acyl-CoA and processed by oxidation to produce acetyl-CoA. Therefore, the ultimate product of the oxidation of aliphatic hydrocarbons is acetyl-CoA being catabolized in the Krebs cycle and completely oxidizing the substrate into CO_2 and H_2O. In contrast to aliphatic hydrocarbons, the oxidation of aromatic hydrocarbons produces phenolic intermediates. Their further degradations are via central pathways catalyzed by intradiol or extradiol dioxygenases leading to the formation of simpler compounds, which are then inserted into the Krebs cycle and utilized (Figure 11.1).

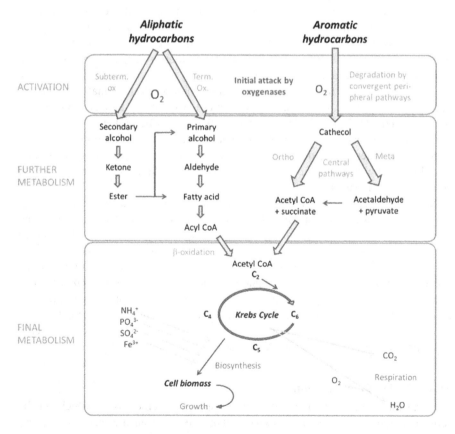

FIGURE 11.1 Degradation of aliphatic and aromatic hydrocarbons by aerobic bacteria. (From: Rohrbacher and St-Arnaud, 2016.)

11.4.2.1.3   Pesticides

Pesticides as organic chemical pollutants are purposely intended to control the insect population and decrease losses of agricultural products resulting in agricultural yield enhancement. Recently, powerful and remedial technologies have appeared to degrade or remove pesticides by using native or genetically modified organisms. There are many reports of various organisms such as bacteria, algae, yeasts, fungi, and plants which are specified in terms of their genome and enzymes applying for bioremediation of soil and water.

Pesticides could be degraded by different methods in the soil, such as physical degradation, chemical degradation, physical-chemical degradation, and microbial degradation. The latter one has been used more frequently in recent years because the applied pesticides are mostly microbial nutrients, which are finally degraded into small molecules like CO_2 and H_2O in microbial degradation processes. Through the enzymatic reactions, the pesticide compound first enters into a microorganism's body in a certain way. Then they are completely degraded or broken down into less toxic compounds by a series of physiological and biochemical reactions under the action of various enzymes.

Both plants and microorganisms have metabolic pathways that can degrade most of the pesticide compounds released in the environment. However, the long-term persistence of some pesticides, such as organochlorine compounds, limits the ability of microbes and plants to degrade them due to their hydrophobic nature, which prohibits their uptake and translocation.

To degrade most of the pesticide compounds, microbial transformation in the soil must play an essential role in making them more available for uptake and degradation by plants. Thus, various chlorinated pesticides can be broken down better in a vegetated soil than in a non-vegetated soil. A higher microbial population contributes to a faster rate of degradation of lindane, atrazine, metolachlor, trifluralin, butachlor, and aldicarb in the rhizospheric soil of some herbicide-tolerant plants in comparison with bare soil.

Some of the researchers have also indicated the role of endophytes in facilitating the phytoremediation of various pesticides. The decrease in phytotoxicity and an increase in phytoremediation of 2,4-dichlorophenolxyacetic acid (2,4-D) resulted from the colonization of the roots of phytoremedial plant with some endophytes. The removal potential of 2,4-D and the pea plants' biomass were enhanced after successful colonization of its roots with a poplar tree endophyte, *Pseudomonas putida*, having the ability for breaking this compound. Accordingly, it might be proposed that a plant-endophyte partnership could be employed as the main mechanism not only to increase the phytoremediation of pesticide-contaminated soils but also to decrease the concentrations of toxic herbicide residues in crop plants.

11.4.2.2   Phytodegradation

The transformation of organic pollutants is a species specific process which may be restricted in the given plant only to a particular tissue or organ or even in one particular developmental stage. The degradation processes of pollutants can be carried out in three phases: transformation, conjugation, and elimination reactions. In the first stage, the reactions are oxidation, reduction, and hydrolysis. These reactions may modify compounds which enable them to convert into more polar and water-soluble forms through enzymes like cytochrome P450 and carboxylesterases. It is assumed that these processes are localized inside mitochondria or the endoplasmic reticulum. These reactions may modify compounds so that they can conjugate with amines, acids, and alcohols in the second phase of detoxification. The second stage of pollutant degradation is the biotransformation phase, including specific conjugation processes with saccharide, glutathione, or amino acids, sulphatizing, and methylation. These processes may be marked as a group of synthetic metabolic reactions when a xenobiotic reacts with either endogenous compound. The reactions are catalyzed by relatively few specific enzymes such as glycosyl transferases and glutathione S-transferases, transferring them into the different parts of the cell, such as into the vacuole or cell wall.

The third stage of biotransformation is detoxification, in which the organic pollutants are essentially removed from metabolic tissues. This process is also defined as sequestration/compartmentalization, where conjugates are isolated by catabolic and anabolic mechanisms in plants. Created conjugates are transferred from the cytoplasm to the vacuole by trans-membrane ATP-dependent transporters (ABC-transporters) for accumulation. Alternatively, they can be additionally processed by hydrolytic reactions within the vacuole and then re-exported into the cytoplasm for more metabolisms.

Plants can also accumulate soluble conjugates in the vacuole and non-soluble conjugates in the cell wall for further metabolism. The organic pollutants such as aromatic compounds with hydroxyl, carboxyl, amino, or sulfhydryl groups are deposited into lignin or to other cell wall components such as protein, hemicellulose, and cellulose. A study on organic pollutants has shown that in maize, phenanthrene can be metabolized into more polar products, and in another study anthracene and its derived-compounds are bound to cell wall components such as pectin, lignin, hemicellulose, and cellulose.

11.5   INORGANIC POLLUTANTS

Inorganic pollutants are naturally present in the earth's crust and atmosphere. They are also contributed by human activities such as industry, mining, motorized traffic, agriculture, and military actions promoting their release and concentration in the environment leading to toxicity. These pollutants comprise metals or metalloids such as, Cd, Cu, Hg, Mn, Se, and Zn, radionuclides such as Cs, P, and U, and plant fertilizers such as nitrate and phosphate. The inorganic pollutant mainly constitutes heavy metals, making them more persistent than organic contaminants are. Although they cannot be broken down, they can be modified by reduction or oxidation. They can also migrate into different plant parts and be accumulated and volatilized.

11.5.1   Uptake and Sequestration of Inorganic Pollutants

The uptake of inorganic pollutants such as HMs consists of ion uptake from the soil and into the root cells. The metal must mobilize into the soil solution for uptake and accumulation by plants. Siderophores, including mugineic acids and avenic acids, are set free by some plant species (grass species) to increase the bioavailability of HMs for root uptake. The transportation and accumulation of HMs are carried out by different compounds such as organic acids (OAs) within tissues and other compartments. OAs (citrate, malate, and oxalate) also perform a substitutive role of excluding metals from plants. For example, Al absorption is prevented by the exudation of OAs, making the OAs-Al complex in wheat. Similarly, Cu uptake in *Populus tremula* root is prevented by the exudation of formate, malate, and oxalate, while Zn absorption is prevented through the exudation of formate.

After uptake from the root, HMs enter into the cytosol by transporter protein, mainly ZRT-IRT-like proteins (ZIP) family members. Into the cytosol, HMs provokes phytochelatin synthase (PCS) an enzyme by which the reaction of phytochelatin (PC) production through glutathione is catalyzed. HMS phytochelatin complexes (HM-PC) (low-molecular-weight complexes) move to the vacuole through ABC-transporters. This complex collects within the vacuole, and as more HMs are added, they convert into a high-molecular-weight (HMW) complex. HMS is also transferred into the vacuole through cation/proton exchanger (CAX) transporters exchanging the HMs directly with protons, metal tolerance proteins (MTP) as HMs-proton exchanger transporters, and natural resistance associated macrophage proteins (NRAMPs). These transporter proteins are established into the tonoplast and mediate the pathway of metal ions for compartmentation or remobilization.

Further, metallothioneins (MT) are low-molecular-weight, cysteine-rich, and metal-binding proteins applying for HMs remediation through intracellular transference or sequestration. These chelators joined to the metals and make a complex move into vacuole like Zn transportation within the vacuole by MTP (Figure 11.2).

11.5.2 PLANT RESPONSE TO INORGANIC POLLUTANTS

Plants present various mechanisms for removing metals indicating the rate of metal uptake by the plants. In general, phytoremediation approaches for inorganic pollutants are phytostabilization (immobilization), phytoextraction (rhizofiltration), and phytovolatilization.

11.5.2.1 Phytostabilization

Phytostabilization, also named as phytosequestration or phytodeposition, is an approach dealing with stabilizing or moving pollutants in the soil near the root to forbid the movement of heavy metals to either groundwater or food. In recent years, two kinds of grass—*Agrostis* species and *Festuca* species—were applied in the phytostabilization of soil polluted with Cu, Zn, and Pb. Thus, phytostabilization is considered as a distinct approach to restrict migrations of HMs and to inactivate toxins.

11.5.2.2 Phytoextraction

Phytoextraction, also named as phytoaccumulation, is used to absorb pollutants from soil and water through roots and transferring the pollutants to the shoot and leaves. Hyperaccumulator plants are generally small and slow growing. But they are able to

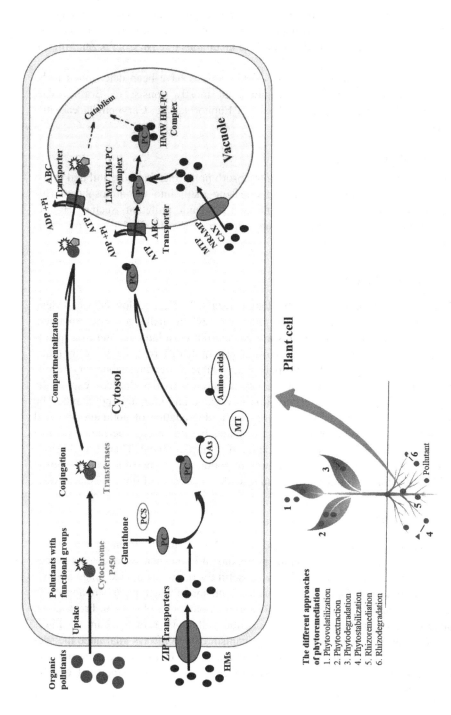

FIGURE 11.2 Metabolism of organic and inorganic pollutants in plants cells. (From: Modified from Yang and Chu, 2011; Yadav et al., 2015; Chandra et al., 2017.)

take up HMs 50–500 times more than non-hyperaccumulator plants can without any negative influence on their growth and development. Accordingly, they can accumulate and detoxify exceptionally high levels of metal ions like Ni, Co, Pb, Zn, Mn, Cd, and more in their shoots.

Recently, nearly 450 plant species of 45 families have been determined to be hyperaccumulators. These hyperaccumulators mostly consist of Brassicaceae, Asteraceae, Fabaceae, Lamiaceae, Poaceae, Euphorbiaceae, Caryophyllaceae, and Violaceae.

11.5.2.3 Phytovolatization

Phytovolatilization means using plants to absorb pollutants from the soil and transforming them into volatile form and releasing them into the atmosphere. This approach is used for removal of organic and inorganic pollutants such as Hg, Se, and As from soil. For example, Se changes into volatile methyl selenate form and is released into the atmosphere.

11.6 FUTURE PROSPECTS

Although the application of biotechnology to establish effective, low-cost, and clean bioremediation technologies is much promised, several challenges have remained. Most phytoremediation experiments are carried out on a labscale and in a hydroponic environment, while the soil environment is quite different, and there are many metals in insoluble forms that are poorly accessible. Extending laboratory results to the real environment is one of the key research activities that can exploit the real potential of plants in the elimination of organic and inorganic pollutants. The efficiency and capability of many plants in the degradation of pollutants are still unknown, and the production of transgenic plants by genetic engineering is one possible future role of plants in the purification of the environment. Therefore, illustrating how genes are involved in the pathway of pollutant decomposition and detecting suitable genes for phytoremediation can provide other cases of future perspectives.

11.7 CONCLUSION

Phytoremediation is an innovative and safe technique for the management of polluted soil with organic and inorganic pollutants; it is inexpensive and solar energy driven, and naturally clean technology is useful for removing a wide range of environmental pollutions. There are various mechanisms in which plants are applied to remove or remediate polluted sites. To remove pollutants, plants can break down organic contaminants and stabilize inorganic pollutants through acting as filters or traps. Using only one remediation mechanism cannot always guarantee the success of phytoremediation. Thus, a combination of mechanisms, including uptake, transformation, stabilization, and rhizosphere degradation, may be required to break down pollutants.

12 Conservation of Biodiversity by Biotechnology

Alka Rani, Rajni Kashyap and Wamik Azmi

CONTENTS

DOI: 10.1201/9781003131427-12

12.1 INTRODUCTION

Biodiversity, or biological diversity, is defined as the variety and variability of all species on earth. It includes plants, animals, microorganisms and their genes, marine ecosystems, and terrestrial and water ecosystems. Basically, humans are dependent on services provided by biodiversity. Since the beginning of human civilization, biodiversity is the basis of agriculture and is the source of crops, food products, timber, fiber, and fresh air and water, which are all necessary to lead a healthy life. Three different levels of biodiversity exist in nature: genetic, species, and ecosystem. Continuous actions of human to expand agricultural land for crop production and livestock, human habitat, and industrialization have destroyed the habitat of animals, causing them to move into smaller territories. Further, with the increase in the human population, the areas of habitat for both wild animals and humans become limited. Organisms or animals facing extinction are referred to as endangered species. At present, degradation of species and groups of species and a decrease in the number of organisms are taking place very rapidly. The major causes of increased degradation and extinction of genetic diversity are habitat loss, change in climate, pollution, invasive species, pressure of human population, change in lifestyle, industrialization, and an increase in agricultural pressure and practices. The conservation of biodiversity is a global concern.

Biotechnology is used for the enhancement of biodiversity. It is defined as any technology that uses living organisms, or substances, to make a product or to improve plants or animals. Biotechnology includes various techniques that are applied to change and improve living organisms to provide various benefits for human use. It has been used for the conservation and utilization of biodiversity, particularly for important crops. Molecular biology is essential for the characterization and conservation of biodiversity. Molecular markers help to identify the genes of interest to breeders and establish the extent of diversity within a species. Molecular markers have also been used for the mapping of genes of crops and selecting favorable traits to develop better germplasm for growers. Such techniques can also help in establishing the priorities for conservation. Biotechnology already helps in the conservation of plant and animal genetic resources through new methods. These methods include the collection and storage of genes (as seed and tissue culture), identification of useful genes, detection of genes responsible for disease and elimination of those diseases by gene therapy, improved techniques for long-term storage, and safer and more efficient distribution of germplasm to users.

DNA technologies are very effective in conservation strategies, and the DNA bank is an efficient, very simple, and long-term method used for the conservation of genetic resources for biodiversity. Entire plants cannot be obtained from DNA; the stored genetic material must be introduced by using genetic techniques. *In vitro* techniques are also useful for conserving plant biodiversity for the short, medium, and long term. Such techniques involve three basic steps: culture initiation, culture maintenance and multiplication, and storage. Medium-term storage allows the conservation from several months to few years with the use of a slow-growth strategy. The temperature reported for medium-term conservation usually ranges from 4 °C to room temperature. However, tropical plant species are often sensitive to cold and have to be stored in the range of 15–20 °C or even higher, depending on their sensitivity. For an undefined time of storage, cryopreservation has extensively been applied. Cryopreservation is a very beneficial method, and it allows 20% increases in process of regeneration compared to other conservation methods. The great advantage of storage of biological material at such a low temperature is that both metabolic processes and biological deterioration are considerably slowed or even halted.

Recent advances in genomics, proteomics, and metabolomics supply distinctive opportunities for the identification and industrial utilization of biological merchandise and molecules within the pharmaceutical, nutraceutical, agricultural, and environmental sectors. There is a wide range of biotech products that are highly profitable for farmers. Biotechnology applications offer opportunities to make substantial advances in our knowledge of the diversity of some of the most important crops.

12.2 BIODIVERSITY IN ECONOMIC DEVELOPMENT

Biodiversity is obligatory for our existence because it provides the elementary building blocks for many things that provide a healthy environment for us. For human survival and economic development, biodiversity provides food, clothing, housing, industrial raw materials, medicine, and much more (Figure 12.1). Many industrial products such as dyes, waxes, oils, resins, lubricants, perfumes, fragrances, paper, rubber, latex, poisons, and cork have been derived from various plant species.

A variety of wild plant species have been used for medicinal applications from the beginning of history. Many plant species contain thousands of important compounds with medicinal properties isolated from leaves, roots, bark, herbs, and so on. Pharmaceutical companies have produced many medicines from these chemicals, but the original formulas come from plants. For example, quinine derived from the bark of the Amazonian cinchona tree, morphine from the poppy plant, and digitalis from the foxglove plant are used, respectively, to treat malaria, relieve pain, and treat chronic heart trouble. According to the National Cancer Institute in the United States, over 70% of the potent anti-tumor drugs derived from plants in the tropical rainforests area, and out of 250,000 known plant species about 5,000 have been explored for possible medical applications. Most of the world's populations (about 80%) depend on the medicines obtained from natural resources, either directly or indirectly (Table 12.1). Further, 200–250 herbal species are exported in the Mediterranean region for the treatment of human illness.

Supplies based on animal origin include wool, fur, leather, silk, lubricants, and waxes. Animals may also be used as a mode of transportation. Biodiversity also produces economic wealth in many areas, such as parks and forests, where wild nature and animals are a source of attraction for many people. Ecotourism, in particular, is an emerging recreational outdoor activity. A variety of natural ecosystem services and processes such as atmospheric regulation, purification of water and air, pollination, prevention of soil erosion, and recycling of nutrients are supported by

TABLE 12.1
Estimated Number of Plant Species Used Medicinally in Different Countries

Sr. No.	Name of Country	Estimated Number	References
1	India	7500	Shiva, 1996
2	China	10,000–11,250	He and Gu, 1997; Pie, 2002; Xiao and Yong, 1998
3	Mexico	2237	Toledo, 1995
4	North America	2572	Moerman, 1998

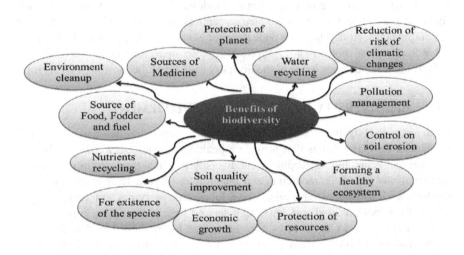

FIGURE 12.1   Benefits of biodiversity.

biodiversity. Each species performs a specific function within an ecosystem. They can capture and store energy, manufacture organic material, decompose organic material, facilitate the cycling of nutrients and water throughout the scheme, control soil erosion and pests, fix atmospheric gas, and facilitate climate regulation. The more diverse an ecosystem is, the better it can cope with environmental stress and the more productive it is. Biodiversity is the reflector of our relationships with the other living species. An ethical view of biodiversity considers the rights of others species, the human duty to those species and education about the topic. Humans cannot voluntarily cause their own extinction if they consider that all species have a right to exist. Every day, the degradation of species increases because of the ecosystem's decreased ability to maintain itself or to recover from damage. Highly complex mechanisms underlie these ecological effects.

About 1.75 million species of animals, plants, and microorganisms have been identified out of the 13 million total species estimated by scientists. The United Nations Convention on Biological Diversity notes, "at least 40 per cent of the world's economy and 80 per cent of the needs of the poor are derived from biological resources".

12.3 CAUSES OF BIODIVERSITY DEGRADATION

In recent years, the degradation of biodiversity has increased significantly, which poses a high risk to the survival of everyone. The earth's atmosphere is changing on all scales from local to global due to human activities. Biodiversity degradation is a global crisis. The major causes of loss of biodiversity is shown in Figure 12.2.

12.3.1 HABITAT LOSS AND FRAGMENTATION

Destruction of habitat is the process by which native habitat becomes incapable of supporting its natural species. The organisms that previously used the site are displaced or destroyed, which ultimately leads to their extinction. Increase in population size and acceleration in demand for resources lead to the conversion of natural habitats to agricultural land, grassland, plantations, infrastructure, industrial production, and urbanization.

Other causes include logging, mining, trawling, and urban sprawl. Currently, habitat destruction is the primary cause of the disappearance of many species worldwide. More than 90% of the original area of the Mesopotamian marshes of Iraq was lost due to water diversion projects during the 1990s. It is obvious that populations and species will experience hardship when their habitat becomes degraded or is lost entirely. Loss of habitats leaves many of species to continuously decline and become extinct. Table 12.2 lists some critically endangered, endangered, and vulnerable medicinal plants.

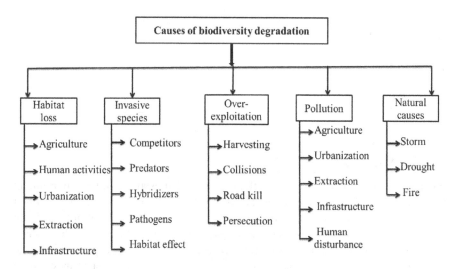

FIGURE 12.2   Causes of biodiversity loss.

TABLE 12.2
IUCN 2015 Critically Endangered, Endangered, and Vulnerable Medicinal Plants

Sr. No.	Vulnerable	S. No.	Endangered	S. No.	Critically Endangered
1	*Hydnocarpus pentandrus*	1	*Cinnamomum wightii*	1	*Aconitum chasmanthum*
2	*Magnolia nilagirica*	2	*Coptis teeta*	2	*Chlorophytum borivilianum*
3	*Malaxis muscifera*	3	*Decalepis hamiltonii*	3	*Gentiana kurroo*
4	*Phyllanthus indofischeri*	4	*Dysoxylum malabaricum*	4	*Gymnocladus assamicus*
5	*Nilgirianthus ciliates*	5	*Gymnema khandalense*	5	*Lilium polyphyllum*
6	*Garcinia indica*	6	*Humboldtia vahliana*	6	*Nardostachys jatamansi*
7	*Piper pedicellatum*	7	*Illicium griffithii*	7	*Saussurea costus*
8	*Salacia oblonga*	8	*Iphigenia stellata*	8	*Tribulus rajasthanensis*
9	*Terminalia pallid*	9	*Lamprachaenium Microcephalum*	9	*Valeriana leschenaultia*
10	*Myristica dactyloides*	10	*Nepenthes khasiana*	10	*Commiphora wightii*
11	*Aconitum violaceum*	11	*Pimpinella tirupatiensis*		
12	*Boswellia ovalifoliolata*	12	*Piper barberi*		
13	*Calophyllum apetalum*	13	*Syzygium alternifolium*		
14	*Cayratia pedata*	14	*Shorea tumbuggaia*		
15	*Cinnamomum macrocarpum*	15	*Aconitum heterophyllum*		
16	*Cinnamomum sulphuratum*	16	*Angelica glauca*		
17	*Diospyros candolleana*				
18	*Diospyros paniculata*				

Source: Dhyani and Dhyani, 2016.

12.3.2 POLLUTION

Pollution is the introduction into or the presence of substances in the environment that have harmful or poisonous effects. All types of pollution cause a serious threat to our biological diversity (Figure 12.3). Transport and agriculture are the main sources of pollution. International Union for Conservation of Nature (IUCN) classification distinguishes among pollution threats according to primary activities (such as agriculture and forestry) in generating pollutants and media (such as effluent) through which mixtures of pollutants are introduced to ecosystems. Pollution has

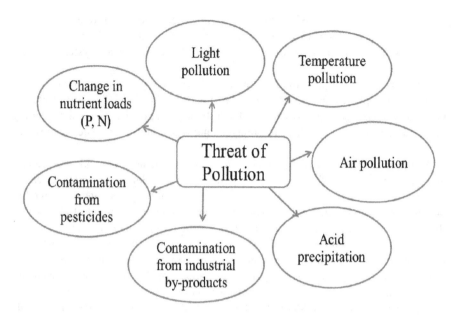

FIGURE 12.3   Various threats of pollution.

become a great threat to global biodiversity due to industrialization. Many species cannot deal with the speedy changes in physical parameters that are occurring in the environment. An increased level of pollution results in toxicity. Toxicity of any substance depends on the level at which it occurs in the environment and an organism's capacity to tolerate that substance. Different studies distinguishes pollutant-generating activities—agriculture, urbanization, extraction, infrastructure, and human disturbance. Another increasing cause of biodiversity degradation and ecosystem dysfunction is nutrient loading, primarily of nitrogen and phosphorus. In addition to this, nitrogen compounds can lead to the eutrophication of ecosystems. Eutrophication is defined as an increase in concentration of nutrient content to the extent that it leads to an increase in primary productivity or a great increase of phytoplankton in a water body.

Pollution from burning fossil fuels like oil, coal, and gas will stay within the air as particle pollutants. Sulfur and nitrogen are released from fossil fuel combustion as oxides. These gaseous oxides in combination with atmospheric water transform to acidic particles and acid precipitation in the atmosphere. Acid rain mainly consists of sulfuric and nitric acid. These acids cause the acidification of lakes, rivers, and sensitive forest soils. Acid rain contributes to slower forest growth and damage of trees at high altitude. The gas and particles may be deposited directly on plants and soil surfaces in a process known as dry deposition, or they may be incorporated into cloud droplets, raindrops, or snowflakes to increase the acidity of precipitation. Sulfur oxides are released primarily from coal combustion, whereas any combustion can produce N oxides. Thus, the contribution of motor vehicles to the pollution problem is greater for N oxides than for S oxides. Sulfate and nitrate ions can be readily leached from soil to surface water. Deposition of these pollutants to terrestrial ecosystems may cause a cascade effect that includes freshwater systems, estuaries, and ocean coasts.

Ozone is a well-studied pollutant known to be toxic to plants and animals. There is reduction in photosynthesis of plants due to ozone. This leads to slower growth of plants and in very severe cases causes death. In animals, ozone effects have mainly been studied in humans, where it damages lung tissue and exacerbates respiratory problems such as asthma.

Mercury is ranked third by the US Government Agency for Toxic Substances and Disease Registry of the most toxic elements on the planet. Arsenic and lead is dumped into our waterways and soil, spilled into our atmosphere, and consumed in our food and water. Human activities have nearly tripled the amount of mercury in the atmosphere. The atmospheric burden of the mercury is increasing by 1.5% per year. As mercury contaminates the soil and water, it has the potential to enter the food chain through plant and livestock. Once mercury enters the food chain, it will bioaccumulate, inflicting adverse effects on human health. The exact mechanism by which mercury enters organic organisms remains mostly unknown and probably varies among ecosystems. Mercury is known to accumulate in soil, but studies of its effects have primarily targeted aquatic ecosystems wherever anaerobic conditions facilitate the production of a form called methyl mercury. Exposure of humans and animals is largely associated with methyl mercury. Previously, terrestrial organisms considered to be in danger due to mercury contamination were animals that feed on other animals from the aquatic food web, like birds feeding on aquatic insects or raccoons that eat aquatic invertebrates.

12.3.3 OVEREXPLOITATION

Overexploitation is also known as overharvesting, which refers to the exploitation of natural sources to the point of diminishing returns. Human have always depended on nature for food and shelter, but when "need" turns to "greed", it leads to the overexploitation of natural resources. Continued overexploitation can lead to the destruction of the resource, including extinction. The term applies to natural resources such as wild healthful plants, grazing pastures, game animals, fish stocks, forests, and water aquifers. Overexploitation is one of the five main activities threatening global biodiversity. Overexploitation is not an activity limited to humans. Introduced predators and herbivores, for example, can also overexploit native flora and fauna. The population of the world now stands at seven billion and is supposed to reach nine billion by 2045. Overpopulation causes the overconsumption or overexploitation of natural resources. The threat of overexploitation includes intentional (for example, harvesting and persecution) and unintentional (for example, by catch and road kill) activities of humans. While the IUCN classification does not include the threats from unintentional activities, it distinguishes among harvesting activities under the category of biological resource use. Consumptive use of terrestrial animals, terrestrial plants, logging, wood harvesting activity, and fishing are distinguished in the IUCN classification (Figure 12.4). Overuse does not necessarily cause the destruction of

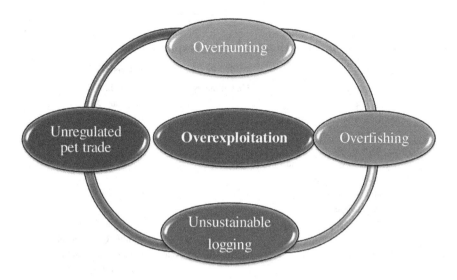

FIGURE 12.4 Factors responsible for overexploitation of biodiversity.

the resource, nor is it necessarily unsustainable. However, a decrease in numbers or quantity of the resource can modify its quality.

Forests are considered overexploited when they are logged at a rate quicker than rehabilitation takes place. Replantation competes with different land uses like food production, livestock grazing, and living space for more economic processes. Historically, utilization of forest products, including timber and fuel wood, have played a key role in human societies, comparable to the roles of water and cultivable land. Today, developed countries still utilize timber for building their homes and pulp for paper. In developing countries, almost three billion people depend on wood for heating and cooking. Short-term economic benefits created by conversion of forest to agriculture, or overuse of wood product, usually result in loss of long-term financial gain and biological productivity. West Africa, Madagascar, Southeast Asia, and numerous other regions have experienced lower revenue because of overuse and the successive decline of timber harvests. In some areas, meat is in such less quantity that animals are hunted for food or similar livelihood activities. Unsustainable hunting for consumption and trade of wild meat (also known as bush meat) by humans represents a significant extinction threat to wild terrestrial mammal populations, perhaps most notably in parts of Asia, Africa, and South America. Tropical species are especially in danger. In African nations, between one and five million tons of bush meat is collected annually, a figure thought to exceed sustainable levels. People try to hunt additional animals for the extra sale or income of their meat, hides, organs, or different body parts, and some species are acquired illegally and sold on the black market. Populations of enormous primates such as gorillas and chimpanzees, hoofed animals and other mammals have been reduced by 80% or more by hunting, and certain species may be eliminated altogether. This decline has been called the bush meat crisis.

TABLE 12.3
List of Some of the Species Affected by Overexploitation

Sr. No.	Species	Native Region	References
1	Carolina parakeet	United States	Vie et al., 2009
2	Great auk		Vie et al., 2009
3	*Achantinella* (15 sp. listed as extinct and 24 critically endangered)	Hawaii	IUCN, 2003
4	*Partula* (60 sp. extinct and 14 critically endangered)	Polynesia	IUCN, 2003
5	Asian black bear	Asia	Collin, 2012
6	Saiga antelope		
7	Steller's sea cow	Eastern Russia	Domning, 2016
8	Passenger pigeon	North America	Birdlife International, 2016
9	Atlantic bluefin tuna	North Atlantic Ocean	Endangered Atlantic bluefin tuna formally recommended for international trade ban, 2009
10	Cod		Frank et al., 2005

The footstool palm is a wild palm tree found in Southeast Asia. The leaves of this plant are used for thatching and wrapping of food. The overharvesting of the leaves has affected its leaf size, making it smaller. As another example, once the common or garden hedgehog was imported to the Scottish island of Uist. By the mid-1990s, hedgehogs were rich throughout the fertile areas of South Uist, Benbecula, and the extreme south of North Uist but were absent from the northern and western parts of North Uist. Surveys done in the 1990s, which indicated that the populations of some shorebird species were declining, and research revealed that there are high levels predation of eggs by hedgehogs. Twelve species of avi fauna are affected, with some species numbers being reduced by 39%. Many species have become extinct within the last 500 years (Steller's sea cow, passenger pigeon) because of overexploitation by humans. Table 12.3 lists some of the species that have been affected by overexploitation.

12.3.4 SPECIES INVASION

Great damage to biodiversity has been done by the introduction or spread of non-native species. Invasive species can cause disturbances to native species, competing with them for food, eating them, spreading diseases, causing genetic changes through interbreeding with them, and disrupting various aspects of the food web and the physical environment. Not all the species that enter a new place become invasive, in fact most do not. Many garden plants have been imported from different places, and they do not achieve big populations even if they grow wildly and do not have a large effect on native species' survival. These are just called non-native or introduced species. Imported non-native species that become invasive, however, cause big problems to native species. Species invasions have contributed to the extinction of many species worldwide, particularly on islands. Predation by introduced species has caused the extinction of various native animal species on islands, and competition of exotic plants with native plants causes their extinction. These extinctions have occurred disproportionately among taxonomic groups. For example, there has been a loss of several species of birds, both in absolute terms and relative to their total number of species, whereas plants have lost only a few species. Humans have caused or contributed to the extinction of several plants and animals. Over the past 15,000 years, humans have contributed to the extinction of large fauna on most continents of the world. The total range of recent extinction is unknown; as a result, several species have seemingly gone extinct before ever being recorded by science.

The extinction of invasive species has been well documented across a wide variety of islands and for a number of taxonomic groups. In general, many species of plants, vertebrates, and invertebrates have been imported to the islands. Many of those introduced species have become naturalized, that is, they have formed independent populations capable of perpetuating themselves. The loss of several native species occurs on islands. Among vertebrates, bird species have been extinted greatly, mostly because most other vertebrate groups are relatively depauperate (weak) on islands. People, and also the merchandise we use, travel around the world so quickly and typically carry uninvited species with them. Ships can carry aquatic organisms in their ballast water, whereas smaller boats can carry them on their propellers. Insects can get into wood, shipping palettes, and crates that are shipped around the world. Some decorative plants can escape into the wild and become invasive, and some invasive species are purposely or accidentally discharged pets. In the Florida Everglades, Burmese pythons are becoming a big problem. These snakes prey on a wide variety of birds, mammals, and crocodilian species occupying the Everglades.

Over the past few decades, globalization has increased the movement of people and goods around the world. This leads to a rise in the number of species introduced to areas outside their natural ranges. It has been reported that over one-third of all introductions in the past 200 years occurred after 1970. The rate of introductions is showing no sign of slowing down. The impacts of invasive alien species (IAS) can be compounded by climate change—the change in the earth's climate due to rising greenhouse gas emissions. Extreme climatic events resulting from climate change, such as hurricanes, floods, and droughts, can transport IAS to new areas and decrease the resistance of habitats to invasions. Climate change is additionally forming new pathways of introduction of IAS. The emerging Arctic shipping passages due to melting ice caps will greatly reduce the time taken for ships to travel from Asia to Europe, but there are chances that this will increase the danger of alien species surviving the journey. IAS reduce the resilience of natural habitats, making them more vulnerable to the impacts of climate change. Some grass and trees that have become IAS can considerably alter health regimes, especially in areas that are becoming warmer and drier. This will increase the frequency and severity of wildfire and puts habitats, urban areas, and human lives in danger. IAS may also affect agricultural systems by reducing crop and animal health. Exotic pests include non-native microorganisms, plants, insects, and other animals that cause or transmit diseases, displace native species, or diminish the economic value of a product. These pests produce toxins or act as a vector for plant, animal, or human diseases. These pests cause damage to domestic animals, cultivated crops, ornamentals, forests, wildlife, pets, and their habitats and humans. This will also influence the efficiency of production and quality of product directly. Alien pests may be a nuisance in one environment while being beneficial in another (Asian lady bird beetles invade homes for winter refuge in the midwestern United States but provide biological control of the pecan aphid in Georgia).

The greatest impact is caused by introduced species that change an entire habitat, because many native species flourish only in a certain habitat. Asian chestnut blight fungus nearly eliminated the American chestnut from over 180 million acres of eastern US forests within the first half of the 20th century. It was a disaster for several animals that were extremely adapted to live in forests dominated by this tree species. The ten moth species that would live solely on chestnut trees became extinct. Similarly, the Australian paperbark tree has taken place of native plants, such as sawgrass, in over 400,000 acres of south Florida, because it has a fusion of traits (spongy outer bark and ignitable leaves and litter) that increase hearth frequency and intensity. Many birds and mammals adapted to the native plant community declined in abundance as paperbark took over. In similar fashion, aquatic plants like the South American *Eichhornia crassipes* (water hyacinth) and marine algae like the Australian *Caulerpa* in Texas and Louisiana were introduced in the United States as an ornamental plant from the Mediterranean Sea, and they changed the vast expanses of habitat by replacing formerly dominant native plants. Water hyacinth forms a thick sheet on the surface of freshwater bodies. Its populations are very much known to double in as little as 12 days, leading to the blockage of waterways, limiting boat traffic, and affecting fishing and trade. In Victoria Lake in eastern Africa, the plant can grow to such extent that ships are unable to move from the dock. Ship rats (*Rattus rattus*) are an invasive species native to the Indian subcontinent but occurring throughout the world. They are frequently identified with catastrophic declines of birds on islands, and they transmit plague bacterium via fleas in certain areas of the world. The zebra mussel (*Dreissena polymorpha*) is a small freshwater mussel, accidentally brought to the United States from southern Russia. It transforms aquatic habitats by filtering prodigious amounts of water (thereby lowering densities of planktonic organisms) and settling in dense masses over vast areas. Treatment plants are most affected because the water intakes bring the microscopic free-swimming larvae directly into the facilities. Zebra mussels also cling to pipes under the water and clog them. At least 30 freshwater mussel species are threatened with extinction by the zebra mussel. They are also accountable for the near extinction of several species in the Great Lakes system by outcompeting native species for food and by growing on top of and suffocating the native clams and mussels. The numbers of introduced species are increasing within the US and elsewhere because of enhanced trade and travel, but the situation is not hopeless. Internationally, the Rio Convention on Biological Diversity (1992) identified the threat and called for action to limit it. Further, a Global Invasive Species Program, formed by the United Nations and other international organizations, is beginning to answer this call with a series of programs designed to deal with particular sorts of introduced species.

12.4 CONSERVATION OF BIODIVERSITY

Conservation of biodiversity means the shielding of valuable natural resources for future generations as well as for the well-being of ecosystem functions. Conservation of biological diversity results in conservation of essential ecological diversity to preserve the continuity of food chains. The conservation of biodiversity is an important issue concerning the human population worldwide. Human-induced pressure, introduction of alien species, as well as domesticated species and chronic weed invasion have the worst effects on biodiversity. All these factors lead to increasing the number of threatening species. The concept of conservation biology grew in the mid-20th century as ecologists, naturalists, and other scientists worked together to address the major issue of global biodiversity decline.

12.4.1 BIODIVERSITY CONSERVATION TRENDS IN BIOTECHNOLOGY

The UN Convention on Biological Diversity (Article 2) defined biotechnology as "any technological application that uses biological systems, living organisms or their derivatives, to make or alter products or processes for a defined purpose" (https://www.cbd.int/convention/articles/?a=cbd-02). Biotechnology covers a variety of techniques and applications. It ranges from widely used and long-established techniques of traditional biotechnology to novel and advanced biotechniques of cell and tissue culture and transgenic methodology. The ultimate purpose satisfied by these techniques is to allow changes and improvements in living organisms so as to provide desirable products for human use. At present, different parts of the world are facing the loss and/or endangerment of distinct species and the mass extinction of a group of species. In this regard, biotechnology plays a crucial role in the conservation, evaluation, and utilization of biodiversity. Biodiversity conservation techniques can be divided in two general categories viz. *in situ* and *ex situ* conservation techniques. Both techniques contribute equally for biodiversity conservation. In order to restore the overall biodiversity at a given eco-site, conservation strategies generally involve both preservation techniques. Although working to achieve the same goal, the principal difference between the two is that *in situ* conservation deals with the maintenance of viable populations in their natural surroundings; it is a dynamic system. Further, *in situ* techniques allow the biological resources to evolve and change over time through natural or human-driven selection processes. In contrast, *ex situ* conservation implies the maintenance of genetic materials outside of the "normal" environment where the species has evolved. The *ex situ* technique aims at maintaining the genetic integrity of the material at the time of collection. Recognizing the complementary role of both techniques (*in situ* and *ex situ*), the Convention on Biological Diversity (CBD, 1992) set out guidelines (Articles 8 and 9) for their use. Appropriate strategy should be selected based on a number of criteria, including the biological nature of the species and the viability of applying the chosen methods.

12.4.2 IN SITU CONSERVATION

In situ conservation can be defined as the conservation of natural habitats and ecosystems and therefore the maintenance and rehabilitation of viable populations of species within their natural habitats where they have developed their distinctive properties (CBD, 1992). The existing genetic diversity and viable population of wild taxa can be managed and maintained using *in situ* conservation techniques. The *in situ* conservation technique helps in the maintenance of biological interactions, ecological processes, and functions under natural conditions. This in turn allows natural evolution such as mutation, migration, recombination, and selection of a viable population. Variation is considered an important component for the long-term survival of biodiversity in natural settings and can only be met by adopting *in situ* conservation policy. Conservation of wildlife in national parks, sanctuaries, and conservation reserves comes under the category of *in situ* conservation. In addition to these *in situ* conservation techniques, biotechnology contributes to *in situ* conservation of biodiversity by bioremediation of polluted landscapes using an eco-friendly, cost-effective, and versatile technology such as microbial bioremediation that will rescue the biodiversity heritage and restore the destroyed ecosystems naturally. However, *in situ* methods alone are insufficient for saving endangered species in cases where heavy loss of species, populations, and ecosystem composition would ultimately lead to complete loss of biodiversity. In that case, the *in situ* technique must be complemented with the *ex situ* conservation technique.

12.4.2.1   Limitations of *In Situ* Conservation

Many risks and uncertainties are linked with *in situ* conservation, as the establishment of a conservation area does not mean complete protection of biodiversity. The near extermination of the entire habitat of the golden lion tamarin (*Leontopithecus r. rosalia*) in 1992 by fire is an extreme example of the risk associated with this technique (Castro, 1995). The requirement of large areas of the earth's surface to preserve the full complement of biodiversity of a region can be considered the most important drawback of the *in situ* conservation technique. To ensure the survivability of endangered species, the conserved area may not be large enough because of the fragmentation of habitat. The genetic diversity of the organism in question may have already been dramatically diminished as the conditions that threatened the organisms in the area may still be there, ascribed to interspecific competition or the invasion of pathogenic organisms.

12.4.3   *Ex Situ* Conservation Technique

The *ex situ* conservation technique of biodiversity conservation targets all levels of biodiversity such as genetic, species, and ecosystems outside their native habitats. *Ex situ* employs a range of *in vitro* techniques and advanced molecular techniques, especially transgenic technology that leads to the production of a new class of germplasm, genetically transformed material, and cell lines with special attributes. Generally, *ex situ* conservation is applied supplementary to *in situ* conservation; however, in some instances *ex situ* plays the central role and is the only viable method for saving biodiversity from extinction by overcoming the problems faced during *in situ* conservation. *Ex situ* conservation techniques are generally used for a species exhibiting one or more of the subsequent characteristics: endangered species, species of ethno botanical interest, species having past, present, or future local importance, species of interest for the restoration of a local ecosystem, taxonomically isolated species, symbolic local species, and monotypic and oligotypic genera. In general, *ex situ* conservation covers a broad range of conservation goals such as managing captive populations, raising awareness about biodiversity conservation, supporting research initiatives, and cooperating with *in situ* attempts.

12.5 BIOTECHNOLOGICAL TRENDS IN PLANT BIODIVERSITY CONSERVATION

As stated in *Global Strategy for Plant Conservation* report, 2009 (https://www.cbd. int/doc/publications/plant-conservation-report-en.pdf), 60% of threatened plant species (particularly critically endangered species) should be accessible in an *ex situ* collection center by the year 2010 in their native country. The *ex situ* method of plant biodiversity conservation involves the propagation of plant species or specific clones either by following the classical method or by using biotechnological methods. The classical method includes using seeds, rhizomes, corms, or cutting. The biotechnological methods involves *in vitro* techniques such as micropropagation from cell, tissue, and organ cultures, cryopreservation, germplasm banking, gene banking, and advanced molecular techniques to introduce new genetic modification in the existing population, augment existing populations, and reintroduce populations into wild but controlled environments.

12.5.1 *IN VITRO* TECHNIQUES FOR PLANT CONSERVATION

In vitro techniques play an invaluable role in conserving plant biodiversity. Culture initiation, maintenance, and storage are the three basic steps of *in vitro* techniques. Medium-term storage may vary from a few months to a few years, and slow-growth storage strategy works on the principle of lengthening the period between subcultures by reducing growth rates of concerned plant species. Growth reduction can be attained by modifying the environmental conditions by modification in gaseous environment, desiccation, encapsulation, covering with paraffin or mineral oil, and/or by changing the culture medium by diluting mineral elements, reducing sugar content, or altering the concentration or nature of growth regulators. Further, by adopting slow-growth strategy, *Gladiolus imbricatus*, which is a rare wild species and an important resistant gene pool of its genus was stored. This plant is resistant to many biotic and abiotic stresses on Murashige and Skoog medium in the dark at low temperature up to one year. However, for the storage of undefined time, cryopreservation is the technique of choice. A detailed account of all the *in vitro* techniques currently used for plant biodiversity conservation is discussed in the following sections.

12.5.2 PLANT TISSUE CULTURE

In order to accomplish sustainable resource conservation, management, and improvement of plant biodiversity, plant tissue culture (PTC) forms the basic and integral part of any plant biotechnological activity. PTC is an efficient, season-independent, and quick *in vitro* propagation technique for plants under aseptic microenvironment. PTC techniques are of eminent interest for the collection, multiplication at high rates, and storage of disease-free clean plant stock. The PTC technique relies on the idea of culturing isolated cells to exploit cellular totipotency to develop a whole new plant, originally developed by German botanist Haberlandt in 1902. The different types of culture method are applied using different organs satisfying different objectives. However, whatever may be the technique, the most important objective of all PTC techniques is to raise the rate of plant production. A million genetically identical plants can be obtained from a single bud by using these techniques. PTC techniques provide some conclusive advantages over traditional methods of plant propagation. Quick plant regeneration in the absence of seeds, germplasm conservation of medicinally important plants, and preservation of somatic embryos for medium- and long-term conservation of plants in a relatively short time period and small space are a few advantages associated with PTC techniques.

12.5.2.1 Micropropagation

Micropropagation or clonal propagation is a distinctive method used to protect different endangered plants. It deals with the rapid multiplication of stock plant material to produce large numbers of plant progeny by using plant tissue culture techniques. Micropropagation is a method of asexual reproduction for the mass production of plant propagules from any cell or part of a plant. Micropropagation involves utilization of apical shoots, meristems, and axillary buds. The newly formed roots and shoots served as explants for repeating plant proliferation. Micropropagation includes many steps, starting from shoot initiation either directly from the nodal part of explants or indirectly by callus mediated dedifferentiation of shoot initials. Further, the shoot initials undergo a stage of elongation and development whereby the well-developed plantlets (having shoot and root system) will be generated for the transplantation and hardening in the soil following acclimatization. Table 12.4 lists some critically endangered, endangered, and threatened plants of medical importance propagated and conserved through micropropagation.

12.5.2.2 Somatic Embryogenesis and Organogenesis

Organ development by way of organogenesis and somatic embryogenesis from different cultures of explants is another frequently used technique applied to regenerate many endangered plants. Somatic embryogenesis can be outlined as an induced cellular totipotency in which a non-zygotic cell gives rise to bipolar structure, resembling a zygotic embryo with no vascular connection with the original tissue. Somatic embryos are generally used to assess the regulation of embryo development, but also are an important tool for vegetative propagation on a large scale. There are many steps in somatic embryogenesis initiating with the development of pro-embryogenic masses, subsequently leading to somatic embryo formation, maturation, desiccation, and finally plant regeneration. The whole plant can be regenerated by culturing explants on a suitable media either by way of direct or indirect somatic embryogenesis. Direct somatic embryogenesis involves plant regeneration without the intervening step of callus (mass of unorganized cells) induction. In contrast, indirect somatic

TABLE 12.4
Critically Endangered, Endangered, Threatened Medicinal Plants Conserved through Micropropagation

Plant Species	Plant Type	Importance	Explant Used	Multiplication	Reference
Hypericum gaitii Haines	Endangered	Wound healing, bactericidal, anti-inflammatory	Apical axillary meristems	Micropropagation	Swain et al., 2016
Dendrobium thyrsiflarum	Threatened	Smooth muscle relaxant, vasodilation, anticoagulant	Nodal explants	Micropropagation	Bhattacharya et al., 2015
Nardostachys jatamansi	Critically endangered	Hepatoprotective, cardioprotective, neuroprotective, antidiabetic, anticancerous, antibacterial, antifungal	Leaf and petiole	Micropropagation	Bose et al., 2016
Thymus persieus	Endangered	Anti-inflammatory, hepatoprotective, anti-tumor, anti-HIV, antimicrobial, anti-ulcer, antihyperlipidic	Leaf and internode	Micropropagation	Bakhtiar, 2016
Picrorhiza kurroa	Endangered	Asthma, wound healing, hepatoprotective	Nodal explant	Micropropagation	Patial et al., 2016
Saussurea lappa	Endangered	Anti-inflammatory and analgesic effect, anticancer, antifatigue, cardiotonic	Apical shoot explants	Micropropagation	Zaib-un-Nisa et al., 2019
Acorus calamus Linn	Endangered	Anticancerous, repatoprotective, anti-arthritis	Rhizome	Micropropagation	Quraishi et al., 2017
Aristdochiaindica Linn	Endangered	Anticancerous, analgesic, anti-inflammatory	Nodal explant	Micropropagation	Shah et al., 2013
Hemides indicus	Endangered	Antioxidant, repatoprotective, anti-ulcer, antimicrobial, hypoglycemic, antihyperlipidic, analgesic, anti-inflammatory, anticancerous	Nodal segment	Micropropagation	Shekhawat and Monokari, 2016

TABLE 12.5
Critically Endangered, Endangered, Threatened Medicinal Plants Conserved through Somatic-Embryogenesis and Organogenesis

Plant Species	Plant Type	Importance	Explant Used	Multiplication	References
Ceropegia barnesii	Endangered	Medicinal	Nodes, internodes, leaves, root	Somatic embryogenesis	Ananthan et al., 2018
Urginea altissima	Threatened	Medicinal	Leaf	Organogenesis	Baskaran et al., 2017
Angelica glauca	Endangered	Medicinal	Rhizomes	Direct organogenesis	Mishra et al., 2018
Anoectochilus elatus	Endangered	Medicinal	Nodes	Somatic embryogenesis	Sherif et al., 2017
Crytanthus mackenii	Threatened	Medicinal	Leaf	Organogenesis and somatic embryogenesis	Kumari et al., 2017
Pterocarpus marsupium	Threatened	Medicinal	Green pods	Somatic embryogenesis	Tippani et al., 2018
Malaxis wallichii	Threatened	Medicinal	Pseudostem and pseudobulbs	Organogenesis	Bose et al., 2017
Nardostachys jatamansi	Endangered	Medicinal	Leaf and petiole	Direct shoot organogenesis	Bose et al., 2015
Ledebouria revolute	Endangered	Medicinal	Bulbs	Indirect somatic embryogenesis	Haque et al., 2016
Nilgirianthus ciliates	Endangered	Medicinal	Nodes	*In vitro* propagation	Rameshkumar et al., 2016
Thymus persicus	Endangered	Medicinal	Leaf and internodes	*In vitro* propagation	Bakhtiar et al., 2016

embryogenesis involves formation of undifferentiated, or partially differentiated, cells (often referred to as callus) from explants and dedifferentiation of callus into an organized plant growth. Several endangered plant species are regenerated through somatic embryogenesis and organogenesis process as described in Table 12.5.

12.5.2.3   Gene Banks

Genome resource banking is an invaluable management approach for the conservation of maximum possible genetic diversity of particular genetic stock. It includes different types of genes banks depending on the type of material used for the purpose of conservation. Gene banks can broadly be divided into seed banks (stores seeds to preserve genetic diversity), field gene banks (preserves live plants), *in vitro* gene banks (preserves plant cells and tissues; often used when the germplasm is either difficult or impossible to conserve as seeds), and pollen and DNA banks (embryos, tissues chromosomes, or DNA), which are maintained in short- and long-term storage laboratories generally by cryopreservation and freeze drying. The Genebank

Standards for Plant Genetic Resources for Food and Agriculture (PGRFA) impart international standards for *ex situ* conservation in seed banks and field gene banks. Currently, there are about 7.5 million PGRFA accessions conserved worldwide in gene banks (FAO, 2014). The principle objective of gene banking is to maintain genetic diversity as long as possible and to decrease the frequency of regeneration that may cause the loss of genetic diversity.

Seed banking is the preferred method for plant conservation in gene banks due to the easy handling during collection and maintenance in a viable state for long duration. Temperature, moisture, and relative humidity are the most important factors that affect seed longevity, hence in seed banks the dried seeds of threatened plant species are stored at low temperature (4 °C for a short period and −18 to −20 °C for long-term storage) with moisture content 3%–7%. However, there is variability among different species stored under the same conditions. Moreover, some drawbacks are associated with seed banks, for example many plants cannot be protected as seeds ascribed to their recalcitrant nature, as such type of seeds cannot be dried to a sufficiently low moisture level to allow their storage at low temperatures.

Further, with the rapid development in molecular biology and increasing demand for professionally preserved samples yielding high-molecular-weight DNA or RNA, DNA banks offer several advantages over traditional seed banks such as stable nature and small sample size, and they reduce the risk of vulnerability of genetic information to natural surroundings. Even though a single gene or small numbers of genes could be subsequently utilized, the regeneration of whole organisms from DNA cannot be predicted at present. Moreover, there are many issues associated with this technique such as gene isolation, cloning, and transfer. Some organizations in the world such as Kew Royal Botanic Garden, US Missouri Botanical Garden, Japan National Institute of Agrobiological Sciences, Australian Plant DNA bank, and South Africa Leslie Hill Molecular Systematic Laboratory of the National Botanical Institute are managing plant-derived DNA in DNA banks. With the ever-increasing advancement in information technology, geographical information system (GIS), and DNA marker technology, there is a marked enhancement in the gene bank documentation. Information derived on DNA assessment of variation can be useful for the identification and search of important genes using these technologies. Online information from DNA collections is available on Global Biodiversity Information (www.gbif.net), Inter-American Biodiversity Network (www.ukbiodiversity.net), and Species 2000 (www.species2000.org).

12.5.2.4 Cryopreservation

Cryopreservation is the slow freezing of viable cells, tissues, and organs at ultra-low temperature (usually −196 °C) to preserve their current state by arresting their mitotic and metabolic activities for future use. Cryopreservation is the most efficient technique of biotechnology enabling safe and cost efficient long-term preservation of various categories of plant species. Plant species producing non-orthodox seeds (recalcitrant or intermediate, i.e. those seeds that cannot stand desiccation below a relatively high critical water content value of 10%–12% on fresh weight basis, and cold storage without losing viability as in case of *Hevea brasiliensis*, and *Elaeis guineensis*), plants that can only be propagated vegetatively (*Dioscorea* sp., *Solanum* sp., *Musa* sp., *Manihot* sp., *Colocasia esculentum*, and *Ipomoea batatas*), and rare and endangered plant species can be successfully conserved through cryopreservation. In comparison to other conservation methods, the regeneration process increases by 20% if cryopreservation technique is followed. 14% and 100% survival of shoot tips of *Fragaria* and *Pisum* stored in liquid nitrogen for 28 years, respectively. A high amount of water content is present in all the experimental systems (calli, shoot tips, cell suspension, and embryo) used in cryopreservation, therefore water removal plays a key role in preventing freezing injury and to maintain post-thaw viability of cryopreserved plant materials. There are two protocols of cryopreservation: classical and new cryopreservation techniques (Figure 12.5) differing in their physical mechanism. In classical cryopreservation, freeze-induced dehydration, that is, cooling in the presence of ice, is performed. Classical cryopreservation involves various steps such as pre-growth of sample, cryoprotection by using different combinations of cryoprotectant, slow cooling (0.5–2 °C/min) to a determined pre-freezing temperature (around 40 °C), rapid immersion of plant sample in liquid nitrogen, followed by storage and rapid thawing

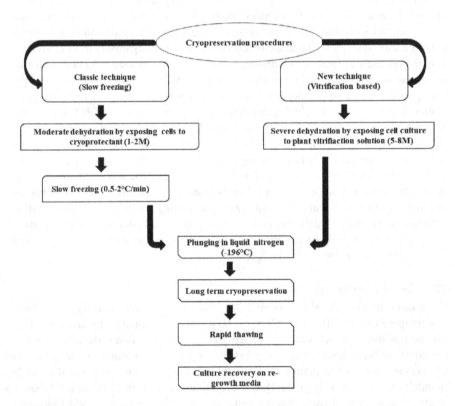

FIGURE 12.5 Flow diagram of cryopreservation techniques.

for recovery against cryogenic injuries. Preservation of *Manihot esculenta* using the classical cryopreservation method is an exceptional example. However, the classical cryopreservation method is operationally complex, as it requires technical skills and expensive programmable freezers.

In contrast, new cryopreservation techniques are based on vitrification, which involves cell dehydration prior to cooling, and cooling takes place without ice formation by exposing samples to highly concentrated media named plant vitrification solution (PVS). The vitrification process can be defined as the transition from the liquid phase to a glassy amorphous solid at glass transition temperature (T_g). It is believed that this glass helps in preventing tissue collapse and pH and solute concentration alteration. In the vitrification process, the slow freezing regime characteristic of classical cryopreservation is replaced by an ultra-rapid cooling process. The highest survival of up to 80% of shoot tips was achieved for the endangered Australian species *Androcalva perlaria* by preculturing on 1.2 M glycerol for 48 hours, incubation in PVS media at 0 °C for 30 minutes, followed by rapid immersion in liquid nitrogen and recovery. Many new vitrification-based protocols have been developed, such as encapsulation dehydration, encapsulation vitrification, droplet vitrification, and cryo-plate in order to increase the rate of cooling from the temperature at which cryoprotective treatments are performed.

12.6 BIOTECHNOLOGY TRENDS IN CONSERVATION OF ANIMAL BIODIVERSITY

About half of the world's biodiversity has disappeared since the 1970s, and tens of thousands of animal species are becoming extinct every year. This calls for the development and application of some innovative approaches to safeguard animal diversity. Biotechnological techniques offer great opportunities in this regard, as biotechnology is advancing at a rapid pace. For example, the cost of genome sequencing has reduced to $1000 today from $100 million in 2001. There are two main and interlinked objectives of conserving animal biodiversity: conservation of genes and conservation of population or breed. Gene conservation refers to ensuring the survival of each genetically controlled trait inherent within a population or breed, while population or breed conservation ensures the survival of a population as defined by the genetically inherent traits that it possess. The development and the utilization of genetic resources are the most efficient strategy for animal conservation. *Ex situ* or *in vitro* biotechnological approaches for animal diversity conservation include cryopreservation of DNA segments, gametes, embryos, semen, and ova, as well as reproductive biotechnological or assisted reproductive techniques (ART) viz. embryo transfer, artificial insemination, *in vitro* fertilization, embryo/gamete micromanipulation, embryo sexing, and genome resource banking.

12.6.1 ANIMAL CELL OR TISSUE CULTURE BANKING

Animal cell culture banking by preserving viable cell culture *in vitro* is an option for animal biodiversity conservation. Vertebrate cells were propagated successfully first in the 1990s. Fibroblast cells are usually used for viable cell culture banking. Therefore, it is common to use skin or muscle tissues (obtained either directly from fresh tissues or from frozen tissues) as a primary source material of somatic cell lines for different applications. Cell and tissue culture banking focuses on preserving the innate functions of an organism for an extended period of time through cryopreservation in liquid nitrogen. A bank of cells and tissues was generated from a Przewalski's horse, or Mongolian wild horse (*Equus przewalskii*), after its death. This is the only species of wild horse in existence. Fibroblast monolayer cell lines were successively developed from some biopsy samples, while other samples were stored in liquid nitrogen. A frozen zoo, which was initiated in 1975 by Kurt Benirschke at the San Diego Zoo Institute for Conservation Research, is the most longstanding and continuous active program of viable cell culture banking of wild animals. The viable cells derived from the northern white rhinoceros (*Ceratotherium simumcottoni*) (the last male of which, named Sudan, died on 19 March 2018, leaving only two females alive) and many other critically endangered species have been successfully maintained at the frozen zoo. However, establishment of *in vitro* culture is associated with high cost, susceptibility to contamination, and phenotypic and genotypic drift.

12.6.1.1   Tissue Graft Banking

The recent advancement in the autografting and xenografting of testes and ovaries witnessed the potential value of cryopreservation and subsequent use of gonadal tissues. The objective of testicular and ovarian tissue cryopreservation is to preserve spermatogonial and primordial follicles, respectively. Ovaries from large mammals, including humans, monkeys, marsupials, cows, pigs, and dogs, can be transplanted to recipient mice for the development of antral follicles and even matured oocytes. Ovarian tissues of common wombats were used (*Vombatus ursinus* and *Lasiorhinus latifrons*) as a model system to develop a method for the conservation of the highly endangered marsupial species *Lasiorhinus kreffti*. The study demonstrated the successful survival and functional development of ovarian tissues following grafting into an immunocompromised rat, from which the ovarian tissues could even be collected and developed in an ovarian xenograft.

12.6.1.2   Genome Resource Banking

Genome resource banking (GRB) can be defined as the organized collection, storage, and redispersal of animal origin biomaterials in an organized, logistical, and secured way. Genome resource banking offers an important opportunity to assist in the research and management of wildlife, including endangered species. Genome resource banks provide a safeguard against gene loss by preserving cryopreserved germplasm in the form of several cell types and tissues, including oocytes, sperm, embryos, embryonic or adult stem cells, induced pluripotent stem cells, and gonadal tissues. The cryopreserved germplasm can be stored for generations and can further be used for assisted reproductive programs after thawing.

The frozen ark, or the field gamete rescue program, has accumulated a large collection of gametes, embryos, and gonadal tissues at the Leibniz Institute for Zoo and Wildlife Research (IZW).

12.6.1.3 Animal Embryo Culture

The *in vitro* embryo culture technique involves the splitting and cloning of embryos, marker-based selection, embryo sexing, and transfer of new genes into an embryo. To transfer fertilized embryos into surrogate mothers, embryo culture and transfer techniques are used. Sometimes, in order to produce offspring of an endangered species, a closely related species can be used. For example, an interspecies nucleus transfer was made to clone a gaur bull (*Bos gaurus*), a large wild ox on the verge of extinction (Species Survival Plan < 100 animals) by electrofusing somatic cells from it with enucleated oocyte from domestic cows. The *gaurus* origin of the offspring's nuclear genome was confirmed by microsatellite marker and cytogenetic analyses. Normal fetal development was observed in gaur nuclei, even if the mitochondrial DNA (mtDNA) within all the tissue types was derived exclusively from the recipient cow's oocytes.

12.6.2 REPRODUCTIVE BIOTECHNOLOGICAL TOOL

Reproductive biotechnological tools or assisted reproductive tools (ART) are widely applied to overcome the breeding difficulties in captivity and expanding the gene pool for *in situ* and *ex situ* conservation of animal diversity. The ART methodology includes hormonal monitoring of cyclic ovarian activity, artificial insemination, oocyte aspiration, spermatozoa collection, *in vitro* embryo production or fertilization, and embryo transfer.

12.6.2.1 Cloning

Although surrounded by controversy, the consideration of the potential role of cloning in the conservation of endangered animals begins with the successful cloning of Dolly by Ian Wilmut in 1997. Moreover, by cloning of a mouflon (a species of wild sheep) in a domestic surrogate sheep, pointed towards the potential conservation opportunities and additional challenges in the evaluation of appropriate technologies for present and future efforts to conserve gene pools of endangered species. The cloning process involves somatic cell nuclear transfer (SCNT), in which the nucleus is transmitted from a donor cell to an enucleated recipient cell to create an exact genetic copy of the donor (same species or another subpopulation or closely related species). If this process results into a viable embryo that proceeds to term, the offspring produced will have similar genetic makeup to that of the donor, except the mitochondrial DNA will be without any genetic dilution (Figure 12.6). SCNT can help in the preservation and propagation of endangered animals in captivity until their natural habitats are restored and the populations have been reintroduced to their native habitats.

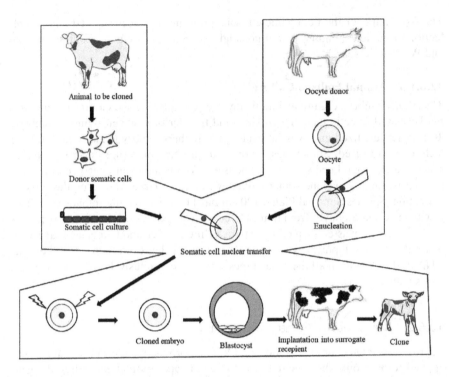

FIGURE 12.6 An overview of cloning through somatic cell nuclear transfer.

12.7 GENOMICS

Genomics (a term coined by Thomas Roderick in 1986) refers to the mapping, sequencing, and analysis of the genome. Genomics is generally divided into two parts: functional genomics and structural genomics. Functional genomics refers to the study of expression and function of the genome, while structural genomics is the study of structure, organization, and evolution. Functional genomics requires the assistance of various fields such as structural genomics, mathematics, computer sciences, computational biology, and other areas of biology. With the rapid advancement and falling cost of genomics, genome analysis is not limited to model organisms only, and the genomes of thousands of organisms including plant, vertebrates, and invertebrates have been sequenced. Moreover, now it is possible to attain genomes of extinct species such as the passenger pigeon and woolly mammoth. The genomes of extinct species from DNA are left in the environment (eDNA). Today, population genomic approaches have made it very easy to analyze the population structure, genetic variations, and recent demographic events in endangered species. The most commonly used genetic techniques for the identification of genetic variations in plants and animals include random amplification of polymorphic DNA (RAPD), random fragment length polymorphism (RFLP), amplified fragment length polymorphism (AFLP), single strand conformation polymorphism (SSCP), mini-satellites, microsatellites, single nucleotide polymorphisms (SNPs), DNA and RNA sequence analysis, and DNA fingerprinting. All these tools provide genomic insight to improve monitoring, management, and restoration of endangered species. The recent application of such studies include identification of unrecognized lineages deserving conservation attention, delineated conservation units, estimation of gene flow, identification of source populations for translocations, detection of inbreeding in small populations, management of captive populations, and detection of hybridization with invasive species.

12.7.1 Metabarcoding

Metabarcoding is a rapid method of species composition evaluation by amplification using universal primers, sequencing, and analyzing target genomic regions of environmental DNA samples. Metabarcoding utilizes two technologies: DNA-based identification and high-throughput DNA sequencing. This technique enables the DNA sequencing and hence biodiversity assessment of mixed and bulk samples by bypassing the sample sorting step, which is difficult to assess by conventional methods. However, the reliability of metabarcoding is still to be resolved, for example the PCR amplification biases that affect species identification assessment of a relative abundance of species.

12.7.2 Genetic Modification

Production of transgenic animals or plants by transferring adaptive traits within and between species offers another potential of biotechnological technique for the conservation of diversity. Several conservation difficulties such as threats from invasive species, pests and diseases, inbreeding problems, and temperature fluctuations can be overcome by adopting this strategy. Several lines of transgenic farm animals have been generated, however, due to government and public concern, conservationists are wary of releasing transgenic species into the wild.

12.7.3 Genome Editing

Genome editing is the newest reproductive biotechnological tool that can modify genes of interest without the insertion of foreign DNA. The recent approach of genome editing by using CRISPR/Cas9 has almost replaced the other alternative tools for genome editing, such as zinc-finger nucleases (ZFNs) and transcription-activator like effector nucleases (ALENs), due to its simplicity and cost efficiency. Endangered species can be conserved by performing gene editing for disease resistance from infectious disease and for drought resistance in order to survive in a changing climate.

12.7.4 Gene Drives

A new gene-editing tool called gene drive has made it possible to carry out successful eradication of harmful invasive pest species (rodents, feral pigs, and insects). The CRISPR-based gene drive is a process through which a gene variant is inherited in such a way that all the offspring of an edited parent have two copies of the edited gene so as to spread the deleterious gene evenly in the population. The resulting genetic transformation breaks the 50–50 sex chance rule that ultimately controls harmful non-native species and increases survival of threatened species. The possibility to target drive to any gene has led scientists to speculate on their potential use for biodiversity conservation. For example, applying gene drive to alter the sex ratio of the rat population on an island may result in all-male offspring who will fail to breed further. However, despite the immense possibilities of gene drive, there are risks associated with their field application due to high inheritance, and there is need for innovative safe and specific localization.

12.8   MICROBIAL CONSERVATION TECHNIQUES

The microbial world, which includes bacteria, archaea, fungi, algae, and protozoans and protists, is the largest unexplored and unnoticed reservoir of biodiversity. Microbes perform various functions essential for biosphere management such as nutrient cycling and environmental detoxification as well as for sustainable development with commercial and industrial applications. The enormous possibility of environmental heterogeneity and the ability to provide novel antibiotics, novel enzymes (thermophiles/alkaliphiles), novel pharmaceutical products, biofertilizers for agriculture, new strains for crop improvement, and land reclamation by way of biotechnology is a large untapped resource. In spite of the obvious environmental and economic value, the conservation and management of microbial diversity is largely an ignored issue. The reasons for this ignorance include the invisibility of microorganisms, a general perception of their association only with diseases, and a lack of knowledge about their potential role in the improvement of the quality of life. Therefore, there is a great mismatch between the knowledge of microbial diversity and their importance in economic and environmental development.

Conservation of microbial diversity demands some specialized techniques for the reclamation of degraded habitat because microbes have only few morphological and metabolic characteristics that better describe them. As with the case of plant and animal conservation techniques, both *ex situ* and *in situ* microbial conservation techniques are applied. *In situ* conservation of microbial diversity involves onsite maintenance of microbial diversity by avoiding the washing out of surface soil containing diverse microflora by torrential rain through plantation and avoiding deforestation. *Ex situ* microbial conservation involves microbial culture collection and microbial resource centers forming the repository of microbial isolates and microbial gene banks.

12.9   ROLE OF INDIA IN BIODIVERSITY CONSERVATION

India has been recognized as one among 17 identified megadiverse countries. Being enriched with approximately 91,000 animal species and 45,000 plant species, India accounts for about 6.4% of all known floral and faunal species. India is acknowledged as a center of origin of many crops and their wild relatives. India was among the few countries that have mapped biodiversity-rich areas and have developed biogeographic classification for implementation of conservation planning. Four of the 34 biodiversity hotspots around the world are present in India, including the Himalayas, the Western Ghats, the Northeast, and the Nicobar Islands. Approximately 5% of total geographical area in India is safeguarded as national parks, wildlife sanctuaries, ecologically fragile areas, and biosphere reserves.

India was one of the first countries to have enacted legislation in 2002 in order to implement the provisions of Article 15 of the CBD, which allows equitable sharing of benefits arising out of the use of biological resources and traditional knowledge. The biodiversity rules were publicized by the government in 2004. A three-tier structure, namely the National Biodiversity Authority (NBA), State Biodiversity Boards (SBBs), and Biodiversity Management Committees (BMCs), works at the national level, provisional level, and local level, respectively, to implement the act. NBA is a statutory body under the Government of India and was established in 2003. NBA headquarters is situated at Chennai (Tamil Nadu) and performs facilitative, regulatory, and advisory functions for the Government of India on issues of conservation, sustainable use of biological resources, and equitable sharing of benefits arising out of the use of biological resources. Since its establishment, NBA has supported the creation of SBBs in 29 states and facilitated establishment of around 155,868 BMCs.

In addition, India is also endowed with billions of diverse microbes, including agriculturally important microorganisms, many of which are only found in India. Despite the well-recognized benefits of microorganisms in the environment and economic development, only less than 5% of the world's microorganisms have been described. The National Bureau of Agriculturally Important Microorganisms (NBAIM) is a microbial bioresource center developed by the Indian Council of Agricultural Research (ICAR) for the collection, conservation, supply, and preservation of microorganisms. The National Agriculturally Important Microbial Culture Collection (NAIMCC) facility at NBAIM has an archive of 4668 cultures, which includes 4644 endemic and 24 exotic accessions. NBAIM has successfully implemented a research program for the analysis of the microbial diversity of extreme regions and a program for the conservation of microorganisms during the 11th plan period (2007–2012). In order to increase crop production, agro-waste utilization, and abiotic stress management related to microbial-based technologies, NBAIM has implemented an ICAR network project called "Application of Microorganisms in Agriculture and Allied Sectors."

12.10 CONCLUSION

Conservation of biodiversity is a global concern, hence all the member states of CBD are actively participating to preserve indigenous biodiversity. Though biotechnological tools and techniques are proving valuable for efficient and effective conservation attempts and offer many advantages over conventional procedures, there are some drawbacks associated with biotechnological methods leading to genetic pollution and ethical controversies. Therefore, biotechnological conservation tools must be integrated with conventional conservation strategies. Moreover, a consistent level of financial, human resource, and appropriate policy development is needed for the successful generation, adaptation, and adoption of biotechnological tools for biodiversity conservation.

13 Environmental Risks and Concerns of Biotechnology

Tarun Sharma and Neetu Sharma

CONTENTS

13.1 INTRODUCTION

The influence of human activity in any sphere supersedes that of any other species. Hence, we have an inherent moral responsibility to assess the risks associated with every event that has a severe global impact. Out of all the emerging sciences in the last few decades, the most lucrative and complex is the application of biotechnology. On the surface, the concepts and practices offer viable and even successful solutions to the modern problems of the planet. This industry is, however, not without its new environmental concerns and risks. When we use the word environment here, it concerns our immediate surroundings and other ecosystems of various sizes. We are a part of some of these systems (for example, food chains and webs) while some exist without our direct involvement (for example, microbial biomes).

With every significant development in biotechnology, there is also a wave of new challenges. One might wonder why these challenges exist. The simplest explanation is that any innovation or change in this field has a direct or indirect impact on human lives and the environment. Thus, any development has great potential to improve our lives or, in some instances, harm us directly.

For example, if a mathematical proof turns out to be incorrect, we can erase the blackboard and continue working on it, or if a particular physical equation is proven wrong, it is discarded and cannot cause any significant harm. Even in practical chemistry, we get to see quick results, and we can test the products as we wish immediately. There is but a small chance of something going very wrong. However, a problem or threat on the same level in biotechnology could mean an antibiotic-resistant microbe

DOI: 10.1201/9781003131427-13

released in public, or perhaps a genetically modified super plant takes over wild variety crops. Since the world we occupy is set in a very delicate balance, even small changes can cause increasingly more significant problems as we move up in this network of organisms which rely on each other. Furthermore, biotechnology directly affects consumers. The risks and concerns in the present and the ones perceived to happen in the near future will be discussed briefly.

We now know that biotechnology will enhance our species and the environment just as physics has boosted IT (information technology) and space technology. Many systematic measures and studies have been conducted in an effort to scale the potential rewards and risks associated with the impacts of biotechnology. European government councils which deal with the regulation of GMOs have proposed that we find a system where we can balance the risks and rewards from each individual technology. We use various mathematical and computational models for this. However, we keep in mind that these methods may not be accurate in the field of biology. This is because the margin for randomness in biology is much higher than it is in other sciences. The main reason for this margin of error lies in its essential unpredictability.

In various publications and papers, different statistical methods have been used to calculate values of risk and how it works across multiple subcategories. Thus, detailed estimation of various chemicals in nature and samples taken from the natural environment can be carried out employing such methods. These methods can provide a mathematical index of the respective risks and using these observations, we can engineer bioprocesses beforehand instead of waiting for the damage to happen.

As the complexity of the natural environment increases, the balance of the environment tends to get more unstable. Biotechnology possesses a unique potential to repair a lot of environmental damage that has been done, but to develop such processes, we require sophisticated and dedicated research. There are still concerns that we may not be able to see potential risks because of the exponential growth model. The final most impactful threats may happen so fast that it will not be possible for us to try to repair.

Much like Albert 2 aboard a V2 rocket, the first monkey sent to space, any early research requires a lot of sacrifices. It's all about venturing into the unknown with no maps and trying to solve the significant problems that concern us. Since the processes in biotechnology need the sacrifice from an existing domain such as test organisms, plants and microbes, progress is halted or significantly inhibited. This is sometimes for a good reason.

13.2   CONCERNS AND RISKS THAT HAVE EMERGED AND ARE BEING MANAGED

The following concerns and risks have developed over the years:

- Unpredictable impact of GM species on wild varieties
- Financial risks
- Environmental accumulation of metabolites from processes
- Antibiotic resistance and superbugs
- Risks associated with recombinant therapeutics

13.2.1   Unpredictable Impact of GM Species on Native Varieties

The creation of any genetically modified species comes with a unique set of challenges. One of these is the choice of habitat for it, the ideal option being the wild and a more feasible option being a controlled environment in the lab. However, an organism that is designed to carry out bioremediation or repair ecological damage can't be confined to controlled conditions in the laboratory.

The principle problem that we face in such a scenario is that we do not know how the modified species will interact with the wild type. This issue may sound simple, but it has gravity because it ranges from the simplest of bacteria to complex organisms. One example from history highlights the problem: a small population of goats brought to an island managed to change the plant life of the place by eating through most of the plants very quickly, turning the landscape into a grassland. Another famous example is the problem of hippopotamuses in the nation of Colombia, which started because a drug lord named Pablo Escobar brought in a few of the animals from Africa for his private zoo. The few escaped, and with an abundance of food and no natural threats, their populations has now grown to about 700, posing a severe threat to the ecosystem and the daily life of the people since they are powerful, aggressive predators. If the displacement of normal and even domesticated animals has the potential to change whole countries, we should be careful as to what may happen when superior species are released into unprepared ecosystems. Another more specific example concerns human evolution: the development of cranial capacity, posture and behavioural traits were non-uniform around the continents. It has often been theorized that through selection or perhaps the flow of genes, whenever a more evolved humanoid appeared in a specific inhabited area, its arrival was followed by the immediate extinction of the less evolved kin. This coincides well with the theory of natural selection, explaining how the more adaptive species survive while the less adaptive eventually become extinct.

The issue here is not related to the speculation of extinction of wild-type species. Instead, it is the fact that the changes that we experiment within biotechnology at a molecular level take thousands of generations to happen naturally. Substitution of one nucleotide or one gene naturally would involve adaptation across generations, for example a chance infection of a particular virus on a germ cell can result in a quick change in the next generation or a mutation caused by severe abiotic stress. In the lab, it is possible to edit genomes using new-age technology such as CRISPR. While the time taken by the natural route gives the ecosystem time to adjust to such changes, this grace period is removed when GMOs are released into the wild.

McClintock's studies on maize discovered and confirmed the presence of transposons, also called "jumping genes". This also validated her idea that the genome of an organism is not a stationary set of molecules, but may be modified or altered under certain conditions. The mechanisms by which these genes are modified and the methods by which we choose to edit them are very different. This is often subject to the scrutiny of the most important genes selected for editing with respect to the environmental impact.

With the release of any genetically modified organism into the environment, the primary concern is how competitive it will be against the pre-existing wild-type organisms. For example, the modified salmon dubbed as "AquAdvantage" is much bigger than a regular salmon. The research points to the fact that it will quickly dominate the wild type if released in the ocean. This fish is a famous example in any biotechnology discussion since it might have all kinds of consequences, including disrupting the fisheries industry.

Bt cotton, a genetically altered strain of the cotton plant which was designed to have a natural resistance to pests, received a mixed response in India. This cotton plant has genes from a bacterium *Bacillus thuringiensis* which allows it to produce a toxin. This toxin is only harmful to pest species such as bollworms, which makes it extremely useful. In the most recent ecological backlash, researchers claim that this has caused the emergence of a more powerful type of parasite in the absence of competition. The central issues of Bt cotton are

1. Relatively expensive seeds
2. No change in fertilizer usage
3. Hybrid strain usage is ignored
4. Incorrect production statistics

Many times in theology and science, the notion that this aspect of biotechnology disrupts the natural order of things has been put forward. While a large portion of the scientific community stresses the potential of GMOs to be a force for good, they are still taboo for some scientists.

There is a special organization under the Government of India called the Genetic Engineering Appraisal Committee under the Ministry of Environment, Forest and Climate Change. The main purpose it serves is to regulate research and the widespread application of recombinant species at any scale and of any type. The approval of this committee is crucial for the introduction of any product related to GMOs in the market. There are a total of 24 members who conduct monthly meetings to review the applications they receive, deciding whether they should be approved or not.

Some speculated and observed risks known are as follows:

- Changes induced in the local and ecosystem-wide food chains and food webs due to competition
- The destruction of wild-type variants and diversity by modified organisms
- The rejection of modified organisms by wild varieties and hence the failure of the project as a whole
- The overconsumption of resources by these species causing habitat destruction
- Transposons and the mutations induced in the species that the GMOs interact with
- Unchecked mutations happening more often due to uncontrolled natural resources available to the organisms

In a reversal, GMOs have the potential to cause significant environmental issues, but this does not change the fact that bioremediation is the most rapid and cost-effective way to repair the damage already caused to the environment. As we innovate and come up with better ways to assess risk and rewards, perhaps in the near future we can come to equilibrium between wild types and GMOs.

13.2.2 FINANCIAL ASPECTS

In the 21st century, we have seen perhaps the most significant union of finance and life science technology. For any industry to be able to find success in the very fast-changing world, the critical factors required are innovation in the industry, competent professionals in research and a capable workforce, finance and resources.

Since national and international markets have shown unexpected fluctuations, investors have been cautious as to where they invest their money. Even with technology specially designed to predict market flow, it is often said in trading circles that nobody understands how money works in the world of stocks.

In modern times, science and finance are like purines and pyrimidines. They do exist separately, but when together, they have unprecedented potential. The global market share of biotechnology has rapidly grown for the past decade, but most of it is in pharmaceuticals. Biotechnology is usually considered a high-risk business for a variety of reasons.

The business itself is costly to start up, while an ordinary tech start-up asks for anywhere between 1 to 5 lakhs in initial funding, Biotech businesses estimate first-round funding to reach 50 lakhs and above. This is because the reagents and machinery are expensive. Laboratories have to be spacious and air-conditioned, with uninterrupted electricity and away from dense settlements. Special sections have to be built for incubation of plants and cell cultures. Everything from essential equipment to waste disposal methods has to be of as high quality as possible to yield useful results. The scientists and support teams are highly qualified individuals who expect salaries which justify their knowledge and skills. Most private companies shy away from investing such high amounts of money in research that may or may not yield any profitable technology. The risks are also built around the fact that a biotech business takes a considerable amount of time to pay back revenue. It takes substantially longer to bring a product to market. The first step, putting plans on paper, is perhaps the one that takes the smallest amount of time. The steps following this one are as follows:

1. Formulating a business model
2. Getting academic approval of the technical parts of the plan
3. Arranging the raw materials from media to organism used
4. Getting government permits
5. Building prototypes
6. Getting government approval
7. Finding investors
8. Developing the product
9. Filing patents
10. Branding and sales
11. Profit

12. The initial public offering of stock
13. Dividend generated
14. Licensing after the patent is expired
15. Research and development to maintain a position in the industry

Even if the product is studied and projected to be valuable, there are lists of approvals at both national and international levels that one must have before bringing the finished product to the market. These cost both time and money. Government policies on almost all biotechnology-based products, except pharmaceuticals and nutraceuticals, are stringent, which delays the issues of permits and grants. In the aforementioned steps, one might think that these are the same for any business, but they take significantly longer and are more expensive to get through. People who do invest in the high-risk sector know that biotech is a gold mine, but not without its limitations.

In the minds of the common people, there is still a lot of doubt regarding the viability of processes and developed products. This is because biotechnology has been under scrutiny for very long and is difficult to experiment as compared to other forms of technology. It also takes longer for a product to be declared entirely safe for use since we cannot predict long-term effects.

In the world of investors, a simple concept is always mentioned by the greatest of the investment giants of all time, Mr Warren Buffet. He insists that people should only invest in businesses that they fully understand. Since people studying biotechnology rarely have the time to study economics too, hard-core companies lack expertise in this area, leading to their fall in the competition. This pushes biotechnology-based companies further and further down the watchlists of capital giants.

The consumer end of the financial risks is that although modified plant or animal products are cheaper and of better quality, consumers often hesitate to buy them due to a lot of media propaganda against them. More affordable food products mean better quality and more resilient products which can solve hunger problems in countries with such crises. One simple, inevitable change we all hope to see is for the agriculture sector to progress without extensive mechanization. The existence of better crops and chemical-free methods to grow food will empower the farmer, general public and, through cause and effect, every organism on this planet.

Studies and surveys have often shown how the public is unaware of the workings of the biotechnology industry. There is a lot of ignorance towards the products that are in the market. People are unaware of whether the products they use are GMOs, not realizing that in some cases there is little choice. This is because even minimal-scale improvements through biotechnology have made significant changes in multiple industries that require plant or animal products as raw material. As we make progress and take better safety measures, the market size will grow, and investments from capital giants will allow this discipline to establish itself in the world firmly.

13.2.3 ENVIRONMENTAL ACCUMULATION OF METABOLITES FROM PROCESSES

All industries inherently produce secondary waste products that need to be disposed of properly. In the recent past, the disposal of such products sustainably and safely had been a very controversial and urgent issue. In bio-industrial processes, the waste products are of many types. Even though most of it is biodegradable, certain waste products are hazardous. Taking the example of electrophoretic gels, they contain ethidium bromide, which is a toxic reagent. By-products such as these require proper biomedical disposal carried out by trained professionals. Even concerning disposal, one primary concern is the accumulation of said products in the environment and bio-magnification of these through various trophic levels.

Cellular machinery maintains a very delicate balance, in terms of both the breaking down and the building up of biomolecules. Even a slight change can result in totally different products. In terms of quantity, substances produced by living organisms such as microbes, plants and animals are present in large numbers. These substances are called metabolites and are an integral part of the ecological balance.

Any changes in metabolites are a significant source of concern for the environment. What is more dangerous is that it is much easier to estimate the nature of changes by organisms since they have habits and observable behaviour. But one can only understand chemical changes once they have already reached a critical point. By this time, it is usually too late to be able to reverse the damage.

The carbon and nitrogen cycle may have some impact due to the biotechnology reforms relating to genetically modified organisms. Whenever the problem of deforestation is discussed, it is usually associated with oxygen problems. Still, very little attention is drawn to the fact that forests are necessarily crucial massive sinks of carbon dioxide.

Another objective lies with the inner mechanisms of a cell. When a cell is modified for maximum yield or to make it into a microscopic industrial unit, the primary aim is to extract out as much metabolite as possible. However, the biochemical arc associated with the metabolite may cause another unwanted chemical residue to be produced. This will either accumulate inside the cell or be excreted. In both cases, the longevity of the cell is significantly decreased. If the unwanted metabolite is excreted, the following impacts are observed:

- It may have a negative ecological impact
- If the immediate environment is closed, it will decrease the overall quality of the target metabolite
- It will cause cytotoxicity to the cell once it exceeds a specific tolerable limit

If the substance accumulates inside the cell, it will merely cause the cell to die or decrease the efficiency of the process significantly. Keeping this in mind, large-scale biological changes can shake up the chemical hotspots which produce organic molecules or act as sinks for them. There is no way of reversing any damage done to the biological portion of primary cycles of our environment, such as the carbon cycle, nitrogen cycle, and so on.

13.2.4 ANTIBIOTIC RESISTANCE AND SUPERBUGS

One of the most significant risks emerging with the development of high-level biotechnology is the way it challenges modern medicine and healthcare. Addressing a significant concern, we generally use antibiotics to treat microbial infections.

Antibiotics like penicillin, ampicillin and kanamycin are substances produced by living organisms that inhibit or stop the growth of other microbes. In some cases, medicines may be able to kill a specific type of microbes. With new biotechnology, we can make highly particular antibiotics that somehow disrupt only the functions of the pathological cells. While this approach has saved millions of lives, it has given birth to a new problem. If the required objective is to be achieved, it leads to have an artificial selection in the microbes against the antibiotics. The ones that are susceptible are killed, and the ones that survive are highly resistant organisms called superbugs.

The concern here is that at one point in time, we will run out of antibiotics and treatment methods for these new pathogens. At a time when we can cure most diseases and because of general economic prosperity, the basal immune tolerance of the masses has decreased, and any one superbug has the potential to start an epidemic. An organism like this raises concerns not only for humans but also for potential zoological infections and problems caused eventually for biodiversity and food webs.

The other concern about superbugs are the mutations induced on exposure to these antibiotics. The supercolonies in the culture grow harder and harder to kill. Their presence hinders aseptic operation and pure cell cultures.

In the last few years, superbugs have become a significant concern in the healthcare industry. MRSA or methicillin-resistant *Staphylococcus aureus* is the most famous name in the superbug family. This new phenomenon has introduced another parameter in drug design at the molecular level. Scientists are now required to effectively study precise mechanisms of action and theorize the level of difficulty the organism will face developing resistance to the said antibiotic.

MRSA is a significant risk factor in situations where the general factors associated with the infection have already been considered and neutralized. These include surgical procedures, tests on mammals and production of certain products that have an infection risk. Though there are methods to prevent or eradicate MRSA infection, they are rarely viable in such cases.

A ubiquitous term when talking about superbugs is MDR; it is expanded as "multidrug resistance". In instances where the strain of microbial infection is diverse with respect to the intensity of disorder, doctors often prescribe multi-drug therapy (MDT) to ensure that the patient is cured, irrespective of the strain. However, MDT is the biggest reason for the existence of superbugs.

Another case study that can be used to understand superbugs and their evolution better is that of tuberculosis. This disorder affects multiple organ systems, sometimes separately and sometimes at the same time. Evidence of tuberculosis has been found in mummified corpses in Egypt, which means that it is a disease of great antiquity. One could argue that it is one of the most widespread diseases in the whole world, both demographically and chronologically. When antibiotics like streptomycin and isoniazid were used as potent and fast treatments for tuberculosis, the bacterial strains developed resistance. The mechanism of development of resistance in this particular pathogen was found to be a spontaneous mutation. The famed writer George Orwell, author of the great dystopian novel *1984*, apparently suffered from a resistant strain of tuberculosis. As the treatments have evolved since then, the pathogen has made amends to carry out the infection still. We saw the emergence of XDR (extensively drug-resistant) strains which were resistant to multi-drug treatments. They came into being because of mismanagement and overuse of the MDT techniques. Now we have strains called TDR (Totally drug resistance) strains which are resistant to any drug.

There is widespread speculation that the very phenomenon of drug resistance came into being the same time as biotechnology and recombinant DNA technology gained popularity.

The best approach to tackle emergence of drug-resistant tuberculosis is through prevention and discipline (Figure 13.1).

13.2.5   RISKS ASSOCIATED WITH RECOMBINANT THERAPEUTICS

Ever since Eli Lilly successfully produced insulin from the animal pancreas, a new industry was born. The potential of living organisms as industrially viable units was known, but this was the step that made it clear that such a practice could be safe. However, this whole idea has two major pitfalls

1. We have not been able to use it to its potential
2. Since the usage is very less, there is no way to know the long-term effects of such medicines and substances

The risks associated with this kind of therapeutics can be better understood by studying the workflow of a product such as this one into the market.

1. Need for certain therapeutic seen as not being met through traditional production methods
2. Research and sequencing to find the gene sequence responsible for the production of the particular biomolecule
3. Finding the ideal vector and a target model organism for production of said therapeutic
4. Carrying out the infection and testing of the target organism and the product hence obtained
5. Purifying said product and cellular-level testing
6. If approved, testing on the mammalian specimen, mice and chimpanzees
7. Publishing results
8. If not approved, then back to step 3
9. If approved further, human testing and FDA and WHO documentation and approval
10. Publishing results
11. Applying for government trade approvals
12. Applying for patents
13. Market research and release
14. Marketing and sales
15. If any discrepancy then the recollection of sold items and payment of damages
16. Modifying production and selling costs to suit the expired patents and to compete against generic branded alternatives

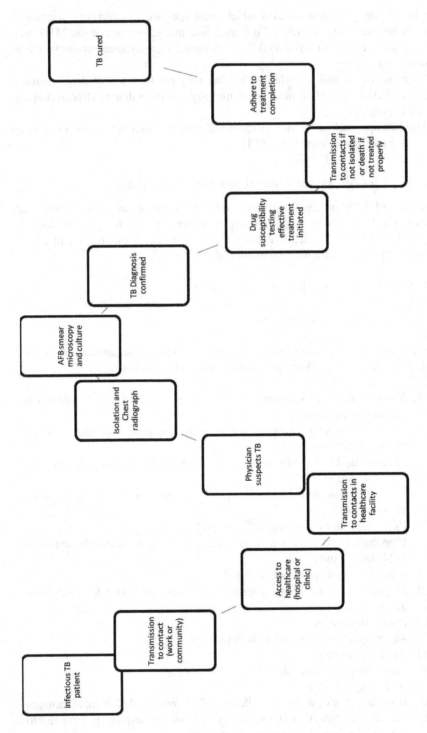

FIGURE 13.1 Prevention of development and transmission of drug-resistant tuberculosis.

This simple step-by-step process has many financial aspects as well, but we will focus on the risks concerned with the keyword. Some of the Government-approved recombinant products to be used in treatment are summarized in Table 13.1. Recombinant technology-based therapeutics are much more challenging to work with as compared to traditional pharmaceuticals. The beliefs around these products are also very volatile, with even reputed scientists at times making radical claims.

A famous rumour has been doing rounds in the scientific community for a few years now. We are talking about the great vaccination scare. In the last 10 years or so, societies, especially in the West, have started to become opposed to the idea of infant vaccination because of a very controversial study that linked vaccination to autism in infants.

Between the 1980s and the early 21st century, over 90 recombinant therapeutic proteins alone were approved for consumption by the FDA. However, even in highly developed countries, the actual sale and usage are far below this figure.

One of the most widely used recombinant therapeutics is interferon. Interferons are a family of proteins which are produced by cells when the organism or cellular system faces an immunity threat. These threats may be related to viral infections, cell death, both natural and cancerous, and regulation of general immunity. They have applications as cancer drugs and hence attract a lot of industrial interest. They are currently widely in use and will most likely be so because of the urgency of their use. The primary concern here is a hyperacute immune rejection of interferon by the host body. We do not have a very accurate system of being able to predict or process such risks, but animal testing has revealed that there is a great potential danger. It is a common practice for hospitals to use high risk and experimental treatments for terminal illness. The concern here is, where exactly do we draw the line on experimental therapies? How do we decide which treatments have a stable risk-reward equation? From the patient's perspective, isn't it better to have a chance than to face imminent death?

TABLE 13.1
Some Government-Approved Recombinant Proteins, Their Source Organisms/Cells and Associated Risks

Approved Recombinant Protein	Microorganism/Cells	Associated Risks
Engerix B (Hep B Vaccine)	Hepatitis B virus	Speculated increased risk of multiple sclerosis
IGF (insulin-like growth factor) 1	*E. coli*	Carcinogenic risk
TPA	Cucumber, *Cucumis sativus*	Haemorrhagic concerns
Erythropoietin	Mammalian cells	Heart disease, venous thromboembolism risks
Granulocyte colony-stimulating factor	*E. coli*	Malignancy risk
Glucocerebrosidase	Rice	None as of yet except not being as effective as expected
Interferon	*E. coli*	None yet

Recombinant products like human growth factor were subject to scrutiny for concerns pertaining to toxicity, uncontrolled stimulation of cells and even hypoglycaemic activity.

13.2.6 ETHICAL CONCERNS

The definition of life has been debated, as has the age-old argument of biotechnology, whether it is ethical to use living and sentient beings such as animals as factory machines. As we find more techniques for industrial production in all four significant biotechnology domains (healthcare, food, fuel and environment), often ignored questions come to mind.

What is the limit that we will go to in an exploration of the potential of this discipline?

What next, if and when the damage is all repaired and all is right with the ecosystem?

How does biotechnology come into play in tandem with our place in the environment, not as productive species but as a species capable of forming and modifying the environment?

There is one significant fact that we often come across while assessing the environmental risks in biotechnology. The human aspect is what causes these substantial changes in the ecological balance; some people argue that this is, in fact, a part of organically happening evolution. The reason these changes are justified is that the same forces of nature have made us capable of causing this change. Some futurists claim that this is the next step of human evolution and the evolution of life on our planet.

As we progress in research and pursuit of a sustainable life, we find capable people and organizations doing their part. In light of recent warning studies which tell us how urgent it is for us to be aware of our environment, people are now more concerned than they have been at any point in history. In time, we will find a way to balance risks and rewards. To eradicate any negative consequence, humans should work in harmony with nature towards a sustainable future.

14 Biofuel
Production, Applications and Challenges

Neetu Sharma and Abhinashi Singh Sodhi

CONTENTS

14.1 INTRODUCTION

Different fossil fuels across the world fulfill the majority of energy needs. Fossil fuels like petroleum, oil, coal, and natural gas produce energy through combustion. With the increase in population, the gap between energy production and demand has deepened. The available resources are depleting at a faster rate than their availability and replenishment. The other drawback associated with the prolonged use of fossil fuels is the adverse impact on the environment due to the toxic by-products produced by them during combustion. The by-products produced result in air pollution and land pollution. These factors lead to the search for alternative sources of energy generated through renewable resources. Such energy resources in current use include solar energy, wind energy, nuclear fission, and geothermal energy. But

DOI: 10.1201/9781003131427-14

these energy sources have their limitations too in terms of availability in different geographical locations across the world. In this context, biofuel has emerged as an alternative source of energy in the last two decades due to its environmental remuneration, which includes reduction in the release of various harmful gases like SO_x, NO_x, CO_2, unburnt hydrocarbons, RPM (respirable particulate matter: PM_{10}, $PM_{2.5}$), and SPM (suspended particulate matter), with subsequent decline in greenhouse effects.

The use of biomass for the production of biofuels holds a great future due to increasing research and interest in the field. The biomass currently employed for biofuel production includes agricultural crops, agricultural waste, nonfood crops, agroindustrial and municipal soild waste, and so on. The use of plants for the production of ethanol dates back to the World War I era. But the research in that era was limited due to lack of resources and the abundance of fossil fuels.

14.2   BIOFUEL PROCESSING TECHNOLOGY

The processing of biomass into biofuel involves chemical conversion processes based on thermal and biochemical approaches. The production of biofuels—biodiesel and bioethanol—is based on the biochemical method. The production of biodiesel is based on the transesterification principle, which involves the transformation of triglycerides of raw materials into fatty acid esters. Bioethanol processing involves hydrolysis or fermentation of raw materials through a microbial approach. The transesterification approach is divided into two categories: catalytic and non-catalytic methods. The former is further subdivided into two subtypes based on the nature of catalyst: homogeneous and heterogeneous. The homogenous types are alkaline (KOH, NaOH) and acidic (HCl, H_2SO_4). The heterogeneous type involves ionic resins, enzymes, and alkaline earth metal compounds. The process is carried out by dissolving the raw material to be esterified and the catalyst in a reactor. Transesterification will produce two phases: crude glycerine and ester. The non-catalytic transesterification involves the use of different solvents, including propanol, ethanol, and supercritical methanol. Supercritical methanol is the most widely used approach due to simultaneous transesterification of triglycerides in raw materials and esterification of fatty acids. The other advantage of the process is a more straightforward approach and high yield as compared to other available methods. The return and production rate of biodiesel is governed by several factors, such as type of raw material, catalyst, temperature, pressure, and free fatty acid content (see Figure 14.1).

14.3   RAW MATERIALS FOR THE SYNTHESIS OF BIOFUELS

The selection of raw materials is mainly governed by the presence of a primary substrate that is capable of being converted into biofuel. Organic material consisting of different types of starch and cellulose is the primary choice of raw material. Some examples include wheat, corn, barley, beetroot, sugarcane, and a variety of trees. The choice of raw materials for biodiesel differs from that of ethanol, as the requirement for production for the former is the presence of fats and oils in the raw materials.

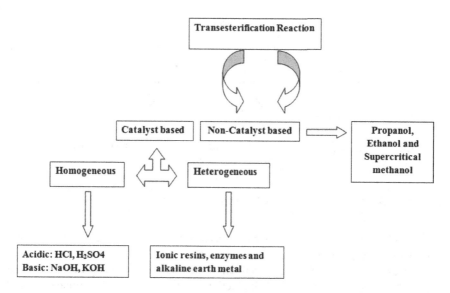

FIGURE 14.1   Schematic representation of biofuel processing technology.

Some widely used sources across the world are rapeseed, *Jatropha curcas*, sunflower, soybean, African palm, used vegetable oils (from the hotel industry and households), and animal fat (from slaughterhouses). The use of hydrogen as a biofuel is gaining interest because it can oxidize to water without emitting any carbon dioxide. But the production is currently dependent on chemical-based raw materials under high temperature and pressure. The raw materials mainly include coal and oil. The alternative for the aforementioned process is in the form of microbial-dependent hydrogen production, that is, biohydrogen, which holds a great future. Current research on biohydrogen production involves the use of:

a. Hydrogenases in photosynthetic microbes, for example, blue-green algae and cyanobacteria.
b. Nitrogenases in anoxygenic conditions in photoheterotrophic bacteria, for example, *Rhodopseudomonas palustris*.
c. Use of fermentative bacteria such as *E. coli* and *Enterobacter aerogenes*.

14.4   BIOHYDROGEN

Hydrogen is being viewed as the future of the fuel industry due to its environmentally friendly nature. Hydrogen can be produced from a range of fossil fuels and biological materials using different approaches. Some of the commonly used methods are steam reformation and the thermal cracking of natural gas and coal gasification. But these methods mainly utilize nonrenewable raw materials. The production of hydrogen using water involves photolysis, electrolysis, and thermolysis. The biomass-based approach is primarily based on the principle of gasification or

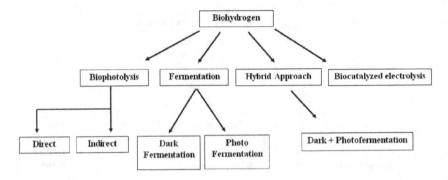

FIGURE 14.2 Different methods for biohydrogen production.

pyrolysis but suffers from the major drawback of producing a mixture of gases like carbon dioxide, nitrogen, methane, and carbon monoxide due to weak efficiency of the process. Biological hydrogen production methods are currently being explored across the globe. The methods of production of biohydrogen are summarized next (Figure 14.2).

14.4.1 BIOPHOTOLYSIS

The direct biophotolysis method is based on the principle of hydrolysis of water to produce hydrogen as the main product. Based on the type of approach used, it is subdivided into two categories: direct and indirect photolysis. In the former type, hydrogenase activity of algae is used to produce hydrogen via photosynthesis.

$$2H_2O + \text{light energy} \rightarrow 2H_2 + O_2$$

Hydrogenase activity was reported in *Chlorococcum littorale*, *Chlorella fusca*, and *Chlamydomonas reinhardtii*. The sensitivity of the hydrogenase enzyme towards oxygen is a major rate-limiting step. The other limiting factors are the cost of operation and photosynthetic efficiency of the system, which will limit the yield and needs to be targeted for achieving high production yield. In indirect biophotolysis, the problem of oxygen sensitivity is resolved by separating the stages of hydrogen and oxygen production. The hydrogen production in indirect biophotolysis is summarized as follows:

$$12H_2O + 6CO_2 + \text{"light energy"} \rightarrow C_6H_{12}O_6 + 6O_2$$
$$C_6H_{12}O_6 + 12H_2O + \text{"light energy"} \rightarrow 12H_2 + 6CO_2$$

Carbon dioxide fixation, along with nitrogen fixation capability, is reported in many algae and cyanobacteria. In this process, CO_2 is fixed and used to manufacture carbohydrates, as discussed earlier. In the next step, the product splits to release hydrogen via the action of hydrogenase. In the case of cyanobacteria, the hydrogenase was present in heterocysts in the oxygen-free environment.

14.4.2   PHOTO FERMENTATION

Purple non-sulfur bacteria were reported to carry out the production of hydrogen using light as the source of energy under anaerobic conditions. Due to the presence of photosystem-I and the absence of photosystem-II, nitrogenase can carry out hydrogen production. Purple non-sulfur bacteria are associated with the major drawback of photosynthetic efficiency, which falls in the range of 2%–10%. This problem can be overcomed by the construction of bioreactors with a large surface area to receive more light energy, but it will increase the cost of the process. Another approach for maintaining an optimal nitrogenase to hydrogenase ratio is to increase the yield of hydrogen in a reactor. Purple non-sulfur bacteria are reported to make use of a range of organic wastes as carbon source for the production of hydrogen.

14.4.3   DARK FERMENTATION

A number of anaerobic bacteria, including *Clostridium* and *Enterobacter,* were reported to produce hydrogen as a by-product along with acids and methane as the other important components. Depending on the type of carbohydrates used as the carbon source, bacteria produced a different type of acids with the release of hydrogen as the end product. Ideally, 1 mole of glucose will be broken down to 12 moles of hydrogen gas under standard conditions, but the actual yield is low.

$$C_6H_{12}O_6 + 2H_2O \rightarrow 2CH_3COOH + 2CO_2 + 4H_2$$

Apart from different process parameters, the partial pressure of hydrogen plays an important role in the production of hydrogen. With the increase in pressure of hydrogen, the overall yield of hydrogen decreases due to end-product inhibition.

14.4.4   HYBRID TECHNOLOGY

As discussed previously, under ideal conditions, 1 mole of glucose will produce 12 moles of hydrogen, but the actual yield is low. However, in the absence of light as a source of energy, the breakdown of glucose by anaerobic bacteria leads to the production of other by-products with a yield of 4 moles of hydrogen per glucose. As discussed in the following reaction, the acetate produced would be further broken down into hydrogen and carbon dioxide by photosynthetic bacteria using light energy. Hence, dark and photo fermentation techniques can be assimilated to increase the yield of hydrogen.

$$C_6H_{12}O_6 + 2H_2O \rightarrow 4H_2 + 2CO_2 + 2CH_3COOH$$
$$CH_3COOH + 2H_2O + \text{"light energy"} \rightarrow 4H_2 + 2CO_2$$

14.4.5   BIOCATALYZED ELECTROLYSIS

Acetate can be split by using an alternate source of external energy Another way of oxidizing the acetate (or the effluent of the dark fermentation process) to produce hydrogen is to provide external energy in the form of electrical energy instead of solar energy. In this approach, the bioreactor containing acetate forms the anodic compartment of an electrolyzer cell, and protons and electrons produced by bacteria are collected at the cathode (a platinum electrode catalyzing a hydrogen evolution reaction). Anodic and cathodic reactions are as follows:

Anode: $2CH_3COOH + 2H_2O \rightarrow 2CO_2 + 8H + + 8e$

Cathode: $8H + + 8e \rightarrow 4H_2$

It can be concluded that an external supply of around 100 mV is required to produce hydrogen at the cathode. However, because of over potential at the electrodes, a voltage higher than 100 mV is required to produce hydrogen.

14.5 BIOETHANOL

In the recent past, bioethanol has emerged as a better alternative to conventional fossil fuels. Bioethanol offers several advantages over conventional fuels in terms of reduced greenhouse gas emissions, high octane number, low boiling point, and ease of production and transportation.

Raw materials for bioethanol production mainly consist of lignocellulose biomass, sugars, and microalgae (Table 14.1). Based on raw materials, three different kinds of bioethanol can be produced.

14.5.1 FIRST-GENERATION BIOETHANOL

First-generation bioethanol is mainly produced by direct fermentation of sugarcane or starch. Major producers of first-generation bioethanol are the United States and Brazil. This class of fuel is dependent upon agricultural crops and thus directly affects the human food supply chain and also requires a large area of cultivable land for the necessary amount of crop production. Therefore, the whole idea is economically not viable.

14.5.2 SECOND-GENERATION BIOETHANOL

Second-generation bioethanol is produced by lignocellulose feedstock, such as agricultural residues and forest residues. This has an advantage over first-generation bioethanol as they are produced using lignocellulosic feedstock and agricultural residues, thus not affecting the human food supply chain and resulting in a low environmental impact.

14.5.3 THIRD-GENERATION BIOETHANOL

Because of the total dependency of first- and second-generation bioethanol production upon agricultural crops and related by-products, an alternative technology has been developed. For the production of third-generation bioethanol, algal biomass

TABLE 14.1

Brief Comparison of First-Generation, Second-Generation, Third-Generation, and Fourth-Generation Bioethanol

	1st-Generation Bioethanol	2nd-Generation Bioethanol	3rd-Generation Bioethanol	4th-Generation Bioethanol
Type of feedstock	Agricultural crops like sugarcane, sorghum, etc.	Lignocelluloses residue, agricultural residue	Algal biomass	Genetically modified microalgae
Type of land requirement	Agricultural land	Agricultural land, forest area	N/A	N/A
Impact on food supply chain	Large impact due to dependency on agricultural crops	Low impact	No impact	No impact
Water requirement	Potable water required	Potable water required	Potable, non-potable, saline water	Potable, non-potable, saline water
Environmental impact	Low GHG mitigation	Low GHG mitigation	Moderate carbon dioxide fixation	High carbon dioxide fixation
Economic viability	Low initial cost High product recovery cost	Low initial cost High product recovery cost	High initial cost	High initial cost
Product yield	Low in relation to biomass	Low in relation to biomass	High	High

is used. The use of algal species such as *Chaetocero scalcitrans*, *Botryococcus braunni*, and *Chlorella* species has been reported for the production of biofuel.

Production using algal biomass is advantageous in many aspects.

a. Low operating cost and easy cultivation.
b. Non-requirement of a large area of land.
c. The food supply chain is not interfered.
d. A high amount of lipid content in comparison to first- and second-generation biofuels.

14.5.4 FOURTH-GENERATION BIOETHANOL

Fourth-generation biofuels will be based upon genetically modified version of microalgae. The success of microalgae biomass in the synthesis of third-generation biofuel shifted the focus of biofuel synthesis from conventional crops and related by-products towards the use of improved algal strains. Microalgae-based production of bioethanol can be enhanced by increasing microalgae's exposure to light and thus improving its photosynthetic ability. One approach to increase the penetration of light is by using a truncated chlorophyll antenna and even by reducing light absorption in between. *Chlorella vulgaris, Chlorella sorokiniana, Spirulina platensis, Spirulina maxima, Botryococcus braunii,* and *Phaeodactylum tricornutum* are some of the selected microalgae strains being used for biofuel production.

Bioethanol production from microalgae can be done using two systems: contained and uncontained. In contained systems, closed systems are used for microalgae cultivation, and better protection towards contamination is offered due to highly controlled systems. However, the operating cost of the contained system is comparatively higher than the uncontained system. In uncontained systems, the cultivation of genetically modified microalgae is more natural, but the practice is discouraged because of the associated risk of GM strain release into the environment.

14.6   ENVIRONMENTAL IMPACT OF BIOFUELS

Large-scale production and application of biofuels is one of the key features for achieving the goal of sustainable development. Being a fundamental part of future energy sources and the type of production technology involved, the environmental impact of biofuels is of great importance. We all are well aware of the negative impact of conventional fuels, ranging from pollution to substantial scale contributions to global warming. Therefore, measuring and monitoring the environmental impact of biofuels at all stages of processing and application is of great significance.

1. *Impact on land*

   To meet the global biofuel demand in the future, an enormous amount of raw material will be required in the form of agricultural crops. Therefore, a large area of land will be required for the cultivation of feedstock. Also, the suitability of the soil for feedstock cultivation is another factor for consideration. With limited land availability a major threat in the coming future due to a continuous rise in population, feedstock cultivation will remain in jeopardy.

2. *Impact on water resources*

   Production of biofuels requires a large amount of water. Agriculture crops being used for biofuel production such as sugarcane, sorghum, corn, and maize requires, for example, approximately 3900 liters of water for processing of 1 ton of sugarcane crop for bioethanol production. Such an extensive amount of water requirement will put an extra burden on countries at both the environment and the economic front. With climate shift, reduced rainfall, and ever-decreasing water resources, water availability will be a critical limiting factor for biofuel production, especially in countries already facing the problem of water scarcity.

3. *Greenhouse gas emissions*

   Production of biofuel can lead to the emission of GHGs at different stages of the process. During the cultivation of feedstock, the use of fertilizers will result in the release of a large amount of nitrous oxide, which is a massive contributor to global warming. In addition to fertilizers, the nature of the soil used for cultivation is another factor to be considered.

Soil rich in peat has a large amount of organic carbon stored in it, which will be released into the environment during feedstock cultivation in the form of carbon dioxide, thus increasing GHG emission. Moreover, land conversion for feedstock cultivation will also contribute towards the carbon footprint due to the reduction in forest cover.

Thus, the idea of cutting down GHG emissions using biofuels and slowing down climate change still needs to be analyzed more precisely. More data must be gathered and evaluated for GHG emission during energy input for fuel generation to the reduction obtained during the use of biofuel.

4. *Impact on biodiversity*

Expansion of land for agricultural crop production will destroy the natural habitat and severely affect the flora and fauna of a particular area. The concept of mono cropping will affect the natural diversity of the area and encourages a narrow range of habitat. Invasion of foreign species is another primary concern as the use of genetically modified crop varieties, and microalgae strains can result in competition with wild types. Therefore, a regulatory approach must be followed while selecting the area and species for biofuel production to prevent harm to biodiversity.

14.7   BIOFUELS AND FOOD SECURITY

In the past few decades, with the advent of technological development in the agricultural field, food production around the globe has increased. But with an enormous increase in population, reduction in arable land, lack of proper food storage facilities, and a high amount of food wastage, an imbalance has been created, resulting in major food shortage issues. In this scenario, cultivating major agriculture crops for the sole purpose of biofuel generation is not fruitful; food prices will increase, and the whole food supply chain will break down. For example, biofuel feedstock contributed to the rise in food and fertilizer prices in the years 2002–2007. With a mere target of 7% global biofuel production, a 20% increase in cereal prices was observed. Underdeveloped nations such as African nations where crops like sugarcane, maize are used as staple food items, and malnutrition rate is very high; these crops can't be used for biofuel production. Increased biofuel production in such places will affect human food consumption and result in an increase in the hunger index. An increase in the global share of biofuel from 8% in 2020 to 10% in 2030 would directly affect the food requirement of 200 million people.

Thus, innovative strategies and better policies are required to make biofuel generation an income-generating market both economically and environmentally. The negative impacts of biofuel generation on food security can be reduced if policymakers first ensure surplus agricultural productivity, which can be used for biofuels.

14.8   ADVANTAGES OF BIOFUELS

1. One major advantage of biofuel use is reduced greenhouse gas emissions, thus contributing less towards carbon content of the planet compared to conventional fuels.

2. Biofuels cause less air pollution and thus are eco-friendly.
3. Biofuels have a high octane number; therefore, high energy content is obtained.

14.9 DISADVANTAGES OF BIOFUELS

1. High processing cost is still a major constraint.
2. Biofuels, specifically biodiesel, releases a high amount of nitrogen oxides (NO_x) comparing to diesel.

14.10 CONCLUSION

To make biofuels like bioethanol, biodiesel, and biohydrogen a mainstream energy source, the associated environmental impacts must be strongly considered and overcome. Only after addressing the biofuel-associated environmental pros and cons and framing adequate national and international policies, sustainable development can be achieved.

15 Bioplastics
Origin, Types and Applications

Damanjeet Kaur and Saurabh Gupta

CONTENTS

15.1 INTRODUCTION

The word plastic is a derivative of a Greek word *plastikos* meaning "can be moulded into various shapes". Plastics are a wide range of high-molecular-weight organic substances derived from nonrenewable resources like hydrocarbons, petroleum, and allied compounds. These long-chain, polymeric, synthetic compounds are extensively used for numerous purposes such as food packaging, pharmaceutical and cosmetic products, transportation, biomedical, structural, and electrical components as well as other consumer products. Further, these synthetic polymers serve as one of the highly valuable compounds due to their astonishing versatility and low cost. Plastic is widely used in many different forms such as polyethylene (PE), polyvinyl chloride (PVC), polycarbonate (PC), polypropylene (PP), polyethylene terephthalate (PET), polybutylene terephthalate (PBT), and polystyrene (PS).

Some forms of plastic are reusable, while others produce hazardous substances after several uses. Some are easily recyclable, whereas others require a sophisticated handling process during recycling. In 1988, the Society of Plastic Industry (SPI) had classified plastics into seven categories depending upon their use and properties, allowing consumers and recyclers to differentiate among them:

Type 1—Polyethylene terephthalate: PET is highly flexible, thermoplastic polymer belonging to the polyester family. These semi-crystalline resins have excellent thermal, mechanical, and chemical properties. These are

DOI: 10.1201/9781003131427-15

one of the most widely used recyclable polymers with enormous applications in clothing and textile fibres, soft drink and water bottles, medicine jars, films, electronic devices, packaging sheets, other household items, and more. The PET polymers contain antimony trioxide—a carcinogenic substance.

$$(-CH_2-CH_2-O-\overset{\overset{\displaystyle O}{\|}}{C}-\langle\!\bigcirc\!\rangle-C-O-)_n$$

Type 2—High-density polyethylene: HDPE is a flexible, translucent thermoplastic polymer made by the polymerization of ethylene. These polymers have a high melting temperature and excellent moisture and chemical resistance properties. HDPE plastics are inexpensive, recyclable, more stable than PET, and used for the manufacturing of storage containers, plastic bags, toys, plant pots, buckets, crates, wires and cables, fuel tanks and more. Polyethylene exposure can result in several defects in the immune system, respiratory system, skeletal system, and muscular system.

$$(-CH_2-CH_2-CH_2-CH_2-CH_2-)_n$$

Type 3—Polyvinyl chloride: PVC is the second most widely used thermoplastic made from polymers of vinyl chloride. It is produced in two forms: rigid plastic and flexible plastic. The addition of phthalates prepares flexible plastics. PVC is used for the manufacturing of sliding windows, pipes and fittings, cables and wire insulation, floppy-disk covers, roof linings, automobile seat covers, raincoats, toys, housewares, and so on. PVC is considered as hazardous plastic because it releases various toxic compounds such as bisphenol A, lead, cadmium, mercury, and phthalates. It also poses serious health hazards as it emits hydrogen chloride when burnt.

$$\overset{\overset{\displaystyle Cl}{|}}{(-CH_2-CH-)_n}$$

Type 4—Low-density polyethylene: LDPE is semi-rigid, translucent, branched polymer having properties similar to HDPE. Due to its low density, LDPE is highly flexible and durable but quite difficult to recycle. LDPE is used as cling films, squeeze bottles, grocery bags, food packaging films, frozen food bags, caps and closures, lamination films, and so on. Chemicals associated with LDPE result in hormonal defects in humans. When inhaled, it causes proximal scleroderma, joint involvement, and pulmonary manifestations.

$$CH_3$$
$$CH_2$$
$$CH_2$$
$$CH_2$$
$$(-CH_2-CH_2-CH-CH_2-CH_2-CH-CH_2-CH_2-)_n$$
$$CH_2$$
$$CH_3$$

Type 5—Polypropylene: PP is a rigid, tough, crystalline thermoplastic produced by the polymerization of monomer propylene. PP exhibits excellent electrical and mechanical properties, even at elevated temperatures. It is most widely used as plastics and fibres in the automotive industry, furniture market, and various consumer goods. PP can cause asthma, hormonal imbalance, and even cancer in some cases.

$$CH_3$$
$$(-CH_2-CH-)_n$$

Type 6—Polystyrene: PS is a hard, transparent, synthetic hydrocarbon resin made by the polymerization of styrene. PS is widely employed as disposable cups, packaging foam, food boxes, egg cartons, plastic cutlery and so on. Polystyrene contains neurotoxic and carcinogenic substances that pose a severe health hazard in living organisms. The burning of polystyrene causes the release of carbon monoxide and styrene monomers, which create a harsh, hazardous environment and are a danger to human health.

$$(-CH_2-CH-)_n$$

Type 7—Others: This category includes polylactide, polycarbonate, acrylonitrile butadiene styrene, acrylic, fibreglass, and nylon. These polymers offer a wide range of uses in different sectors.

15.2   NEED FOR BIOPOLYMERS

Several properties such as stability, durability and mechanical and thermal strength have considerably contributed to the increased use of plastics. Due to extensive applications and properties, the demand and production of these synthetic polymers have increased several fold in the past few decades. Considerable quantities of petroleum-derived synthetic polymers are synthesized every year and dumped into the environment as waste after use. This plastic waste contaminates a wide range of natural terrestrial, freshwater and marine ecosystems.

Further petroleum-based plastics are not readily biodegraded and thus tend to persist in the surroundings for hundreds of years. Environmental contamination because of these synthetic polymers is increasing daily in significant proportions. Accumulation of plastic waste and by-products in the marine ecosystem had hazardous and detrimental effect on aquatic organisms, including both plants and animals. A significant impact from plastic debris on marine birds, fur seals, cetaceans, sea turtles, filter feeders, and sharks has been reported. Further, the chemicals associated with and released along with these synthetic polymers hinder the biological functions and activities of living organisms.

In addition to environmental issues, the production of plastics is also a major concern due to the fast depletion and increasing cost of nonrenewable fossil resources. Due to urbanization and economic growth, plastic usage and consumption in developing countries such as China, Indonesia, the Philippines, Sri Lanka, and Vietnam have been reported much higher than the world's average. Approximately 4% of fossil resources are utilized every year for the production of around 300 million tons of plastics; however, if the usage continues at the same rate, the plastics sector will account for more than 20% of fossil resource consumption by 2050 (World Economic Forum, 2016). Moreover, the incineration of fossil-based polymers leads to depletion of the ozone layer and generation of greenhouse gases, which subsequently results in a rise in temperature and thus contributes to world climate change. The growing scarcity and increasing cost of fossil resources have lead to the development of plastics from renewable resources. Renewable biopolymers were first introduced in the 1980s to overcome the environmental and fossil-related issues associated with the use of synthetic polymers. Biopolymers are considered sustainable because they can be reused or reprocessed indefinitely after their intended use. In recent years, bioplastics are attracting considerable interest due to their remarkable properties of being renewable, environmentally friendly, sustainable, and biodegradable, thus significantly reducing the dependence on fossil resources and the accumulation of plastic waste in the environment.

15.3 BIOPLASTICS

Bioplastics are considered as organic plastic made up partly or entirely of substances derived from renewable raw materials. Bioplastics are not a single type of material but constitute a group of elements with different characteristics. As per the European Bioplastics Association, bioplastics are polymers that are bio-based, biodegradable, or both (European Bioplastics, 2019). Due to such features, these organic polymers serve as a convincing substitute for conventional petrochemical-based plastics.

Bio-based: "Bio-based" means that the substances are human-made organic macromolecules derived from renewable carbon sources or biomass. Further, these biological plastics are made to diminish greenhouse gas emissions and are regarded as environmentally friendly, depending upon the conception of "carbon neutrality" because when these plastics are burnt, the CO_2 produced is again transformed into

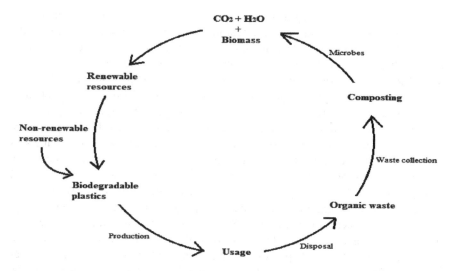

FIGURE 15.1   Life cycle of biodegradable plastics.

biomass by the process of photosynthesis. Bio-based polymers reduce the dependence upon nonrenewable fossil resources and also diminish the carbon footprint.

Biodegradable: Biodegradable plastics are the polymeric substances that are readily decomposed by microorganisms into carbon dioxide, water, biomass, and methane, depending, upon the polymeric structure and environmental factors (Figure 15.1). Biodegradation is a biological process of degradation of complex organic materials into simpler forms by the enzymes produced by microbes. The biodegradation of polymers mainly involves three major steps:

i. *Biodeterioration:* This is the process of altering the physical, chemical, and mechanical properties of the polymers by the augmentation of microbes inside or on the surface of the polymer.
ii. *Biofragmentation:* This involves the conversion of the complex polymeric chain into oligomer and monomer form by the action of microbial enzymes.
iii. *Assimilation:* Microbes utilizes the essential nutrients from the fragmented polymers in the way of carbon and energy and convert them into carbon dioxide, water and biomass.

15.3.1   NATURE OF THE BIOPLASTICS AND INTERNATIONAL LEGISLATION

The terms bio-based and biodegradable are frequently confused as meaning environmentally friendly, even though these are not similar in original perception. Biodegradable plastics have been synthesized to meet the objective of degradability, while bio-based plastics have been developed from biomass rather than fossil resources. However, there is a general conviction that if a product is derived from biomass, it will also be readily degradable. But the utilization of biomass as a raw material does not indicate the degradability of the end product.

Furthermore, product degradability depends not only on its raw constituents but also on the chemical nature and configuration of the final product along with the environmental conditions in which the product is likely to be degraded. Some bio-based polymers are biologically degradable while others are not, depending upon their definite polymeric composition. It is important to understand that bio-based polymers are not necessarily degradable, and conversely, biodegradable polymers are not always made from biomass. The rate of biodegradation of polymeric substances depends upon the temperature, microbial strain, oxygen, humidity, and the environment such as soil, marine, or landfill. Microbes utilize these polymers as the sole carbon source for their life processes and release carbon dioxide (in an aerobic environment) or carbon dioxide and methane (in an anaerobic environment) as end products. The amount of gas involved determines the biodegradation percentage by the microorganism in the target disposal environment. Various international agencies, such as the American Society for Testing and Materials (ASTM), International Organization for Standardization (ISO), and European Committee for Standardization (CEN), have proposed several standards, test methods and specifications to evaluate the biodegradability percentage in various disposal environments. ASTM standards are referred to as European (EN) in Europe, Australian standard (AS) in Australia, while ISO is the worldwide standard. The most biodegradable polymers are said to be compostable, and the definition of compostability has been laid down in different international standards. Several specification standards proposed by global organizations for compostable plastics include ASTM D6400 and D6868, ISO 17055, and EN 13432. Further, for a polymer to be characterized as compostable, it must fulfil the following criteria:

- *Chemical characteristics*: The polymer should be composed of a minimum of 50% organic components.
- *Biodegradation*: The developed product should have the ability to degrade at least 90% within the timeframe of six months under controlled conditions.
- *Disintegration*: Manufactured goods, in a form that enters the marketplace, should be able to disintegrate adequately into visually undetectable particles having a size less than 2 mm within three months under the controlled composting environment.
- *Ecotoxicity*: The manure obtained after the completion of the biopolymer composting experiment eventually contains a huge amount of non-degradable residues. These residues should not cause any harmful hazard or have a significant effect on the ecosystem or the growth of vegetation.

Thus, bioplastics are considered as a large group of polymers having characteristics similar to fossil-based plastics along with the additional properties of being renewable or biodegradable and significantly reducing the harmful impact on the ecosystem. The worldwide production and usage of biodegradable polymers are growing tremendously and continuously in various industrial fields. An extensive array of bioplastics is commercially available, and new ones are constantly emerging. The biopolymer market is expanding at a rate of 8% to 10% every year and constitutes around 10% to 15% of fossil-based plastics. The market share of bioplastics is expected to increase up to 21.7% by 2025.

15.3.2 Classification

Depending on the raw materials and their biodegradability, plastics have been classified into four categories (Table 15.1) (European Bioplastics 2018):

- *Conventional plastics*: Plastics derived from fossil resources and are non-biodegradable are known as traditional or conventional plastics. These consist of PE, PP, PET, PS, PVC, and others.
- *Fossil-based biodegradable plastics*: These include a distinct group of second-generation fossil-based synthetic polymers that are genuinely biodegradable. These are mostly used in amalgamation with starch and other polymeric substances having ester bonds that are degraded by enzymes produced by microbes. These degradable synthetic polymers include aliphatic polyesters, for example, polycaprolactone (PCL), polybutylene succinate (PBS), polybutylene succinate-*co*-adipate (PBSA), and aliphatic and aromatic copolyesters such as polybutylene adipate-*co*-terephthalate (PBAT).
- *Bio-based non-biodegradable plastics*: Plastics derived from biomass but resist biodegradation are known as bio-based non-biodegradable plastics. Although derived from biomass, these polymers are not necessarily biodegradable, for example, bio-based polyethylene (bio-PE), bio-based polyethylene terephthalate (bio-PET), and bio-based polyvinyl chloride (bio-PVC) are derived from renewable raw materials but have non-biodegradable properties. Bio-PE has non-biodegradable properties, although produced from bioethanol. Bio-PET is also synthesized from bio-based ethylene glycol. Similarly, polysaccharide derivatives such as cellulose acetate are extensively used in films and filters but are non-biodegradable due to high degree of substitution.
- *Bio-based biodegradable plastics*: This group includes the polymers that are made from renewable resources such as polysaccharides, lipids, proteins, and plant oil and are truly biodegradable. The significant substrates used for biodegradable plastics include starch, soy protein (SP), cellulose,

TABLE 15.1
Classification of Plastics Based on Raw Material and Their Degradability

	Non-Biodegradable	Biodegradable
Fossil-based	Conventional plastics e.g. PE, PP, PET	Biodegradable plastics e.g. PBAT, PBS, PCL
Bio-based	Bio-based plastics e.g. bio-based PE, PET, PTT (polytrimethylene terethalate), PA (polyamides)	Biodegradable and bio-based plastics e.g. PLA, PHA, starch blends

wheat gluten, zein, rice, whey, casein, and egg albumin. Starch and soy protein blends are thermoplasticized under heat and mechanical agitation and, after that, are mixed with several other thermoplasticized polymeric materials resulting in biodegradable polymer. Similarly, cellulosic fibres are also prepared by dissolving them in appropriate solvents and then moulding them into fibres and sheets; thus, they can be explored in fibre-fortified polymer complexes as fortification fibre. Further, polylactic acid (PLA) and polyhydroxyalkanoates (PHA), for example poly-β-hydroxybutyrate (PHB), are the most imperative thermoplastics prepared from renewable resources. These biodegradable polymers possess certain thermal and mechanical properties and processability that are analogous to that of fossil-derived polymer types.

Furthermore, bioplastics are classified into three categories depending upon their origin, chemical composition, and method of production:

i. *Biomass extracted polymers*: These include the bioplastics that are extracted from marine and agricultural products such as starch, cellulose, soy protein, and chitin. These bioplastics can be used either alone or along with other polymeric substances such as PLA.

ii. *Polymer production from monomer subunit of biomass by chemical synthesis*: For example, polylactic acid synthesized from sugar present in biomass via fermentation.

iii. *Polymer production by naturally occurring or genetically modified microbes*: This consists of microbial polyesters, including PHA and PHB, which accumulate intracellularly in the cytoplasm of a cell as a carbon and energy source under nutrient stress conditions.

Similarly, bioplastics can also be categorized based on their life cycle management:

i. *Compostable bioplastics*: Polylactic acid

ii. *Biodegradable bioplastics*: Starch, polybutylene succinate, and polyhydroxyalkanoates

iii. *Recyclable bioplastics*: Bio-based polyethylene terephthalate, bio-based polyethylene, and bio-based polypropylene

15.3.3 CONSTITUENTS OF BIOPLASTICS

Starch: Starch is a polymeric substance composed of D-glucose subunits linked together by α(1→4) and α(1→6) glycosidic linkages. Starch is the most abundant and frequently used as a renewable polymer obtained from corn, wheat, rice, potato, cassava, and so on. Starch has found a broad range of applications in various fields, for instance in agriculture, adhesives and emulsions, disposable gloves, personal care, glass fibre, construction, and packaging, due to their abundance, low cost, biodegradability, blending with other conventional polymers, and capability of being processed such as by extrusion and injection moulding. Novamont is the major starch biopolymer (Mater-Bi) producing firm based in Italy. In addition, Biome, Cardia, Cereplast, Teknor Apex, Plantic, and Rodenburg are also major producers of starch-based polymers. Plasticized or thermoplastic starch (TPS) is gaining more significance and interest due to its properties and processability similar to thermoplastic. TPS is obtained after disrupting the molecular structure of starch in the presence of high temperatures and plasticizers like water and glycerol. Starch polymers are also blended with other biodegradable polymers like cellulose, polylactic acid, and polyesters to improve their properties like mechanical strength and water resistance. The starch polymers and blends have been used to make compostable trash bags, mulching film, magnetic card, lamination paper, woven net, ropes, and disposable tableware and also for rigid and flexible packaging purposes like extruded bags, thermoformed trays and containers.

Polylactic acid: PLA is the biodegradable, thermoplastic, and recyclable aliphatic polyester derived from renewable raw materials by ring-opening polymerization of cyclic lactide dimer. PLA plastics are derived from the conversion of carbohydrates and agricultural by-products, for instance, starch-rich materials like maize and wheat into dextrose, and subsequently to lactic acid by bacterial fermentation. PLA exists in the form of the racemic mixture—dl, or as two stereoisomers—l and d. The difference in the proportion of l and d forms of lactic acid results in the synthesis of a wide array of PLA. l and d forms are known as poly(l-lactic acid) (PLLA) and poly(d-lactic acid) (PDLA), respectively, and both are optically active substances and semi-crystalline in nature. In contrast, the dl form—poly(dl-lactic acid) (PDLLA)—is optically inactive and completely amorphous. Several microbial species, for instance *Corynebacterium glutamicum, Escherichia coli, Lactobacillus,* and *Cyanobacteria,* have been reported to produce PLA. Under stress conditions, PLA is synthesized and accumulates in the form of granules in the cell cytoplasm. These polyesters have found numerous applications in the packaging and biomedical field and thus serve as a promising substitute to LDPE, HDPE, PET, and PS. However, these polymers can be degradable under certain conditions of composting, including extreme temperature, moisture, and air. The PLA polymers are commercially produced by Biome (Biotec GmbH), Heritage Plastics, Kingfa, FKuR, Cardia, Cereplast, Novamont, Cortec, and others.

Microbial polyesters: Microbial polyesters are of utmost importance over other biodegradable polymers due to the presence of hydrolysable ester bonds. These polyesters are synthesized by numerous bacterial strains such as *Bacillus megaterium, Wautersia eutropha,* and *Pseudomonas* sp. and can accumulate these polyesters up to 80% of their dry weight. The most widely produced and explored microbial polyesters are the polyhydroxyalkanoates, poly-β-hydroxybutyrate and their associated derivatives. These microbial polyesters can substitute for conventional plastics used for coating and packaging purposes.

Polyhydroxyalkanoate: PHAs are the intracellular, aliphatic polyesters synthesized from sugars or lipids in the form of carbon and energy by various microbes. PHAs are biodegradable and biocompatible thermoplastics having properties similar to petrochemical plastics currently in use. These polyesters can range from several hundred to thousands in their molecular weight depending upon microbial strain, growth conditions, and carbon source. Large-scale manufacturers of PHA include

Metabolix, Ecomann, Tianan, Biomer, Tianjin Green/DSM, Kaneka, and Meredian. The GreenBio PHA production plant in China has a production capacity of around 10,000 tons PHA every year. Due to excellent biocompatibility and biodegradability, PHAs are extensively used for biomedical purposes. Further, these polymers have also found application as toners for printing and adhesives for coating purposes. PHA has also been explored in packaging purposes due to its cost-effective and environmentally friendly properties.

Poly-β-hydroxybutyrate: PHB is an intracellular, biodegradable polymer with linear chains of β-hydroxybutyrate (β-OHB). PHB was first discovered in the cytoplasm of the *Bacillus megaterium* strain and later in a wide variety of archaea and eubacteria. PHB accumulate in the bacterial cell in the form of inclusion bodies. Depending upon the function and number of β-OHB units, three types of PHB have been discovered:

   i.   *Storage PHB*: Long-chain polymer with high molecular weight and made up of 10,000 to >1,000,000 β-OHB residues.
   ii.   *Oligo PHB*: Medium-chain polymer with low molecular weight and consisting of 100–300 β-OHB residues.
   iii. *Conjugated PHB (cPHB)*: Short-chain polymer with ≤30 β-OHB residues.

PHB is the most widely used biopolymer for a vast number of applications, having mechanical properties similar to polypropylene except fragility. Numerous microbial strains such as *Burkholderia sacchari*, *Alcaligenes latus*, *Methylocystis* sp., *Microlunatus phosphovorus*, *Bacillus subtilis*, and *Bacillus* sp. have been reported to produce PHB. These bacteria produce PHB as a reserve food material under nutrient-limiting conditions. Zhao and co-workers studied the production of PHB in *Hyphomicrobium zavarzinii* strain by supplementing methanol as a carbon source. Similarly, synthesis of PHB polymers in the moderately halophilic bacterial strain *Halomonas boliviensis* LC1 has been reported by several researchers. Besides these, genetic engineering of *Escherichia coli* has been carried out, and recombinant strains were reported to accumulate PHB granules in the cytoplasm. Being biodegradable and readily available, PHB biopolymers are finding their applications in different fields like packaging, the biomedical industry, and agriculture and serve as a potential substitute for conventional plastics.

15.4   APPLICATIONS OF BIOPLASTICS

For hundreds of years, conventional fossil-derived plastics have found a number of applications in everyday life owing to their versatility, durability, stability, and low cost. The extensively used synthetic plastics with a diverse application are polyethylene (29%), polypropylene (19%), polyvinyl chloride (13%), polystyrene (7%), polyethylene terephthalate (7%), polyurethane (6%), and others. However, their excessive use and numerous disadvantages like accumulation, resistance to degradation, and toxicity after incineration have further limited their use and led to the development of sustainable biopolymers. Reduced CO_2 production during bioplastics synthesis, along with their biodegradability, is the foremost advantage for their usage. Different techniques are used for the preparation of biopolymer blends that can be a substitute for these synthetic polymers. Starch, PLA, PHA, PHB, and so on are the most commonly used and potential biopolymers having full applications. The advantages of bioplastics over synthetic plastic material in being renewable and completely degradable have paved the way for their use in a wide range of industrial sectors like the food industry, cosmetics, agriculture, biomedical and pharmaceutical industries, textiles, and electronics, for packaging purposes, and in numerous other consumer products.

1. *Packaging and coating*

In a global scenario, 35% of the material used for packaging purposes is paper and cartons; polyethylene terephthalate, polyvinyl chloride, and polystyrene accounts for about 30% of the total plastics; and other materials used for packaging include aluminium, steel, and glass. Bioplastics are being explored widely for a large number of packaging purposes due to the drawbacks associated with conventional plastics. Polymers made from starch, cellulose, chitosan, protein, PLA, and their derivatives have found immense applications in packaging of foodstuff for their protection and preservation. Bioplastics are widely used for the preparation of packaging bottles, clamshell cartons, shipper bags, food-grade lamination films, food containers, and so on. Cellulose-based paper and board are widely used as packing material in various industries. Biodegradable polymers are also being used as edible films over food to prevent the moisture loss, oxidation, and migration of lipids. PHB biopolymers are extensively used for packaging as storage containers and as coating films.

Several renowned brands such as Coca-Cola, Volvic, Heinz, Vittel, PepsiCo, Danone, Unilever, P&G, Sainsbury's, Procter & Gamble, and Johnson & Johnson are resorting to bio-based PET bottles for packaging of drinking fluids and other cosmetic products. The Coca-Cola industry first introduced Plant Bottles in 2009 that were made up of 30% plant-based materials and similar in weight and chemical structure to conventional PET bottles. Heinz licensed the Plant Bottle technology from Coca-Cola for their ketchup bottles in 2011. Similarly, in 2011, PepsiCo developed the world's first 100% bio-based PET bottle from pine bark, corn husks, and switchgrass with properties similar to traditional PET. The biodegradable polymers Solaplast1723 and Solaplast1223 derived from algae have found immense application in the packaging industry as storage containers. In 2012, over 2.5 billion Plant Bottles were produced, and these bottles are being used for packaging purposes in Denmark, the United States, Canada, Japan, Brazil, Mexico, Norway, Sweden, and Chile.

2. *Biomedical*

In the biomedical industry, bioplastics have been used in implantations as bone screws, bone plates, and absorbable skin staples and sutures and as vehicles in drug delivery, membranes for tissue renaissance, and multifilament meshes or porous structures for tissue engineering. Bioplastics prepared from chitin and chitosan material are highly explored as wound plaster, as scaffold preparation for tissue engineering, in drug delivery, for sensor development, implant, antibacterial coatings, and in membrane preparations. Bioplastics such as PLA, PHA, and associated copolymers are widely used as a carrier for implanted active substances, stents, surgical threads, adhesion barriers, nerve guides, bone marrow scaffolds, and dissolvable inserts like screws, orthopaedic pins, and plates, which are readily degradable in the body. PHA blended with hydroxyapatite (HA) has been used as a bioactive and biodegradable composite for tissue replacement and renaissance. Thermoplastic starch is used as a coating substance for pills and capsules, and thus serve as a substitute for gelatin.

3. *Domestic goods*

Several domestic/houseware goods prepared from bioplastics are currently used and available extensively in the market, thus substituting for conventional plastic goods made from polystyrene and polypropylene. Houseware goods such as disposable cutlery, kitchen tools, storage containers, salad cups, plates, drinking cups, bathroom accessories, hangers, hooks, and toys are produced from different biodegradable polymers. Shopping bags made from biopolymers that degrade upon exposure to air, water, and sunlight are also used now.

4. *Agriculture and horticulture*

Starch polymers are extensively used to prepare biodegradable mulching and casing film, plant pots and trays, seeding strips, tapes, that offer various applications in agriculture and horticulture. Biodegradable microbial polyesters like PHA are used for encapsulation of fertilizers for their slow release into the environment and encapsulation of seeds for protection. These biodegradable polymers are also used as films to protect crops and as containers for greenhouse facilities. Biodegradable foils and nets made from biopolymers are used in mushroom farming and coating of the trees. Bioplastics are also used as manure for the crops after disposal, as these can be recycled in composting plants into fertile compost. The compost enhances the fertility of the soil and significantly contributes to the development of crops.

5. *Consumer electronics*

Plastics have been used for a huge number of electronic devices because of their different properties like durability, robustness, light weight, stability, and mechanical strength. A number of electronic devices such as mobile casings, touch screen computer casings, computer mice, keyboards, loudspeakers, vacuum cleaners, headphones, mobiles, laptops, and tablets made from bioplastics are now currently available in the market.

15.5 CONCLUSION

Plastics or synthetic polymers are extensively used in almost every field, but several disadvantages have further restricted their use. To overcome the limitations associated with conventional plastics, bioplastics have been explored in the market. Bioplastics, being bio-based, biodegradable, or both, are significantly contributing towards sustainability. Polymers derived from bio-based resources are renewable, thus reducing the dependence upon depleting fossil resources. Biodegradable polymers do not persist in the environment and reduce the risks associated with conventional plastics. Bioplastics offer numerous advantages over traditional plastics and thus serve as a potential substitute for them. Currently, bioplastics have found numerous applications in different fields like medicine, packaging, agriculture, and electronics, and the application of bioplastics industry is expanding tremendously in other fields as well.

16 Environmental Risk Monitoring Assessment and Management

Vinod Yadav and Mony Thakur

CONTENTS

16.1 INTRODUCTION

Risk is defined as the probability of a harmful outcome of an act. Environmental risk assessment (ERA) involves hazard identification, exposure assessment and risk estimation. It is used for evaluating the impact of human interventions on the environment. It helps in identifying the agent and understanding the interaction of the potential agent with the environment. Based on interaction data, ERA further provides solutions to the issues related to the environment, which will help in its management and will guide policy formulators to take appropriate steps to reduce risks.

16.2 ENVIRONMENTAL RISK ASSESSMENT

Empathetic information about the impact of various stressors, starting from an individual to a population and finally to a group of cooperative species at different geographical scales, is the most challenging task of environmental risk assessment. The proposed model of ecological risk assessment mainly targets a varying number of pollutants in a much fewer number of species under verified conditions in a fixed time period, while retrospective environmental risk assessments are highly focused on the effect of multiple stressors on complex communities with interconnection. Along with chemical stressors, physical stressors like temperature and biological

DOI: 10.1201/9781003131427-16

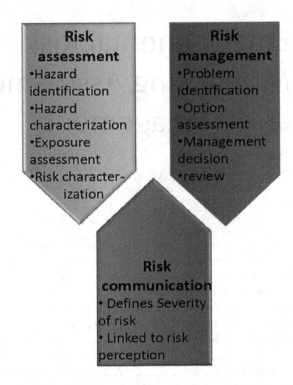

FIGURE 16.1   Approaches for risk analysis.

stressors like parasitism also exist in ecosystems and may positively or negatively affect the impact of chemicals. A number of studies talk about the antagonistic or synergistic effect of stressors with a different way of interaction which may influence bioavailability and uptake of chemicals along with detoxification. The examination of multi-stressor effects requires information about the connection of stressors in the environment as well as field surveys with a full gradient of stressors at different locations. Mechanistic exposure, along with model development and utilization, will be smoothened via environmental scenarios for risk assessment (Figure 16.1).

The distribution of stressors may occur in different habitats at a substantial distance from the initiation point. The risk of the stressors is usually variable and dependent on the context of the habitat's location, type, and quality along with the hazardous effect of pollutants within the locality. Currently, no approach of environmental risk assessment mainly talks about the complexity of the surroundings which is responsible for variation in risk at geographical distribution. The environmental risk assessment in a large distributed geographic area would be beneficial for aiming the interventions with protection necessities, while defining circumstances become costly for overprotection. Simple models for the risk assessment of an environment are unfamiliar with the dynamic approaches of the system and also possess model biases. In contrast, complex models work on the basis of intricate knowledge of the interaction between species and the environment and require various parameters to define the dynamics of the system. An ecotoxic study of a defined area provides a conceptual design for a combination of the mechanical approach with effect modelling.

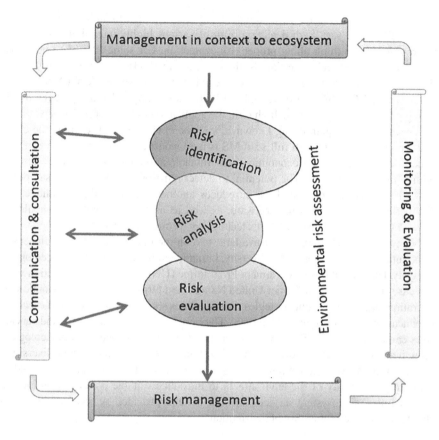

FIGURE 16.2 Standard risk management process.

Synthetic chemicals are demanded for maintaining current living standards, but they are rejected in aspects of environment and human health. The process of risk assessment, as well as management (Figure 16.2), in European countries, is mainly focused on a single chemical and fragmented framework. The combined approach of risk assessment in a similar way along with the life cycle and risk-benefit analysis is necessary for adequate decision-making.

16.3 MONITORING OF WATER AND LAND RESOURCES

A vast amount of emerging pollutants (EPs) are present in water bodies because of point as well as diffuse sources of pollutants. These pollutants, which are not easily monitored, contaminate ecological treasures and harm human health. The NORMAN network, which was initiated in 2005 with an aim to set up permanent research facilities to improve data collection, has identified 700 substances in the water bodies of Europe. Several EPs like hormones, pyrethroids and pesticides adversely affect the surrounding environment, even at a very low concentration. The situation demands urgent and severe action in favour of water resources management. At present, the sampling and analysis approaches are not integrated, but their main focus is on certain classes of EPs. A very high detection limit of various highly hazardous EPs pushes the researchers to think about proper risk evaluation. For some EPs like microplastics, nanomaterials and ionic liquids, the detection methods are at an initial stage of development. To target specific EPs, advanced techniques like GC-APCI-MS/MS and ultrasensitive LC-MS/MS are available. Still, they are barely utilized for regular monitoring purposes. In spite of the high cost, the application of these appliances would be very useful for the quantifying known EPs in the environment. Presently, the chromatographic technique with full scan MS is the favourite option for the screening of generic chemicals. It's an economical and valuable approach for screening samples with a high count of EPs. The potential to detect retrospectively without resampling, as well to detect anonymous EPs, introduces another benefit of this technique. An abstraction of data processing software, as well as associated databases, should be crossed to take full advantage of this technology for routine purposes.

Many assessment strategies are available to monitor and measure the fate and effect of pollutants or chemicals on biodiversity, human health, and indicators of ecological integrity. United Nations Environment Programme (UNEP) has been engaged in various case studies based on former United Nations and Organisation for Economic Cooperation and Development strategies. Such outlines should be modified and executed in education on ecosystem service risks. The operators of ecosystem risk and service risks can be diversified. The defeat of genetic diversity, adverse effects on ecological growth and climate, pollution, and the utilization of land is covered under diversified drivers in the *Atlas of Ecosystem Services*. There is a great and urgent need to understand the complexity of the environment to introduce suitable policies. Considering the dynamics of ecosystem service plans, the benefits provided by ecosystem services should be analyzed for equity of distribution.

16.4   CHARACTERIZATION OF THE EFFECT OF CHEMICALS

Usually, environmental assessment is performed without considering the area's geography. In the age of technology, it is effortless to find ways to determine chemical exposure. Some techniques like remote sensing, high-resolution spatial models, and time-of-flight mass spectrometry may quantify the level of pollution better than present methods can. The advancement in technology is also helpful in monitoring polluted areas and the exposure of humans and ecological treasures to a wide range of chemicals that were difficult to be examined before. However, the utilization of these technologies will also face a number of obstacles such as quality control, social and ethical issues, and more. Also, it is still difficult to quantify a number of contaminant classes such as nanomaterials, plastics and products of unknown and variable composition. The effective toxicity and behaviour of nanomaterials, as well as microplastics, mainly depends on the shape, size, concentration, surface area, charge, aggregation and agglomeration. Therefore, during the examination of nanomaterials and microplastics, physical as well as chemical properties of the pollutants also need to be analyzed along with the composition and concentration. Several analytical techniques like chromatography, filtration, microscopy and spectroscopy approaches may also be utilized. But the utilization of these techniques is restricted to the laboratory scale, as at the environmental level the techniques show less sensitivity and specificity. All these conditions demand the introduction of better methods with the ability to analyze the substances at expected concentrations in the environment.

16.5 ENVIRONMENTAL IMPACT ASSESSMENT

Environmental impact assessment (EIA) is another environmental management program that handles industrial projects to identify, anticipate, evaluate and diminish the environmental as well as a social hazard of a project. It is considered an essential instrument to plan and execute the project and is also utilized for the financial grant and regulatory approval of industrial projects. At present, the International Seabed Authority (ISA) ought to have an EIA to manage deep-sea mining (DSM) in the areas outside the limits of national authorities (Table 16.1). The ISA and its member

TABLE 16.1

EIA Regulatory Steps, Expertise Requirements and Potential Personnel Sources for Deep-Sea Mining (DSM) Enterprises

Regulatory Steps	Skills Required	Potential Personnel Sources
Screening	EIA legislation as well as administration knowledge Some information about the ongoing DSM actions	Various government agencies like environment, mining, fisheries, etc.
Scoping	Information about screening outcomes Knowledge of the marine environment including its flora and fauna	Various government agencies like environment, petroleum, mining, fisheries, etc.
Review of EIA report	Knowledge of scoping phase Identification of risks and impacts including assessments of interrelationships, cumulative and combined effects of activity	Various government agencies like environment, petroleum, mining, fisheries, etc. Regional organizations
Approval or rejection of the proposed project	Basic knowledge about the legal aspects and policies of EIA	Government authorities like a minister of any department
Consent examination of approved project	Information about the DSM activity along with basic monitoring techniques Knowledge about the effect on the environment as well as human health and safety	Various government agencies like environment, petroleum, mining, fisheries, etc.
Execution of approved activities	Familiarity with implementation provisions Information about the threshold for enforcement mechanisms	Government agency which controls EIA

Sources: Bradley & Swaddling, 2016; Durden et al., 2018.

states utilize EIA for the practical purpose of various key responsibilities like taking a precautionary approach and ensuring marine environment protection from the adverse effect of DSM, as obligatory by the United Nations Convention on the Law of the Sea (UNCLOS).

To achieve environmental management plans, EIA permits the description and alleviation of the harmful effects of industrial projects. However, the EIA process has to cross some abstractions for better performance, that is, the EIA process should integrate various tools to point out the uncertainty. ISA utilizes EIAs to apply uniform and consistent environmental standards to all contractors. Along with environmental management by the supporters, it is also essential that the audit of the agreement should be taken care of by the government.

16.6   INTERNATIONAL AGENCIES INVOLVED IN RISK ASSESSMENT

The United Nations have established a clear vision to stamp out poverty, protect our planet and ensure good health for everyone by achieving the development goals of a sustainable environment. A pollution-free and productive nature is the basic need for the achievement of these goals. Knowledge about the adverse effect of toxic chemicals on the environment and on human health becomes essential for better functioning of sustainable development goals. The current situation demands changing the way studies are performed to control the use and adverse effects of chemicals on the environment. The following regulatory authorities play a significant role in achieving the goals of a sustainable environment:

A. *Food and Drug Administration (FDA)*

This is a scientific agency engaged in the assessment of risk for the security and promotion of public wellness via examining and regulating products. The Federal Food, Drug, and Cosmetic Act (FFDCA) is embodied with a set of laws to approbate FDA for the inspection of food, drugs and cosmetics. The FDA also regulates many other products like infant formulas, dietary supplements, animal food, tobacco products and food additives. The approaches used for the management of food safety is discussed in Figure 16.3. Brand name and generic medications come under the authorization of the FDA. The agency also ensures the safety of non-prescription drugs used by human beings. Gene therapy products and allergenic and pharmaceutical products come under the aegis of the FDA. Medical devices like tongue depressors and heart pacemaker as well as electronic appliances such as microwave ovens, laser products and X-ray equipment are also evaluated by the FDA.

B. *US Department of Agriculture (USDA)*

This organization authorizes the safety and the labelling of cultured meat, poultry products, veterinary vaccines and other veterinary pharmaceutical products.

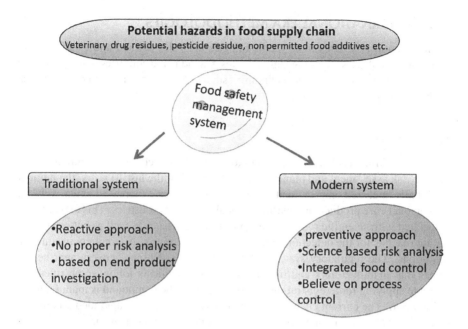

FIGURE 16.3   Food safety management system.

C. *Environmental Protection Agency (EPA)*

Since the introduction of the 1976 Toxic Substance Control Act, the US EPA allows the testing of chemicals only if companies provide evidence of harmful effects from them. The chemical companies lobbied to self-regulate those which are already in the market. The regulatory agencies have to make some assumptions about the authentication of risk and exposure, as ambiguity is the primary feature of every stage of risk assessment. Along with that, risk assessment agencies usually ignore the risks of chemicals that are lacking in reliable toxicological evidence.

D. *Animal and Plant Health Inspection Service (APHIS)*

This multidisciplinary agency comes under the jurisdiction of the USDA. The main aim of this agency is to support the vision of USDA, which is basically to protect and solve issues related to food, agriculture and other natural sources. The responsibilities of this agency include protecting and promoting agriculture, regulating genetically modified organisms, administering the Animal Welfare Act, and managing wildlife damage activities. To achieve these aims, APHIS implements emergency protocols for the detection of pests or disease to immediately eradicate the problem in affected states. This quick and combative strategy of APHIS leads to the successful prevention of pest and disease threats to the crops. APHIS develops advanced science-based tools to promote the health of agriculture at the international level.

16.7 INTERNATIONAL ERA METHODOLOGIES

Environmental risk assessment methodologies are fully defined and established in developed countries such as the United States, Canada, and countries of the European Union. Nations with developing economies such as China and India do not have such methodologies.

16.7.1 CHINA

At present, China has launched a set-up to establish new chemical regulation to utilize the risk assessment and management program. For the last 10 years, the number of globally analyzed environmental studies in China have increased extensively for the personal care product and pharmaceutical chemical industries. This expansion is associated with scientific research on environmental issues as well as better analytical approaches for the detection of contaminants present in different environmental conditions. The regulatory system established by the Chinese government necessitates that manufacturers and importers of new chemicals follow some rules, for example applying for a registration certificate before the compound is manufactured or imported. The current chemical regulation system has been upgraded and undertakes such activities as intensifying the environmental management system, legislating new chemical production and import, obstructing the unregulated practices of new chemicals and enhancing the framework of environmental testing. But the regulatory system still possesses some drawbacks, as it lacks an exposure methodology for the establishment of other quantitative risk evaluations.

16.7.2 EUROPEAN UNION

The Water Framework Directive (WFD) of the European Union (EU) covers the most essential and natural element of human life— water, which is in poor ecological condition in the EU. Inadequate evaluation of chemicals in the water results in the lack of accurate data for the ecological status of water bodies. However, the WFD has initiated several monitoring programs for gathering information about the level of contamination in surface water sources. The adverse impact of chemicals can be directly observed in aquatic life. But it is far more challenging to target the highly toxic chemicals that need to be adequately managed and to have effective solutions for addressing the high risks resulting from pollution. The limited detection ability of analytical instruments is the main problem for studying the negative impacts of pollutants on aquatic as well as human populations. The lack of information available about the sources of pollutants, their transport pathways and their combined effect with other chemicals is another main issue that needs to be solved. Ultimately, the situation demands a better understanding for connecting early biological responses with detectable compounds in bioassays.

A program called SOLUTIONS run by the European Union Seventh Framework Programme (Research and Development Funding Programme) (EU-FP7) addresses these problems via novel approaches for mitigating the negative impact of chemicals in water resources and also for achieving the vision of the WFD. SOLUTIONS is a project formed by a multi- and interdisciplinary consortium of 39 renowned scientific institutions and enterprises from Europe, Brazil, China and Australia and coordinated by the Helmholtz Centre for Environmental Research GmbH (UFZ). Various problem-solving strategies are planned based on connecting hazard identification efforts with available solutions. SOLUTIONS aims to integrate and validate the approaches based on monitoring and modelling and also to figure out the potential and favourable conditions for better interaction between organizations that regulate water quality, such as WFD, and regulations such as REACH should undertake proper chemical product assessment before authorization of its commercial use.

The main objective of SOLUTIONS is to protect water resources via the identification, evaluation and introduction of a conceptual framework for the estimation of growing pollutants which might have a hazardous effect on marine ecosystems as well as on the health of human beings. According to the National Research Council (2009), evaluation of risk is entirely a systematic policy that never leads to the measurement of risk management, even after the characterization of hazardous pollutants. The knowledge gap at any point in the risk assessment procedure may lead to the formation of barriers to the completion of the process. To cross such barriers, SOLUTIONS outlined strategies for identifying regulatory problems at the first stage instead of the final stage. For the eradication of pollutants from water resources, mitigation options will be considered right from the initiation of the process.

Issues related to human health and the environment are generally assessed as separately. However, the approach of SOLUTIONS is to connect both issues on the criteria of information related to modes of action (MoA). The different frameworks will be integrated with defined boundaries as per environmental quality standards (EQS), tolerable daily intake (TDI) along with tipping points towards the management shifts in natural systems. The goal of the SOLUTIONS project is to combine an examination-based framework, especially for the chemicals which are already present in the environment, with the effective designing of new substances and approaches for identifying upcoming trends in pollution (Figure 16.4). The main agenda of SOLUTIONS is to introduce effective screening techniques for various chemicals while ignoring the excessive focus on already known pollutants. The newly introduced approach will use ranking procedures to aid in the adequate consideration of new substances.

The data represent the presence of thousands of compounds in the environment, but experimental findings talk about only a few hazardous compounds. Environmental clearance (EC) focuses on the development of techniques for the identification of hazards or risks associated with chemicals from the mixture of pollutants. SOLUTIONS will examine the effects of chemicals on the diversity of the aquatic ecosystem and on human health from drinking water and eating fish. A focus on evaluating hazardous chemicals brings to light the risk generated by components of the mixture—in general situations along with site-specific risks. Even in the era of advanced technology, several chemicals found in water are still challenging to identify at an (Environmental Quality Standard) level. *In situ* solid-phase extraction, size exclusion devices and high-performance countercurrent chromatography along with passive sampling will all be helpful in solving the problem of low water EQS and low detection limits.

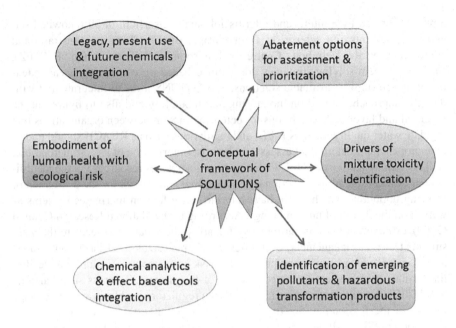

FIGURE 16.4 Scheme of a conceptual framework of SOLUTIONS.

16.8 CONCLUSION

Approaches and strategies such as chemical screening, risk assessment, identification of toxic derivatives, prioritization and so on have been certified under the SOLUTIONS project. Targeted and cost-effective management tools are required to achieve and maintain a proper chemical as well as ecological prominence. Novel environmental monitoring strategies are needed for balancing the need for a healthy environment with the production of chemicals required for modern living. Substantial research is today's requirement for delineating the toxins and chemicals originating from anthropogenic sources. Nations should collaborate for the diversification of information along with the high-level security of human beings and the environment at the global level.

Bibliography

Abbasian, Firouz, Robin Lockington, Megharaj Mallavarapu, and Ravi Naidu. "A comprehensive review of aliphatic hydrocarbon biodegradation by bacteria." *Applied Biochemistry and Biotechnology* 176, no. 3 (2015): 670–699.

Abbasnezhad, Hassan, Murray Gray, and Julia M. Foght. "Influence of adhesion on aerobic biodegradation and bioremediation of liquid hydrocarbons." *Applied Microbiology and Biotechnology* 92, no. 4 (2011): 653.

Abdu, Nafiu, Aisha Abdulkadir, John O. Agbenin, and Andreas Buerkert. "Vertical distribution of heavy metals in wastewater-irrigated vegetable garden soils of three West African cities." *Nutrient Cycling in Agroecosystems* 89, no. 3 (2011): 387–397.

Abed, R. M., S. Al-Kharusi, and M. Al-Hinai. "Effect of biostimulation, temperature and salinity on respiration activities and bacterial community composition in an oil polluted desert soil." *International Biodeterioration & Biodegradation* 98 (2015): 43–52.

Abedi-Koupai, Jahangir, Razieh Mollaei, and Sayed Saeid Eslamian. "The effect of pumice on reduction of cadmium uptake by spinach irrigated with wastewater." *Ecohydrology and Hydrobiology* 15, no. 4 (2015): 208–214.

Abou-Shanab, R. A. I., J. S. Angle, and R. L. Chaney. "Bacterial inoculants affecting nickel uptake by *Alyssum murale* from low, moderate and high Ni soils." *Soil Biol Biochem* 38, no. 9 (2006): 2882–2889.

Abou-Shanab, R. A., J. S. Angle, T. A. Delorme, R. L. Chaney, P. Van Berkum, H. Moawad, . . . and H. A. Ghozlan. "Rhizobacterial effects on nickel extraction from soil and uptake by Alyssum murale." *New Phytologist* 158, no. 1 (2003): 219–224.

Achal, V., A. Mukherjee, and M. S. Reddy. "Characterization of two urease-producing and calcifying *Bacillus* spp. isolated from cement." *Journal of Microbiology and Biotechnology* 20, no. 11 (2010): 1571–1576.

Achal, V., X. Pan, Q. Fu, and D. Zhang. "Iomineralization based remediation of As(III) contaminated soil by *Sporosarcina ginsengisoli*." *J Hazard Mater* 201–202 (2012): 178–184.

Adekola, Folahan A., Rasaq F. Atata, Risikat N. Ahmed, and Sandeep Panda. "Bioleaching of Zn (II) and Pb (II) from Nigerian sphalerite and galena ores by mixed culture of acidophilic bacteria." *Transactions of Nonferrous Metals Society of China* 21, no. 11 (2011): 2535–2541.

Adriano, D. C. *Trace Elements in the Terrestrial Environment.* Springer-Verlag, New York, 1986, p. 533.

Afzal, M., Q. M. Khan, and A. Sessitsch. "Endophytic bacteria: Prospects and applications for the phytoremediation of organic pollutants." *Chemosphere* 117 (2014): 232–242.

Agca, Yuksel. "Genome resource banking of biomedically important laboratory animals." *Theriogenology* 78, no. 8 (2012): 1653–1665.

Ahmed, T., M. Shahid, F. Azeem, I. Rasul, A.A. Shah, M. Noman, A. Hameed, N. Manzoor, I. Manzoor, and S. Muhammad. "Biodegradation of plastics: Current scenario and future prospects for environmental safety." *Environ Sci Pollut Res Int.* 25 (2018): 7287–7298.

Ahsan, Taswar, Nimra Amjad, Adeela Iqbal, and Ansar Javed. "A review: Tissue culturing of important medicinal plants." *Int J Water Res Environ Eng* 2 (2013): 76–79.

Aichner, Bernhard, B. Glaser, and W. Zech. "Polycyclic aromatic hydrocarbons and polychlorinated biphenyls in urban soils from Kathmandu, Nepal." *Organic Geochemistry* 38, no. 4 (2007): 700–715.

Aina, Roberta, Lucio Palin, and Sandra Citterio. "Molecular evidence for benzo[a]pyrene and naphthalene genotoxicity in *Trifolium repens* L." *Chemosphere* 65, no. 4 (2006): 666–673.

Aislabie, Jackie, Asim K. Bej, Janine Ryburn, Nick Lloyd, and Alastair Wilkins. "Characterization of arthrobacter nicotinovorans HIM, an atrazine-degrading bacterium, from agricultural soil New Zealand." *FEMS Microbiology Ecology* 52, no. 2 (2005): 279–286.

Aitken, M. D., W. T. Stringfellow, R. D. Nagel, C. Kazunga, and S.-H. Chen. "Characteristics of phenanthrene-degrading bacteria isolated from soils contaminated with polycyclic aromatic hydrocarbons." *Canadian Journal of Microbiology* 44, no. 8 (1998): 743–752.

Akar, A., E. U. Akkaya, S. K. Yesiladali, G. Celikyilmaz, E. U. Cokgor, C. Tamerler, D. Orhon, and Z. P. Cakar. "Accumulation of polyhydroxyalkanoates by *Microlunatus phosphovorus* under various growth conditions." *J Ind Microbiol Biotechnol* 33 (2006): 215–220.

Akhtar, Shazia, Muhammad Mahmood-ul-Hassan, Rizwan Ahmad, Vishandas Suthor, and Muhammad Yasin. "Metal tolerance potential of filamentous fungi isolated from soils irrigated with untreated municipal effluent." *Soil Environ* 32, no. 1 (2013): 55–62.

Akoto, Osei, Divine Addo, Elvis Baidoo, Eric A. Agyapong, Joseph Apau, and Bernard Fei-Baffoe. "Heavy metal accumulation in untreated wastewater-irrigated soil and lettuce (*Lactuca sativa*)." *Environmental Earth Sciences* 74, no. 7 (2015): 6193–6198.

Albright III, V. C., I. J. Murphy, J. A. Anderson, and J. R. Coats. "Fate of atrazine in switchgrass-soil column system." *Chemosphere* 90, no. 6 (2013): 1847–1853.

Alekhya, M., N. Divya, G. Jyothirmai, and K. Rajashekhar Reddy. "Secured landfills for disposal of municipal solid waste." *Int. J. Eng. Res. Gen. Sci* 1, no. 1 (2013).

Alexander, M. *Biodegradation and Bioremediation*. 2nd ed. Academic Press, San Diego, 1999, pp. 325–327.

Ali, H., E. Khan, and M. A. Sajad. "Phytoremediation of heavy metals-concepts and applications." *Chemosphere* 91 (2013): 869–881.

Ali, Hazrat, Ezzat Khan, and Ikram Ilahi. "Environmental chemistry and ecotoxicology of hazardous heavy metals: Environmental persistence, toxicity, and bioaccumulation." *Journal of Chemistry* 2019 (2019).

Alkio, Merianne, T. M. Tabuchi, Xuchen Wang, and Adán Colon-Carmona. "Stress responses to polycyclic aromatic hydrocarbons in Arabidopsis include growth inhibition and hypersensitive response-like symptoms." *Journal of Experimental Botany* 56, no. 421 (2005): 2983–2994.

Alkorta, I., and C. Garbisu. "Phytoremediation of organic contaminants in soils." *Bioresource Technology* 79, no. 3 (2001): 273–276.

Allen, J., M. Browne, A. Woodburn, and J. Leonardi. "The role of urban consolidation centers in sustainable freight transport." *Transport Rev* 32 (2012): 473–490.

Alloway, Brian J., ed. *Heavy Metals in Soils: Trace Metals and Metalloids in Soils and Their Bioavailability*. Vol. 22. Springer Science & Business Media, 2012.

Altenburger, R., S. Ait-Aissa, P. Antczak, T. Backhaus, D. Barceló, T. B. Seiler, . . . and G. de AragãoUmbuzeiro. "Future water quality monitoring—adapting tools to deal with mixtures of pollutants in water resource management." *Science of the Total Environment* 512 (2015): 540–551.

Al-Thukair, A. A., and K. Malik. "Pyrene metabolism by the novel bacterial strains *Burkholderia fungorum* (T3A13001) and *Caulobacter* sp. (T2A12002) isolated from an oil-polluted site in the Arabian Gulf." *International Biodeterioration and Biodegradation* 110 (2016): 32–37.

Alvarez, A., J. M. Saez, J. S. D. Costa, V. L. Colin, M. S. Fuentes, S. A. Cuozzo, . . . and M. J. Amoroso. "Actinobacteria: Current research and perspectives for bioremediation of pesticides and heavy metals." *Chemosphere* 166 (2017): 41–62.

Amarnath, D., Thamilamudhan R. Joshua, and S. Rajan. "Comparative study on wastewater treatment using activated sludge process and extended aeration sludge process." *Journal of Chemical and Pharmaceutical Research* 7, no. 1 (2015): 798–802.

American Academy of Pediatrics, Committee on Environmental Health. *Handbook of Pediatric Environmental Health.* 2nd ed. American Academy of Pediatrics, Washington, DC, 2003, pp. 311–321.

Ananthan, Rajendran, Remya Mohanraj, and V. Narmatha Bai. "In vitro regeneration, production, and storage of artificial seeds in *Ceropegia barnesii*, an endangered plant." *In Vitro Cellular & Developmental Biology-Plant* 54, no. 5 (2018): 553–563.

Anjum, Muzammil, R. Miandad, Muhammad Waqas, F. Gehany, and M. A. Barakat. "Remediation of wastewater using various nano-materials." *Arabian Journal of Chemistry* (2016).

Ankley, G. T., R. S. Bennett, R. J. Erickson, D. J. Hoff, M. W. Hornung, R. D. Johnson, . . . and J. A. Serrrano. "Adverse outcome pathways: A conceptual framework to support ecotoxicology research and risk assessment." *Environmental Toxicology and Chemistry* 29, no. 3 (2010): 730–741.

Anon. "Understanding the most common sources of noise in the city." New York City Department of Environmental Protection Bureau of Environmental Compliance, 59–17. Junction Blvd, 11th Fl, Flushing, NY, 2010a.

Araujo, K. M., A. Lima, J. N. Silva, L. L. Rodrigues, A. G. N. Amorim, P. V. Quelemes, R. C. Santos, J. C. Rocha, E. O. Andrades, J. R. S. A. Leite, J. Mancini-Filho, and R. A. Trindade. "Identification of phenolic compounds and evaluation of antioxidant and antimicrobial properties of *Euphorbia tirucalli* L." *Antioxidants* 3 (2014): 159–175.

Arena, Umberto. "Process and technological aspects of municipal solid waste gasification: A review." *Waste Management* 32, no. 4 (2012): 625–639.

Arrivault, Stéphanie, Toralf Senger, and Ute Krämer. "The Arabidopsis metal tolerance protein AtMTP3 maintains metal homeostasis by mediating Zn exclusion from the shoot under Fe deficiency and Zn oversupply." *Plant Journal* 46, no. 5 (2006): 861–879.

Arslan, M., A. Imran, Q. M. Khan, and M. Afzal. "Plant-bacteria partnerships for the remediation of persistent organic pollutants." *Environmental Science and Pollution Research* 24, no. 5 (2017): 4322–4336.

Arthur, Ellen L., Pamela J. Rice, Patricia J. Rice, Todd A. Anderson, Sadika M. Baladi, Keri L. D. Henderson, and Joel R. Coats. "Phytoremediation—an overview." *Critical Reviews in Plant Sciences* 24, no. 2 (2005): 109–122.

Arun, A., P. P. Raja, R. Arthi, M. Ananthi, K. S. Kumar, and M. Eyini. "Polycyclic Aromatic Hydrocarbons (PAHs) biodegradation by *Basidiomycetes* fungi, *Pseudomonas* isolate, and their cocultures: Comparative in vivo and in silico approach." *Applied Biochemistry and Biotechnology* 151, no. 2–3 (2008): 132–142.

Aruna Kumara, K. K. I. U., Buddhi Charana Walpola, and Min-Ho Yoon. "Current status of heavy metal contamination in Asia's rice lands." *Reviews in Environmental Science and Biotechnology* 12, no. 4 (2013): 355–377.

Arvind, Kumar Shukla. "An approach for the design of noise barriers on flyovers in urban areas in India." *International Journal for Traffic and Transport Engineering* 1, no. 3 (2011): 158–167.

Ashraf, Muhammad Aqeel. "Persistent organic pollutants (POPs): A global issue, a global challenge." (2017): 4223–4227.

Ashter, S. A. "Types of biodegradable polymers." In: Ashter, S. A. (ed.), *Introduction to Bioplastics Engineering*. William Andrew Publishing, 2016a, pp. 81–151.

Ashter, S. A. "Commercial applications of bioplastics." In: Ashter, S. A. (ed.), *Introduction to Bioplastics Engineering*. William Andrew Publishing, 2016b, pp. 227–249.

Athanasopoulou, A., V. Kollarou, and G. Kollaros. "Soil pollution by transportation projects and operations." Conference Paper, 2014.

Atterby, H., N. Smith, Q. Chaudhry, and D. Stead. "Exploiting microbes and plants to clean up pesticide contaminated environments." *Pesticide Outlook* 13 (2002): 9–13.

Auras, R., B. Harte, and S. Selke. "An overview of polylactides as packaging materials." *Macromol Biosci* 4 (2004): 835–864.

Averous, L., and E. Pollet. *Biodegradable Polymers, Environmental Silicate Nano-Biocomposites, Green Energy and Transport*. Springer-Verlag, London, 2012.

Averous, L., C. Fringant, and L. Moro. "Starch-based biodegradable materials suitable for thermoforming packaging." *Starch* 53 (2001): 368–371.

Ayangbenro, A., and O. Babalola. "A new strategy for heavy metal polluted environments: A review of microbial biosorbents." *International Journal of Environmental Research and Public Health* 14, no. 1 (2017): 94.

Ayodele, T. R., A. S. O. Ogunjuyigbe, and M. A. Alao. "Life cycle assessment of waste-to-energy (WtE) technologies for electricity generation using municipal solid waste in Nigeria." *Applied Energy* 201 (2017): 200–218.

Ayotamuno, J. M., R. B. Kogbara, and O. S. Agoro. "Biostimulation supplemented with phytoremediation in the reclamation of a petroleum contaminated soil." *World Journal of Microbiology and Biotechnology* 25, no. 9 (2009): 1567–1572.

Ayotamuno, J. M., R. B. Kogbara, and O. S. Agoro. "Biostimulation supplemented with phytoremediation in the reclamation of a petroleum contaminated soil." *World Journal of Microbiology and Biotechnology* 25, no. 9 (2009): 1567–1572.

Azzarello, M. Y., and E. S. Vleet. "Marine birds and plastic pollution." *Mar. Ecol. Prog. Ser.* 37 (1987): 295–303.

Babisch, W. "Noise and health." *Environ Health Perspect* 113 (2005): A14–15.

Babu, A. G., J. D. Kim, and B. T. Oh. "Enhancement of heavy metal phytoremediation by Alnus firma with endophytic *Bacillus thuringiensis* GDB-1." *J Hazard Mater* 250–251 (2013): 477–483.

Babu, G. Suresh, M. Farooq, R. S. Ray, P. C. Joshi, P. N. Viswanathan, and R. K. Hans. "DDT and HCH residues in Basmati rice (*Oryza sativa*) cultivated in Dehradun (India)." *Water, Air, and Soil Pollution* 144, no. 1–4 (2003): 149–157.

Baderna, D., S. Maggioni, E. Boriani, S. Gemma, M. Molteni, A. Lombardo, A. Colombo et al. "A combined approach to investigate the toxicity of an industrial landfill's leachate: Chemical analyses, risk assessment and in vitro assays." *Environmental Research* 111, no. 4 (2011): 603–613.

Baird, R. W., and S. K. Hooker. "Ingestion of plastic and unusual prey by a juvenile harbour porpoise." *Mar. Pollut. Bull.* 40 (2000): 719–720.

Baker, D. E., M. C. Amacher, and R. M. Leach. "Sewage sludge as a source of cadmium in soil-plant-animal systems." *Environmental Health Perspectives* 28 (1979): 45–49.

Bakermans, Corien, A. M. Hohnstock-Ashe, S. Padmanabhan, P. Padmanabhan, and E. L. Madsen. "Geochemical and physiological evidence for mixed aerobic and anaerobic field biodegradation of coal tar waste by subsurface microbial communities." *Microbial Ecology* 44, no. 2 (2002): 107–117.

Bakhtiar, Ziba, Mohammad Hossein Mirjalili, and Ali Sonboli. "In vitro callus induction and micropropagation of *Thymus persicus* (Lamiaceae), an endangered medicinal plant." *Crop Breeding and Applied Biotechnology* 16, no. 1 (2016): 48–54.

Baldi, F., M. Gallo, S. Daniele, D. Battistel, C. Faleri, A. Kodre, and I. Arčon. "An extracellular polymeric substance quickly chelates mercury(II) with N-heterocyclic groups." *Chemosphere* 176 (2017): 296–304.

Baldrian, P. "Wood-inhabiting ligninolytic basidiomycetes in soils: Ecology and constraints for applicability in bioremediation." *Fungal Ecology* 1, no. 1 (2008): 4–12.

Bamforth, M. S., and I. Singleton. "Bioremediation of polycyclic aromatic hydrocarbons: Current knowledge and future directions." *Journal of Chemical Technology and Biotechnology* 80, no. 7 (2005): 723–736.

Banerjee, A., A. Roy, S. Dutta, and S. Mondal. "Bioremediation of hydrocarbon." *International Journal of Advanced Research* 4, no. 6 (2016): 1303–1313.

Banger, Kamaljit, Gurpal S. Toor, Tait Chirenje, and Lena Ma. "Polycyclic aromatic hydrocarbons in urban soils of different land uses in Miami, Florida." *Soil and Sediment Contamination* 19, no. 2 (2010): 231–243.

Bank, D. N. A. "Crop genebank knowledge base." (2012).

Bano, S. A., and D. Ashfaq. "Role of mycorrhiza to reduce heavy metal stress." *Natural Science* 12A (2013): 16–20.

Barbara, Griefahn, Anke Marks, and Sibylle Robens. "Noise emitted from road, rail and air traffic and their effects on sleep." *Journal of Sound and Vibration* 295 (2006): 129–140.

Barker, T. "Technical summary in climate change 2007: Mitigation." Contribution of Working Group III to the Fourth Assessment. Report of the Intergovernmental Panel on Climate Change, 2010.

Barnosky, Anthony D. "Megafauna biomass tradeoff as a driver of quaternary and future extinctions." *Proceedings of the National Academy of Sciences* 105, no. Supplement 1 (2008): 11543–11548.

Barreiros, J. P., and J. Barcelos. "Plastic ingestion by a leatherback turtle *Dermochelys coriacea* from the Azores (NE Atlantic)." *Mar. Pollut. Bull.* 42 (2001): 1196–1197.

Basak, B. B., and D. R. Biswas. "Potentiality of Indian rock phosphate as liming material in acid soil." *Geoderma* 263 (2016): 104–109.

Baskaran, Ponnusamy, Aloka Kumari, and Johannes Van Staden. "In vitro propagation via organogenesis and synthetic seeds of *Urginea altissima* (Lf) Baker: A threatened medicinal plant." *3 Biotech* 8, no. 1 (2018): 18.

Basner, M., W. Babisch, A. Davis, M. Brink, C. Clark, S. Janssen et al. "Auditory and non-auditory effects of noise on health." *Lancet* 383, no. 9925 (2014): 1325–1332.

Basner, M., U. Muller, and E. M. Elmenhorst. "Single and combined effects of air, road, and rail traffic noise on sleep and recuperation." *Sleep* 34, no. 1 (2011): 11–23.

Baudouin, C., M. Charveron, R. Tarroux, and Y. Gall. "Environmental pollutants and skin cancer." *Cell Biology and Toxicology* 18, no. 5 (2002): 341–348.

Baumard, P., H. Budzinski, and P. Garrigues. "PAHs in Arcachon Bay, France: Origin and bio-monitoring with caged organisms." *Marine Pollution Bulletin* 36, no. 8 (1998): 577–586.

Baunthiyal, M., and S. Ranghar. "Accumulation of fluoride by plants: Potential for phytoremediation." *Clean–Soil, Air, Water* 43, no. 1 (2015): 127–132.

Bayer, I. S., S. Guzman-Puyol, J. A. Heredia-Guerrero, L. Ceseracciu, F. Pignatelli, R. Ruffilli, R. Cingolani, and A. Athanassiou. "Direct transformation of edible vegetable waste into bioplastics." *Macromolecules* 47 (2014): 5135–5143.

Bedding, R., and A. Molyneux. "Penetration of insect cuticle by infective juveniles of *Heterorhabditis* spp. (Heterorhabditidae: Nematoda)." *Nematologica* 28 (1982): 354–359.

Behera, B. K., A. Das, D. J. Sarkar, P. Weerathunge, P. K. Parida, B. K. Das, . . . and V. Bansal. "Polycyclic Aromatic Hydrocarbons (PAHs) in inland aquatic ecosystems: Perils and remedies through biosensors and bioremediation." *Environmental Pollution* 241 (2018): 212–233.

Beketov, M. A., and M. Liess. "Ecotoxicology and macroecology-time for integration." *Environmental Pollution* 162 (2012): 247–254.

Benito, Maria Elena Gonzalez, Isabel Clavero-Ramírez, and Jose Manuel López-Aranda. "The use of cryopreservation for germplasm conservation of vegetatively propagated crops." *Spanish Journal of Agricultural Research* 2, no. 3 (2004): 341–351.

Benoit, R., K. J. Wilkinson, and S. Sauvé. "Partitioning of silver and chemical speciation of free Ag in soils amended with nanoparticles." *Chemistry Central Journal* 7, no. 1 (2013): 75.

Benson, Erica. *Plant Conservation Biotechnology.* Taylor & Francis, New York, USA, 1999, p. 309.

Beškoski, V. P., G. Gojgić-Cvijović, J. Milić, M. Ilić, S. Miletić, T. Šolević, and M. M. Vrvić. "Ex situ bioremediation of a soil contaminated by mazut (heavy residual fuel oil)–A field experiment." *Chemosphere* 83, no. 1 (2011): 34–40.

Bharagava, Ram N., Diane Purchase, Gaurav Saxena, and Sikandar I. Mulla. "Applications of metagenomics in microbial bioremediation of pollutants: From genomics to environmental cleanup." In *Microbial Diversity in the Genomic Era.* Academic Press, 2019, pp. 459–477.

Bhardwaj, D., M. W. Ansari, R. K. Sahoo, and N. Tuteja. "Biofertilizers function as key player in sustainable agriculture by improving soil fertility, plant tolerance and crop productivity." *Microbial Cell Factories* 13 (2014): 66. doi: 10.1186/1475-2859-13-66

Bhargava, Akshey. "Activated sludge treatment process–concept and system design." *International Journal of Engineering Development and Research* 4 (2016): 890–896.

Bhargava, Atul, Francisco F. Carmona, Meenakshi Bhargava, and Shilpi Srivastava. "Approaches for enhanced phytoextraction of heavy metals." *Journal of Environmental Management* 105 (2012): 103–120.

Bhargava, Gopal. *Development of India's Urban and Regional Planning in the 21st Century.* Gian Publishing House, New Delhi, 2001, pp. 115–116.

Bhatia, Rajiv, Rita Shiau, Myrto Petreas, June M. Weintraub, Lili Farhang, and Brenda Eskenazi. "Organochlorine pesticides and male genital anomalies in the child health and development studies." *Environmental Health Perspectives* 113, no. 2 (2004): 220–224.

Bhattacharyya, Paromik, Suman Kumaria, Nikhil Job, and Pramod Tandon. "Phytomolecular profiling and assessment of antioxidant activity within micropropagated plants of Dendrobium thyrsiflorum: A threatened, medicinal orchid." *Plant Cell, Tissue and Organ Culture (PCTOC)* 122, no. 3 (2015): 535–550.

Biasioli, M., and F. Ajmone-Marsan. "Organic and inorganic diffuse contamination in urban soils: The case of Torino (Italy)." *Journal of Environmental Monitoring* 9, no. 8 (2007): 862–868.

Birdlife International. "*Ectopistes migratorius.*" IUCN Red List of Threatened Species, 2016.

Birolli, Willian G., Darlisson de A. Santos, Natália Alvarenga, Anuska CFS Garcia, Luciane PC Romão, and André L. M. Porto. "Biodegradation of anthracene and several PAHs by the marine-derived fungus *Cladosporium* sp. CBMAI 1237." *Marine Pollution Bulletin* 129, no. 2 (2018): 525–533.

Bisht, Sandeep, Piyush Pandey, Bhavya Bhargava, Shivesh Sharma, Vivek Kumar, and Krishan D. Sharma. "Bioremediation of polyaromatic hydrocarbons (PAHs) using rhizosphere technology." *Brazilian Journal of Microbiology* 46, no. 1 (2015): 7–21.

Blackburn, Tim M., Phillip Cassey, Richard P. Duncan, Karl L. Evans, and Kevin J. Gaston. "Avian extinction and mammalian introductions on oceanic islands." *Science* 305, no. 5692 (2004): 1955–1958.

Blaine, K., S. Kamaldeen, and D. Powell. "Public perceptions of biotechnology." *Journal of Food Science* 67, no. 9 (2002): 3200–3208.

Blight, L. K., and A. E. Burger. "Occurrence of plastic particles in sea-birds from the eastern north pacific." *Mar Pollut Bull.* 34 (1997): 323–325.

Blomkvist, V., C. A. Eriksen, T. Theorell, R. S. Ulrich, and G. Rasmanis. "Acoustics and psychosocial environment in coronary intensive care." *Occupational and Environmental Medicine* 62 (2005): 1–8.

Boiral, Olivier. "Tacit knowledge and environmental management." *Long Range Planning* 35, no. 3 (2002): 291–317.

Bolan, N., A. Kunhikrishnan, R. Thangarajan, J. Kumpiene, J. Park, T. Makino, M. Beth Kirkham, and K. Scheckel. "Remediation of heavy metal(loid)s contaminated soils to mobilize or to immobilize?" *Journal of Hazardous Materials* 266 (2014): 141–166.

Bolan, N. S., J. H. Park, B. Robinson, R. Naidu, and K. Y. Huh. "Phytostabilization: A green approach to contaminant containment." *Adv. Agron* 112 (2011): 145–204.

Bolton, M. "Comparing two remediation alternatives for diesel-contaminated soil in the Arctic using life cycle assessment." (2012).

Bordoloi, Achinta, and Peter A. Gostomski. "Fate of degraded pollutants in waste gas biofiltration: An overview of carbon end-points." *Biotechnology Advances* 37 (2018): 579–588.

Borkar, S. G. *Microbes as Biofertilizers and Their Production Technology.* Wood Head Publishing India Pvt. Ltd., New Delhi, India, 2015, pp. 7–153.

Bose, Biswajit, Suman Kumaria, Hiranjit Choudhury, and Pramod Tandon. "Assessment of genetic homogeneity and analysis of phytomedicinal potential in micropropagated plants of Nardostachys jatamansi, a critically endangered, medicinal plant of alpine Himalayas." *Plant Cell, Tissue and Organ Culture (PCTOC)* 124, no. 2 (2016): 331–349.

Bose, Biswajit, Suman Kumaria, Hiranjit Choudhury, and Pramod Tandon. "Insights into nuclear DNA content, hydrogen peroxide and antioxidative enzyme activities during transverse thin cell layer organogenesis and ex vitro acclimatization of *Malaxis wallichii*, a threatened medicinal orchid." *Physiology and Molecular Biology of Plants* 23, no. 4 (2017): 955–968.

Botkin, Daniel B., and Edward A. Keller. *Environmental Science: Earth as a Living Planet.* 2nd ed. John Wiley & Sons Ltd, 1998.

Bourtoom, T. "Edible protein films: Properties enhancement." *Int Food Res J.* 16 (2009): 1–9.

Brack, W., R. Altenburger, G. Schüürmann, M. Krauss, D. L. Herráez, J. Van Gils, . . . and M. Schriks. "The SOLUTIONS project: Challenges and responses for present and future emerging pollutants in land and water resources management." *Science of the Total Environment* 503 (2015): 22–31.

Braguglia, C. M., A. Coors, A. Gallipoli, A. Gianico, E. Guillon, U. Kunkel, G. Mascolo et al. "Quality assessment of digested sludges produced by advanced stabilization processes." *Environmental Science and Pollution Research* 22, no. 10 (2015): 7216–7235.

Bramer, C. O., P. Vandamme, L. F. D. Silva, J. G. C. Gomez, and A. Steinbuchel. "*Burkholderia sacchari* sp. nov., a polyhydroxyalkanoate-accumulating bacterium isolated from soil of a sugar-cane plantation in Brazil." *Int J Syst Evol Microbiol* 51 (2001): 1709–1713.

Brazier-Hicks, M., M. Kathryn, E. Oliver, D. Cunningham, D. R. W. Hodgson, P. G. Steel, and R. Edwards. "Catabolism of glutathione conjugates in *arabidopsis thaliana* role in metabolic reactivation of the herbicide safener fenclorim." *Journal of Biological Chemistry* 283, no. 30 (2008): 21102–21112.

Breinbauer, H. A., J. L. Anabalón, D. Gutierrez, R. Cárcamo, C. Olivares, and J. Caro. "Output capabilities of personal music players and assessment of preferred listening levels of test subjects: Outlining recommendations for preventing music-induced hearing loss." *Laryngoscope* 122 (2012): 2549–2556.

Briassoulis, D., and C. Dejean. "Critical review of norms and standards for biodegradable agricultural plastics Part I. Biodegradation in soil." *J Polymers Environ* 18 (2010): 384–400.

Brierley, C. L. "How will biomining be applied in future?" *Trans. Nonferrous Met. Soc. China* 18 (2008): 1302–1310.

Bright-Ponte, S. J., T. Zhou, and M. J. Murphy. "Regulatory considerations in veterinary toxicology: An FDA perspective." In *Veterinary Toxicology*. Academic Press, 2018, pp. 89–101.

Brink, Johan A., Bernard Prior, and Edgar J. DaSilva. "Developing biotechnology around the world." *Nature Biotechnology* 17, no. 5 (1999): 434–437.

Bronzaft, A., and G. Van Ryzin. *Neighborhood Noise and Its Consequences: Implications for Tracking Effectiveness of the NYC Revised Noise Code*. Special Report #14. New York: Baruch College/CUNY.

Bronzaft, A. L. "Noise: Combating a ubiquitous and hazardous pollutant." *Noise Health* 2 (2000): 1–8.

Brooks, Thomas M., Russell A. Mittermeier, Cristina G. Mittermeier, Gustavo A. B. Da Fonseca, Anthony B. Rylands, William R. Konstant, Penny Flick et al. "Habitat loss and extinction in the hotspots of biodiversity." *Conservation Biology* 16, no. 4 (2002): 909–923.

Brooks, W. M. "Entomogenous protozoa: Handbook of natural pesticides, microbial insecticides, Part A." In: Ignoffo, C. M., and Mandava, N. B. (eds.), *Entomogenous Protozoa and Fungi*. Vol. 5. CRC Press, Boca Raton, FL, 1988, pp. 1–149.

Brown, Lisa M., Thusitha S. Gunasekera, Richard C. Striebich, and Oscar N. Ruiz. "Draft genome sequence of *Gordonia sihwensis* strain 9, a branched alkane-degrading bacterium." *Genome Announc* 4, no. 3 (2016): e00622–16.

BSM2. "Families affected by metro noise asked to share one apartment." *Copenhagen Post* (Copenhagen, Denmark), 10 October 2012. http://cphpost.dk/news/ families-affected-by-metro-noise-asked-to-share-oneapartment.2987

Bucci, D. Z., and L. B. B. Tavares. "PHB packaging for the storage of food products." *Polymer Testing* 24 (2005): 564–571.

Buhari Muhammad, B. L., R. Sulaiman Babura, N. L. Vyas, Y. Badaru Sulaiman, and Y. Harisu Umar. "Role of biotechnology in phytoremediation." *Journal of Bioremediation and Biodegradation* 7 (2016): 1–6.

Bumpus, J. A. "Biodegradation of polycyclic hydrocarbons by *Phanerochaete chrysosporium*." *Applied and Environmental Microbiology* 55, no. 1 (1989): 154–158.

Bundschuh, J., M. I. Litter, F. Parvez, G. Román-Ross, B. N. Hugo, J. Shuh Jean, Ch. Wuing Liu et al. "One century of arsenic exposure in Latin America: A review of history and occurrence from 14 countries." *Science of the Total Environment* 429 (2012): 2–35.

Bunn, Eric, Shane Turner, Maggie Panaia, and Kingsley W. Dixon. "The contribution of in vitro technology and cryogenic storage to conservation of indigenous plants." *Australian Journal of Botany* 55, no. 3 (2007): 345–355.

Burachevskaya, Marina V., Tatiana M. Minkina, Saglara S. Mandzhieva, Tatiana V. Bauer, Victor A. Chaplygin, Svetlana N. Sushkova, Palma Orlović-Leko, Lyudmila Yu Mashtykova, and Vishnu Rajput. "Comparing two methods of sequential fractionation in the study of copper compounds in Haplic Chernozem under model experimental conditions." *Journal of Soils and Sediments* 18, no. 6 (2017): 2379–2386.

Burken, J. G. "Uptake and metabolism of organic compounds: Green-liver model." *Phytoremediation: Transformation and Control of Contaminants* 59 (2003): 59–84.

Busch-Vishniac, I., J. West, C. Barnhill, T. Hunter, D. Orellana, and R. Chivukula. "Noise levels in Johns Hopkins Hospital." *Journal of the Acoustical Society of America* 118, no. 6 (2005): 3629–3645.

Bwapwa, J. K., A. T. Jaiyeola, and R. Chetty. "Bioremediation of acid mine drainage using algae strains: A review." *South African Journal of Chemical Engineering* 24 (2017): 62–70.

Byun, Y., and Y. T. Kim. "Bioplastics for food packaging: Chemistry and physics." In: Han, J. H. (ed.), *Innovations in Food Packaging*, Academic Press, 2014a, pp. 353–368.

Byun, Y., and Y. T. Kim. "Utilization of bioplastics for food packaging industry." In: Han, J. H. (ed.), *Innovations in Food Packaging*, Academic Press, 2014b, pp. 369–390.

Cabello-Conejo, M. I., C. Becerra-Castro, A. Prieto-Fernández, C. Monterroso, A. Saavedra-Ferro, M. Mench, and P. S. Kidd. "Rhizobacterial inoculants can improve nickel phytoextraction by the hyperaccumulator Alyssum pintodasilvae." *Plant and Soil* 379, no. 1–2 (2014): 35–50.

Cajthaml, T., V. Pacakova, and V. Sasek. "Microbial degradation of polycyclic aromatic hydrocarbons." *Feedback* 91 (1997).

Camenzuli, D., and B. L. Freidman. "On-site and in situ remediation technologies applicable to petroleum hydrocarbon contaminated sites in the Antarctic and Arctic." *Polar Research* 34, no. 1 (2015): 24492.

Campbell, K. C. M. "Oral pharmacologic otoprotective agents to prevent noise-induced hearing loss (Noise-Induced Hearing Loss): When dietary concentration isn't enough. In: Griefahn, B. (ed.), *10th International Congress on Noise as a Public Health Problem of the International Commission on Biological Effects of Noise*, London, UK, 2011. http://www.icben.org/proceedings.html

Campo, P., K. Maguin, S. Gabriel et al. "Combined exposure to noise and ototoxic substances." *European Agency for Safety and Health at Work (EU-OSHA)*, 2009, accessed 11 July 2013.

Cang, L. "Heavy metals pollution in poultry and livestock feeds and manures under intensive farming in Jiangsu Province, China." *Journal of Environmental Sciences* 16, no. 3 (2004): 371–374.

Cardinale, Bradley J., J. Emmett Duffy, Andrew Gonzalez, David U. Hooper, Charles Perrings, Patrick Venail, Anita Narwani et al. "Biodiversity loss and its impact on humanity." *Nature* 486, no. 7401 (2012): 59.

Casati, Stefano, Patrizia Passerini, Maria Rosaria Campise, Giorgio Graziani, Bruno Cesana, Michael Perisic, and Claudio Ponticelli. "Benefits and risks of protracted treatment with human recombinant erythropoietin in patients having haemodialysis." *Br Med J (Clin Res Ed)* 295, no. 6605 (1987): 1017–1020.

Casida, J. E., and G. B. Quistad. *Pyrethrum Flowers: Production, Chemistry, Toxicology and Uses.* Oxford University Press, New York, 1995, ISBN-10: 0195082109, p. 356.

Castelo-Grande, T., P. A. Augusto, P. Monteiro, A. M. Estevez, and D. Barbosa. "Remediation of soils contaminated with pesticides: A review." *International Journal of Environmental and Analytical Chemistry* 90, no. 3–6 (2010): 438–467.

Castilho, L. R., D. A. Mitchell, and D. M. G. Freire. "Production of polyhydroxyalkanoates (PHAs) from waste materials and by-products by submerged and solid-state fermentation." *Biores Technol* 100 (2009): 5996–6009.

Castro, Inés. Smithsonian Institution. Personal communication, 1995.

Caswell, Karen Louise, and Kutty K. Kartha. "Recovery of plants from pea and strawberry meristems cryopreserved for 28 years." *CryoLetters* 30, no. 1 (2009): 41–46.

CBD. "Convention on Biology Diversity, Rio de Janeiro, Brazil." *Convention on Biological Diversity* (1992). http://www.biodiv.org/convention/

Cerniglia, Carl E. "Biodegradation of polycyclic aromatic hydrocarbons." *Current Opinion in Biotechnology* 4, no. 3 (1993): 331–338.

Chaerun, Siti Khodijah, Robi Suryaning Sulistyo, Wahyudin Prawira Minwal, and Mohammad Zaki Mubarok. "Indirect bioleaching of low-grade nickel limonite and saprolite ores using fungal metabolic organic acids generated by Aspergillus niger." *Hydrometallurgy* 174 (2017): 29–37.

Chakraborty, Romy, Cindy H. Wu, and Terry C. Hazen. "Systems biology approach to bioremediation." *Current Opinion in Biotechnology* 23, no. 3 (2012): 483–490.

Chandra, R., and R. Rustogi. "Biodegradable polymers." *Prog Polym Sci* 23 (1998): 1273–1335.

Chandra, R., R. N. Bharagava, S. Yadav, and D. Mohan. "Accumulation and distribution of toxic metals in wheat (Triticum aestivum L.) and Indian mustard (Brassica campestris L.) irrigated with distillery and tannery effluents." *Journal of Hazardous Materials* 162, no. 2–3 (2009): 1514–1521.

Chandra, R., N. K. Dubey, and V. Kumar. *Phytoremediation of Environmental Pollutants.* CRC Press, 2017.

Chattopadhyay, A., N. B. Bhatnagar, and R. Bhatnagar. "Bacterial insecticidal toxins." *Crit Rev Microbiol* 30 (2004): 33–54.

Chaudhry, Q., M. Blom-Zandstra, S. Gupta, and E. J. Joner. "Utilizing the synergy between plants and rhizosphere microorganisms to enhance breakdown of organic pollutants in the environment." *Environmental Science and Pollution Research* 12, no. 1 (2005): 34–48.

Chaudhry, Q., P. Schroeder, D. Werck-Reichhart, W. Grajek, and R. Marecik. "Prospects and limitations of phytoremediation for the removal of persistent pesticides in the environment." *Environmental Science and Pollution Research* 9, no. 1 (2001): 4–17.

Chauhan, Archana, John G. Oakeshott, and Rakesh K. Jain. "Bacterial metabolism of polycyclic aromatic hydrocarbons: Strategies for bioremediation." *Indian Journal of Microbiology* 48, no. 1 (2008): 95–113.

Chauhan, Sippy K., and Anuradha Shukla. "Environmental impacts of production of biodiesel and its use in transportation sector." *Croatia: Environmental Impacts of Biofuels. InTech* (2011): 1–18.

Chekol, T., L. R. Vough, and R. L. Chaney. "Phytoremediation of polychlorinated biphenyl-contaminated soils: The rhizosphere effect." *Environment International* 30, no. 6 (2004): 799–804.

Chen, G. Q., and Wu, Q. "The application of polyhydroxyalkanoates as tissue engineering materials." *Biomaterials* 26 (2005a): 6565–6578.

Chen, W., N. Jongkamonwiwat, L. Abbas et al. "Restoration of auditory evoked responses by human ES-cell-derived otic progenitors." *Nature* 490 (2012): 278–282. [PubMed: 22972191].

Chen, H., G. Gao, P. Liu, R. Pan, X. Liu, and Ch. Lu. "Determination of 16 polycyclic aromatic hydrocarbons in tea by simultaneous dispersive solid-phase extraction and liquid-liquid extraction coupled with gas chromatography-tandem mass spectrometry." *Food Analytical Methods* 9, no. 8 (2016): 2374–2384.

Chen, M., P. Xu, G. Zeng, C. Yang, D. Huang, and J. Zhang. "Bioremediation of soils contaminated with polycyclic aromatic hydrocarbons, petroleum, pesticides, chlorophenols and heavy metals by composting: Applications, microbes and future research needs." *Biotechnology Advances* 33, no. 6 (2015): 745–755.

Chen, S., Q. Hu, M. Hu, J. Luo, Q. Weng, and K. Lai. "Isolation and characterization of a fungus able to degrade pyrethroids and 3-phenoxybenzaldehyde." *Bioresource Technology* 102, no. 17 (2011): 8110–8116.

Chen, W., F. Bruhlmann, R. Richins, and A. Mulchandani. "Engineering of improved microbes and enzymes for bioremediation." *Current Opinion in Biotechnology* 10, no. 2 (1999): 137–141.

Chen, X., and V. Achal. "Biostimulation of carbonate precipitation process in soil for copper immobilization." *Journal of Hazardous Materials* 368 (2019): 705–713.

Cheung, Philip C. W. "A historical review of the benefits and hypothetical risks of disinfecting drinking water by chlorination (Updated and Revised)." *Journal of Environment and Ecology* 8, no. 1 (2017): 73–141.

Chigbo, Ch., and L. Batty. "Phytoremediation for co-contaminated soils of chromium and benzo[a]pyrene using Zea mays L." *Environmental Science and Pollution Research* 21, no. 4 (2014): 3051–3059.

Chinwan, Dipanshu, and Shubhangi Pant. "Waste to energy in India and its management." *Journal of Basic and Applied Engineering Research* 1, no. 10 (2014): 89–94.

Cho-Ruk, K., J. Kurukote, P. Supprung, and S. Vetayasuporn. "Perennial plants in the phytoremediation of lead-contaminated soils." *Biotechnology* 5, no. 1 (2006): 1–4.

Chown, S. L., N. J. M. Gremmen, and K. J. Gaston. "Ecological biogeography of southern ocean islands: Species-area relationships, human impacts, and conservation." *The American Naturalist* 152, no. 4 (1998): 562–575.

Clark, D. A., and P. R. Norris. "*Acidimicrobium ferrooxidans* gen. nov., sp.: Mixed-culture ferrous iron oxidation with Sulfobacillus species." *Microbiology* 142 (1996): 785–790.

Clark, Ross G. "Recombinant human insulin-like growth factor I (IGF-I): Risks and benefits of normalizing blood IGF-I concentrations." *Hormone Research in Paediatrics* 62, no. Suppl. 1 (2004): 93–100.

Clark, Ross G. "Recombinant human insulin-like growth factor I (IGF-I): Risks and benefits of normalizing blood IGF-I concentrations." *Hormone Research in Paediatrics* 62, Suppl. 1 (2004): 93–100.

Clausen, Jay., J. Robb, D. Curry, and N. Korte. "A case study of contaminants on military ranges: Camp Edwards, Massachusetts, USA." *Environmental Pollution* 129, no. 1 (2004): 13–21.

Clavero, M. Milagrosa, and Emili Garcia-Berthou. "Invasive species are a leading cause of animal extinctions." *Trends in Ecology & Evolution* 20, no. 3 (2005): 110.

Clemson, H. G. I. C. *Organic Pesticides and Biopesticides, Clemson Extension, Home and Garden Information Center.* Clemson University, Clemson, 2007.

Clifton II, Jack C. "Mercury exposure and public health." *Pediatric Clinics of North America* 54, no. 2 (2007): 237–269.

Cobbett, Christopher, and Peter Goldsbrough. "Phytochelatins and metallothioneins: Roles in heavy metal detoxification and homeostasis." *Annual Review of Plant Biology* 53, no. 1 (2002): 159–182.

Collie, J. S., L. W. Botsford, A. Hastings, I. C. Kaplan, J. L. Largier, P. A. Livingston, . . . and F. E. Werner. "Ecosystem models for fisheries management: Finding the sweet spot." *Fish and Fisheries* 17, no. 1 (2016): 101–125.

Collins, Ch., M. Fryer, and A. Grosso. "Plant uptake of non-ionic organic chemicals." *Environmental Science and Technology* 40, no. 1 (2006): 45–52.

Collins, Nick. "Chinese medicines contain traces of endangered animals" (2012).

Colozza, N., M. F. Gravina, L. Amendola, M. Rosati, D. E. Akretche, D. Moscone, and F. Arduini. "A miniaturized bismuth-based sensor to evaluate the marine organism Styela plicata bioremediation capacity toward heavy metal polluted seawater." *Science of the Total Environment* 584 (2017): 692–700.

Com, E. "The combination effects of chemicals: Chemical mixtures." *Communication from the Commission to the Council* 252 (2012): 252.

Coman, Simona M., and Vasile I. Parvulescu. "Heterogeneous catalysis for biodiesel production." In: *The Role of Catalysis for the Sustainable Production of Bio-Fuels and Bio-Chemicals*. Elsevier, 2013, pp. 93–136.

Commission to the European Parliament and the Council. *Report from the Commission to the European Parliament and the Council on the Implementation of the Environmental Noise Directive in Accordance with Article 11 of Directive 2002/49/EC*. European Commission, Brussels, 2011.

Convention on biological diversity. (2011).

Copinschi, G. "Metabolic and endocrine effects of sleep deprivation." *Essent Psychopharmacol* 6, no. 6 (2005): 341–347.

Copping, L. G., and J. J. Menn. "Biopesticides: A review of their action, applications and efficacy." *Pest Management Science* 56, no. 8 (2000): 651–676.

Corlett, Richard T. "A bigger toolbox: Biotechnology in biodiversity conservation." *Trends in Biotechnology* 35, no. 1 (2017): 55–65.

Coutris, C., E. J. Joner, and D. H. Oughton. "Aging and soil organic matter content affect the fate of silver nanoparticles in soil." *Science of the Total Environment* 420 (2012): 327–333.

Sharma, Neetu, Abhinashi Singh, and Navneet Batra. "Modern and emerging methods of wastewater treatment." In: *Ecological Wisdom Inspired Restoration Engineering*. Springer, Singapore, 2019, pp. 223–247.

CPCB. Consolidated Annual Report on Implementation of Municipal Solid Waste (Management and Handling) Rules, 2000 (2016).

CPCB. "Water Quality Criteria." 2017. https://cpcb.nic.in/water-quality-criteria/

Cripps, Colleen, John A. Bumpus, and Steven D. Aust. "Biodegradation of azo and heterocyclic dyes by Phanerochaete chrysosporium." *Appl. Environ. Microbiol.* 56, no. 4 (1990): 1114–1118.

Cristaldi, Mauro, Cristiano Foschi, Germana Szpunar, Carlo Brini, Fiorenzo Marinelli, and Lucio Triolo. "Toxic emissions from a military test site in the territory of Sardinia, Italy." *International Journal of Environmental Research and Public Health* 10, no. 4 (2013): 1631–1646.

Cruz-Cruz, Carlos, María González-Arnao, and Florent Engelmann. "Biotechnology and conservation of plant biodiversity." *Resources* 2, no. 2 (2013): 73–95.

Cucchiella, Federica, Idiano D'Adamo, and Massimo Gastaldi. "Sustainable management of waste-to-energy facilities." *Renewable and Sustainable Energy Reviews* 33 (2014): 719–728.

Curtis, T. P., W. T. Sloan, and J. W. Scannell. "Estimating prokaryotic diversity and its limits." *Proceedings of the National Academy of Sciences* 99, no. 16 (2002): 10494–10499.

Daane, L. L., I. Harjono, G. J. Zylstra, and M. M. Häggblom. "Isolation and characterization of polycyclic aromatic hydrocarbon-degrading bacteria associated with the rhizosphere of salt marsh plants." *Applied and Environmental Microbiology* 67, no. 6 (2001): 2683–2691.

Dafforn, K. A., E. L. Johnston, A. Ferguson, C. L. Humphrey, W. Monk, S. J. Nichols, . . . and D. J. Baird. "Big data opportunities and challenges for assessing multiple stressors across scales in aquatic ecosystems." *Marine and Freshwater Research* 67, no. 4 (2016): 393–413.

Dalkmann, Ph., Ch. Siebe, W. Amelung, M. Schloter, and J. Siemens. "Does long-term irrigation with untreated wastewater accelerate the dissipation of pharmaceuticals in soil?" *Environmental Science and Technology* 48, no. 9 (2014): 4963–4970.

Damisa, D., T. S. Oyegoke, U. J. J. Ijah, N. U. Adabara, J. D. Bala, and R. Abdulsalam. "Biodegradation of petroleum by fungi isolated from unpolluted tropical soil." *International J Appl Biol Pharm Technol* 4, no. 2 (2013): 136–140.

Dams, R. I., G. I. Paton, and K. Killham. "Rhizoremediation of pentachlorophenol by Sphingobium chlorophenolicum ATCC 39723." *Chemosphere* 68, no. 5 (2007): 864–870.

Darimont, Chris T., Caroline H. Fox, Heather M. Bryan, and Thomas E. Reimchen. "The unique ecology of human predators." *Science* 349, no. 6250 (2015): 858–860.

Das, D., B. S. Dwivedi, M. C. Meena, V. K. Singh, and K. N. Tiwari. "Integrated nutrient management for improving soil health and crop productivity." *Indian Journal of Fertilizers* 11, no. 4 (2015): 64–83.

Das, N., and P. Chandran. "Microbial degradation of petroleum hydrocarbon contaminants: An overview." *Biotechnology Research International* (2011): 1–13.

Das, Nilanjana, and PreethyChandran. "Microbial degradation of petroleum hydrocarbon contaminants: An overview." *Biotechnology Research International* 2011 (2010): 1–13.

Daud, M. K., Hina Rizvi, Muhammad Farhan Akram, Shafaqat Ali, Muhammad Rizwan, Muhammad Nafees, and Zhu Shui Jin. "Review of upflow anaerobic sludge blanket reactor technology: Effect of different parameters and developments for domestic wastewater treatment." *Journal of Chemistry* 2018 (2018): 1–13.

Davidson, Philip W., Gary J. Myers, and Bernard Weiss. "Mercury exposure and child development outcomes." *PEDIATRICS* 113, no. 4 (2004): 1023–1029.

Davies, Julian, and Dorothy Davies. "Origins and evolution of antibiotic resistance." *Microbiol. Mol. Biol. Rev.* 74, no. 3 (2010): 417–433.

de la Cueva, S. C., C. H. Rodríguez, N. O. S. Cruz, J. A. R. Contreras, and J. L. Miranda. "Changes in bacterial populations during bioremediation of soil contaminated with petroleum hydrocarbons." *Water, Air, and Soil Pollution* 227, no. 3 (2016): 1–12.

De Lacerda, L. "Updating global Hg emissions from small-scale gold mining and assessing its environmental impacts." *Environmental Geology* 43, no. 3 (2003): 308–314.

de Lima, D. P., E. D. A. dos Santos, M. R. Marques, G. C. Giannesi, A. Beatriz, M. K. Yonekawa, and A. D. S. Montanholi. "Fungal bioremediation of pollutant aromatic amines." *Current Opinion in Green and Sustainable Chemistry* 11 (2018): 34–44.

de Moura, I. G., A. V. de Sa, A. S. L. M. Abreu, and A. V. A. Machado. "Bioplastics from agro-wastes for food packaging applications." In: Grumezescu, A. M. (ed.), *Food Packaging*. Academic Press, 2017, pp. 223–263.

de Vileger, J. J. "Green plastics for food packaging." In: Ahvenainen, R. (ed.), *Novel Food Packaging Techniques*. CRC Press, Boca Raton, FL, 2000, pp. 519–534.

Dean, S. M., Y. Jin, D. K. Cha, S. V. Wilson, and M. Radosevich. "Phenanthrene degradation in soils co-inoculated with phenanthrene-degrading and biosurfactant-producing bacteria." *Journal of Environmental Quality* 30, no. 4 (2001): 1126–1133.

Deardorff, T., N. Karch, and S. Holm. "Dioxin levels in ash and soil generated in Southern California fires." *Organohalogen Compd* 70 (2008): 2284–2288.

Deconinck, S., and B. de Wilde. "Benefits and challenges of bio- and oxodegradable plastics, a comparative literature study." Final Study, O. W. S. for Plastics Europe, 2013.

Dedkova, E. N., and L. A. Blatter. "Role of β-hydroxybutyrate, its polymer poly-β-hydroxybutyrate and inorganic polyphosphate in mammalian health and disease." *Front Physiol* 5 (2014): 1–22.

Delhaize, E., P. R. Ryan, and P. J. Randall. "Aluminum tolerance in wheat (*Triticum aestivum* L.) (II. Aluminum-stimulated excretion of malic acid from root apices)." *Plant Physiology* 103, no. 3 (1993): 695–702.

Dettenmaier, E. M., W. J. Doucette, and B. Bugbee. "Chemical hydrophobicity and uptake by plant roots." *Environmental Science & Technology* 43, no. 2 (2008): 324–329.

Dewa, U. N. E. P. "E-waste, the hidden side of IT equipment's manufacturing and use." Early Warning on Emerging Environmental Threats, 2005.

Dhaliwal, G. S., R. Singh, and B. S. Chhillar. In: *Essentials of Agricultural Entomology.* Kalyani Publishers, New Delhi, India, 2006.

Dhankher, Om Parkash, Elizabeth A. H. Pilon-Smits, Richard B. Meagher, and Sharon Doty. "Biotechnological approaches for phytoremediation." In: *Plant Biotechnology and Agriculture.* Academic Press, 2012, pp. 309–328.

Dhyani, A., and D. Dhyani. "IUCN Red List-2015: Indian Medicinal Plants at Risk" (2016).

Dixon, D. P., M. Skipsey, and R. Edwards. "Roles for glutathione transferases in plant secondary metabolism." *Phytochemistry* 71, no. 4 (2010): 338–350.

Domning, D. "Hydrodamalis gigas." IUCN Red List of Threatened Species (2016).

Doran, J. W., and M. R. Zeiss. "Soil health and sustainability: Managing the biotic component of soil quality." *Applied Soil Ecology* 15, no. 1 (2000): 3–11.

dos Santos, J. J., and L. T. Maranho. "Rhizospheric microorganisms as a solution for the recovery of soils contaminated by petroleum: A review." *Journal of Environmental Management* 210 (2018): 104–113.

Dotaniya, M. L. "Impact of various crop residue management practices on nutrient uptake by rice-wheat cropping system." *Curr Adv Agric Sci* 5, no. 2 (2013): 269–271.

Dotaniya, M. L., S. K. Kushwah, S. Rajendiran, M. V. Coumar, S. Kundu, and A. S. Rao. "Rhizosphere effect of kharif crops on phosphatases and dehydrogenase activities in a Typic Haplustert." *Natl Acad Sci Lett* 37, no. 2 (2014): 103–106.

Dotaniya, M. L., V. D. Meena, B. B. Basak, and R. S. Meena. "Potassium uptake by crops as well as microorganisms." In: Meena, V. S., Maurya, B. R., Verma, J. P., and Meena, R. S. (eds.), *Potassium Solubilizing Microorganisms for Sustainable Agriculture.* Springer, New Delhi, 2016, pp. 267–280.

Dotaniya, M. L., Pingoliya, K. K., Meena, H. M., and Prasad, D. "Status and rational use of rock phosphate in agricultural crop production—a review." *Agric Sustain Dev* 1, no. 1 (2013): 103–108.

Dove, Carla J., Ray W. Snow, Michael R. Rochford, and Frank J. Mazzotti. "Birds consumed by the invasive Burmese python (Python molurus bivittatus) in Everglades National Park, Florida, USA." *Wilson Journal of Ornithology* 123, no. 1 (2011): 126–131.

Driscoll, Charles T., Gregory B. Lawrence, Arthur J. Bulger, Thomas J. Butler, Christopher S. Cronan, Christopher Eagar, Kathleen F. Lambert, Gene E. Likens, John L. Stoddard, and Kathleen C. Weathers. "Acidic deposition in the Northeastern United States: Sources and inputs, ecosystem effects, and management strategies." *BioScience* 51, no. 3 (2001): 180–198.

Driscoll, Charles T., Young-Ji Han, Celia Y. Chen, David C. Evers, Kathleen Fallon Lambert, Thomas M. Holsen, Neil C. Kamman, and Ronald K. Munson. "Mercury contamination in forest and freshwater ecosystems in the northeastern United States." *BioScience* 57, no. 1 (2007): 17–28.

Dua, M., A. Singh, N. Sethunathan, and A. Johri. "Biotechnology and bioremediation: Successes and limitations." *Applied Microbiology and Biotechnology* 59, no. 2–3 (2002): 143–152.

Dubey, N. K., R. Shukla, A. Kumar, P. Singh, and B. Prakash. "Global Scenario on the application of Natural products in integrated pest management." In: *Natural Products in Plant Pest Management,* Dubey, N. K., CABI, Oxfordshire, UK. 2011, ISBN: 9781845936716, pp. 1–20.

Dulloo, Mohammad Ehsan, Danny Hunter, and Teresa Borelli. "Ex situ and in situ conservation of agricultural biodiversity: Major advances and research needs." *Notulae Botanicae Horti Agrobotanici Cluj-Napoca* 38, no. 2 (2010): 123–135.

Dunlap, A. "Come on feel the noise: The problem with municipal noise regulation." *U Miami Bus L Rev* 15 (2006): 47–303.

Durden, J. M., L. E. Lallier, K. Murphy, A. Jaeckel, K. Gjerde, and D. O. Jones. "Environmental impact assessment process for deep-sea mining in 'the Area'." *Marine Policy* 87 (2018): 194–202.

Eapen, S., S. Singh, and S. F. D'souza. "Advances in development of transgenic plants for remediation of xenobiotic pollutants." *Biotechnology Advances* 25, no. 5 (2007): 442–451.

Edwards, C. A. *Earthworm Ecology.* CRC Press, 2004.

Edwards, N. T. "Polycyclic Aromatic Hydrocarbons (PAH's) in the terrestrial environment- a review 1." *Journal of Environmental Quality* 12, no. 4 (1983): 427–441.

Edwards, R., D. P. Dixon, I. Cummins, M. Brazier-Hicks, and M. Skipsey. "New perspectives on the metabolism and detoxification of synthetic compounds in plants." In: *Organic Xenobiotics and Plants.* Springer, Dordrecht, 2011, pp. 125–148.

EEA4. Noise in Europe 2014, 2014. (Luxembourg, Contract No.: 10/2014).

EFSA Panel on Plant Protection Products and their Residues (PPR), Colin Ockleford, Paulien Adriaanse, Philippe Berny, Theodorus Brock, Sabine Duquesne, Sandro Grilli et al. "Scientific opinion addressing the state of the science on risk assessment of plant protection products for in-soil organisms." *European Food Safety Authority Journal* 15, no. 2 (2017): e04690.

Ejeian, Fatemeh, Parisa Etedali, Hajar-Alsadat Mansouri-Tehrani, Asieh Soozanipour, Ze-Xian Low, Mohsen Asadnia, Asghar Taheri-Kafrani, and Amir Razmjou. "Biosensors for wastewater monitoring: A review." *Biosensors and Bioelectronics* 118 (2018): 66–79.

Elango, D., M. Pulikesi, P. Baskaralingam, V. Ramamurthi, and S. Sivanesan. "Production of biogas from municipal solid waste with domestic sewage." *Journal of Hazardous Materials* 141, no. 1 (2007): 301–304.

Eldredge, Lucius G., and Scott E. Miller. "Numbers of Hawaiian species: Supplement 2, including a review of freshwater invertebrates." *Bishop Museum Occasional Papers* 48 (1997): 3–22.

El-Kadi, S. *Bioplastic Production from Inexpensive Sources Bacterial Biosynthesis, Cultivation System, Production and Biodegradability.* VDM Publishing House, USA, 2010.

Ellis, Lynda B. M. "Environmental biotechnology informatics." *Current Opinion in Biotechnology* 11, no. 3 (2000): 232–235.

Elwan, Ahmed, Yanuar Z. Arief, Zuraimy Adzis, and Mohd Hafiez Izzwan Saad. "The viability of generating electricity by harnessing household garbage solid waste using life cycle assessment." *Procedia Technology* 11 (2013): 134–140.

Enan, E. E. "Molecular and pharmacological analysis of an octopamine receptor from American cockroach and fruit fly in response to plant essential oils." *Arch. Insect Biochem. Physiol.* 59 (2005): 161–171.

Endangered atlantic bluefin tuna formally recommended for international trade ban. (2009).

Engelmann, Florent. "Germplasm collection, storage, and conservation." In: *Plant Biotechnology and Agriculture.* Academic Press, 2012, pp. 255–267.

Engelmann, Florent. "Importance of cryopreservation for the conservation of plant genetic resources." In: *Cryopreservation of Tropical Plant Germplasm: Current Research Progress and Application: Proceedings of an International Workshop,* Tsukuba, Japan, October 1998. International Plant Genetic Resources Institute (IPGRI), 2000, pp. 8–20.

Engelmann, Florent. "Use of biotechnologies for the conservation of plant biodiversity." *In Vitro Cellular & Developmental Biology-Plant* 47, no. 1 (2011): 5–16.

Ensley B. D., and I. Raskin. "Rationale for use of phytoremediation." In: *Phytoremediation of Toxic Metals: Using Plants to Clean Up the Environment*. Wiley, New York, USA, 1999.

Entwistle, P. F., and H. F. Evans. "Viral control." In: Gilbert, L. I., and Kerkut, G. A. (eds.), *Conprehensive Insect Fisiology: Biochemestry and Farmacology*. Vol. 12. Pergamon Press, Oxford, UK, 1985, pp. 347–412.

ENVIS Centre on Control of Pollution Water, Air and Noise.CPCB. "Water pollution & control." 2016. http://www.cpcbenvis.nic.in/water_pollution_control.html#

Espinoza, Catherine, Ana Panta, and William M. Roca. "Agricultural applications of biotechnology and the potential for biodiversity valorization in Latin America and the Caribbean." (2004).

Etim, E. E. "Phytoremediation and its mechanisms: A review." *Int J Environ Bioenergy* 2, no. 3 (2012): 120–136.

European Bioplastics). "Fact sheet: What are bioplastics? Material types, terminology and labels- an introduction." 2018. www.european-bioplastics.org

European Bioplastics). *Frequently Asked Questions (FAQs) on Bioplastics*, 2019. www.european-bioplastics.org

Evoqua Water Technologies LLC. "Anaerobic vs. Aerobic treatment: Weighing wastewater treatment options." 2019. https://www.evoqua.com/en/brands/adi-systems/Pages/Anaerobic-vs-Aerobic-Treatment.aspx

F. A. O. "Feeding the world in 2050." World Agricultural Summit on Food Security. Food and Agriculture Organization, 16–18 November 2009.

Fa, John E., Carlos A. Peres, and Jessica J. Meeuwig. "Bushmeat exploitation in tropical forests: An intercontinental comparison." *Conservation Biology* 16, no. 1 (2002): 232–237.

Fàbrega, F., V. Kumar, M. Schuhmacher, J. L. Domingo, and M. Nadal. "PBPK modeling for PFOS and PFOA: Validation with human experimental data." *Toxicology Letters* 230, no. 2 (2014): 244–251.

Fahrig, Lenore. "Relative effects of habitat loss and fragmentation on population extinction." *Journal of Wildlife Management* (1997): 603–610.

Falagas, Matthew E., and Drosos E. Karageorgopoulos. Pandrug resistance (PDR), extensive drug resistance (XDR), and multidrug resistance (MDR) among Gram-negative bacilli: Need for international harmonization in.

Falk, Michael C., Bruce M. Chassy, Susan K. Harlander, Thomas J. Hoban IV, Martina N. McGloughlin, and Amin R. Akhlaghi. "Food biotechnology: Benefits and concerns." *Journal of Nutrition* 132, no. 6 (2002): 1384–1390.

Fan, M., Z. Liu, S. Dyer, T. Federle, and X. Wang. "Development of environmental risk assessment framework and methodology for consumer product chemicals in China." *Environmental Toxicology and Chemistry* 38, no. 1 (2019): 250–261.

Fan, Sh., P. Li, Z. Gong, W. Ren, and N. He. "Promotion of pyrene degradation in rhizosphere of alfalfa (*Medicago sativa* L.)." *Chemosphere* 71, no. 8 (2008): 1593–1598.

FAO, ITPS. "Status of the world's soil resources (SWSR)–main report." Food and Agriculture Organization of the United Nations and Intergovernmental Technical Panel on Soils, Rome, Italy 650, 2015.

FAO. "Genebank standards for plant genetic resources for food and agriculture." *Commission on Genetic Resources for Food and Agriculture*, 2014.

Farré, M., J. Sanchís, and D. Barceló. "Analysis and assessment of the occurrence, the fate and the behavior of nanomaterials in the environment." *TrAC Trends in Analytical Chemistry* 30, no. 3 (2011): 517–527.

Faure, Michael G. "Optimal specificity in environmental standard-setting." *Critical Issues in Environmental Taxation: International and Comparative Perspectives* 3 (2012): 730–745.

Favas, P. J. C., J. Pratas, M. Varun, R. D'Souza, and M. S. Paul. "Phytoremediation of soils contaminated with metals and metalloids at mining areas: Potential of native flora." *Environmental Risk Assessment of Soil Contamination* 3 (2014): 485–516.

Fay, P. *Nie Blue-Greens*. Edward Arnold, London, 1983.

Feng, W., and X. Zheng. "Essential oils to control *Alternaria alternata* in vitro and in vivo." *Food Control* 18 (2007): 1126–1130.

Ferrante, M., M. Vassallo, A. Mazzola, M. V. Brundo, R. Pecoraro, A. Grasso, and C. Copat. "In vivo exposure of the marine sponge *Chondrilla nucula* Schmidt, 1862 to cadmium (Cd), copper (Cu) and lead (Pb) and its potential use for bioremediation purposes." *Chemosphere* 193 (2018): 1049–1057.

Feschotte, Cédric, and Ellen J. Pritham. "DNA transposons and the evolution of eukaryotic genomes." *Annu. Rev. Genet.* 41 (2007): 331–368.

Fineman, Stephen. "Enforcing the environment: Regulatory realities." *Business Strategy and the Environment* 9, no. 1 (2000): 62–72.

Fischer, B. B., F. Pomati, and R. I. Eggen. "The toxicity of chemical pollutants in dynamic natural systems: The challenge of integrating environmental factors and biological complexity." *Science of the Total Environment* 449 (2013): 253–259.

Fischer, G., E. Hizsnik, S. Prieler, M. Shah, and H. Van Velthuizen. "Biofuels and food security: Implications of an accelerated biofuels production, summary of the OFID Study prepared by IIASA." (2009). Peplow, M. "Ethanol production harms environment, researchers claim." *Nature* 1 (2005).

Focks, A. "The challenge: Landscape ecotoxicology and spatially explicit risk assessment." *Environmental Toxicology and Chemistry* 33, no. 6 (2014): 1193–1193.

Frank, Kenneth T., Brian Petrie, Jae S. Choi, and William C. Leggett. "Trophic cascades in a formerly cod-dominated ecosystem." *Science* 308, no. 5728 (2005): 1621–1623.

Frascari, D., G. Zanaroli, and A. S. Danko. "In situ aerobic cometabolism of chlorinated solvents: A review." *Journal of Hazardous Materials* 283 (2015): 382–399. doi: 10.1016/j.jhazmat.2014.09.041

Frewer, Lynn J., Richard Shepherd, and Paul Sparks. "Biotechnology and food production: Knowledge and perceived risk." *British Food Journal* 96, no. 9 (1994): 26–32.

Frewer, Lynn J., Richard Shepherd, and Paul Sparks. "Biotechnology and food production: Knowledge and perceived risk." *British Food Journal* 96, no. 9 (1994): 26–32.

Fu, D., R. P. Singh, X. Yang, C. S. P. Ojha, R. Y. Surampalli, and A. J. Kumar. "Sediment in-situ bioremediation by immobilized microbial activated beads: Pilot-scale study." *Journal of Environmental Management* 226 (2018): 62–69.

Fu, Y., and T. Viraraghavan. "Fungal decolorization of dye wastewaters: A review." *Bioresource Technology* 79, no. 3 (2001): 251–262.

Fukuzumi, Toshio, Atsumi Nishida, Kiyowo Aoshima, and Kyoji Minami. "Decolourization of kraft waste liquor with white rot fungi, 1: Screening of the fungi and culturing condition for decolourization of kraft waste liquor." *Journal of the Japan Wood Research Society* (1977).

Fuller, Robert J., and Digger B. Jackson. "Changes in populations of breeding waders on the machair of North Uist, Scotland, 1983–1998." *Bulletin-Wader Study Group* 90 (1999): 47–55.

Furuta, Satoshi, Hiromi Matsuhashi, and Kazushi Arata. "Biodiesel fuel production with solid superacid catalysis in fixed bed reactor under atmospheric pressure." *Catalysis Communications* 5, no. 12 (2004): 721–723.

Galina Vasilyeva, Elena Antonenko, and Vishnu Rajput. "Influence of PAH contamination on soil ecological status." *Journal of Soils and Sediments* (2017): 2368–2378.

Gambhir, Ramandeep Singh, Vinod Kapoor, Ashutosh Nirola, Raman Sohi, and Vikram Bansal. "Water pollution: Impact of pollutants and new promising techniques in purification process." *Journal of Human Ecology* 37, no. 2 (2012): 103–109.

Gandini, A. "Polymers from renewable resources: A challenge for the future of macromolecular materials." *Macromolecules* 41 (2008): 9491–9504.

Gao, Y., and L. Zhu. "Plant uptake, accumulation and translocation of phenanthrene and pyrene in soils." *Chemosphere* 55, no. 9 (2004): 1169–1178.

García, A. *Environmental Urban Noise*. Boston, MA: Wentworth Institute of Technology Press.

García-Pérez, J., E. Boldo, R. Ramis, M. Pollán, B. Pérez-Gómez, N. Aragonés, and G. López-Abente. "Description of industrial pollution in Spain." *BMC Public Health* 7, no. 1 (2007): 40.

Gašić, S., and B. Tanović. "Biopesticide formulations, possibility of application and future trends." *Journal Pesticides and Phytomedicine (Belgrade)* 2 (2013): 97–102.

Gatti, Maria Giulia, Petra Bechtold, Laura Campo, Giovanna Barbieri, Giulia Quattrini, Andrea Ranzi, Sabrina Sucato et al. "Human biomonitoring of polycyclic aromatic hydrocarbonsand metals in the general population residing near the municipal solid waste incinerator of Modena, Italy." *Chemosphere* 186 (2017): 546–557.

Gavrilescu, Maria. "Environmental biotechnology: Achievements, opportunities and challenges." *Dynamic Biochemistry, Process Biotechnology and Molecular Biology* 4, no. 1 (2010): 1–36.

Geissen, V., H. Mol, E. Klumpp, G. Umlauf, M. Nadal, M. van der Ploeg, . . . and C. J. Ritsema. "Emerging pollutants in the environment: A challenge for water resource management." *International Soil and Water Conservation Research* 3, no. 1 (2015): 57–65.

Germaine, K. J., X. Liu, G. G. Cabellos, J. P. Hogan, D. Ryan, and D. N. Dowling. "Bacterial endophyte enhanced phytoremediation of the organochlorine herbicide 2,4-dichlorophenoxyacetic acid." *FEMS Microbiology Ecology* 57, no. 2 (2006): 302–310.

Getachew, A., and F. Woldesenbet. "Production of biodegradable plastic by polyhydroxybutyrate (PHB) accumulating bacteria using low cost agricultural waste material." *BMC Res Notes* 9 (2016): 509.

Ghanbarzadeh, B., and H. Almasi. *Biodegradable Polymers*. InTech Publishers, pp. 141–185.

Ghazaryan, K. A., H. S. Movsesyan, H. E. Khachatryan, N. P. Ghazaryan, T. M. Minkina, S. N. Sushkova, S. S. Mandzhieva, and V. D. Rajput. "Copper phytoextraction and phytostabilization potential of wild plant species growing in the mine polluted areas of Armenia." *Geochemistry: Exploration, Environment, Analysis* (2018): 35. https://doi.org/10.1144/geochem2018-035

Ghazaryan, K. A., H. S. Movsesyan, T. M. Minkina, S. N. Sushkova, and V. D. Rajput. "The identification of phytoextraction potential of *Melilotus officinalis* and *Amaranthus retroflexus* growing on copper- and molybdenum-polluted soils." *Environmental Geochemical Health* (2019). https://doi.org/10.1007/s10653-019-00338-y

Ghosal, Debajyoti, Shreya Ghosh, Tapan K. Dutta, and Youngho Ahn. "Current state of knowledge in microbial degradation of polycyclic aromatic hydrocarbons (PAHs): A review." *Frontiers in Microbiology* 7 (2016): 1369.

Ghosh, N. C., and R. D. Singh. *Groundwater Arsenic Contamination in India: Vulnerability and Scope for Remedy*. National Institute of Hydrology, Roorkee, 2009, pp. 1–24.

Ghosh, Pooja, Mihir Tanay Das, and Indu Shekhar Thakur. "In vitro toxicity evaluation of organic extract of landfill soil and its detoxification by indigenous pyrene-degrading Bacillus sp. ISTPY1." *International Biodeterioration and Biodegradation* 90 (2014): 145–151.

Ghosh, Ruchira, and Arun Kansal. "Urban challenges in India and the mission for a sustainable habitat." *Interdisciplina* 2, no. 2 (2014).

Glazer A. N., and H. Nikaido. *Microbial Biotechnology: Fundamentals of Applied Microbiology.* Freeman, New York, 1995.

Goldman, Lynn R., and Michael W. Shannon. "Committee on environmental health technical report: Mercury in the environment: Implications for pediatricians." *Pediatrics* 108, no. 1 (2001): 197–205.

Gomes, M. V. T., R. R. de Souza, V. S. Teles, and É. Araújo Mendes. "Phytoremediation of water contaminated with mercury using *Typha domingensis* in constructed wetland." *Chemosphere* 103 (2014): 228–233.

Gomez, F., and M. Sartaj. "Field scale ex-situ bioremediation of petroleum contaminated soil under cold climate conditions." *International Biodeterioration & Biodegradation* 85 (2013): 375–382.

Gong, Z. M., J. Cao, X. M. Zhu, Y. H. Cui, L. Q. Guo, S. Tao, W. R. Shen, X. M. Zhao, and L. X. Han. "Organochlorine pesticide residues in agricultural soils from Tianjin." *Agro-Environmental Protection* 21 (2002): 459–461.

Gonzalez-Gutierrez, J., P. Partal, M. Garcia-Morales, and C. Gallegos. "Development of highly-transparent protein/starch-based bioplastics." *Bioresour Technol* 101 (2010): 2007–2013.

Gorovtsov, A., T. M. Minkina, T. Morin, I. V. Zamulina, S. S. Mandzhieva, S. N. Sushkova, and V. D. Rajput. "Ecological evaluation of polymetallic soil quality: The applicability of culture-dependent methods of bacterial communities studying." *Journal of Soils and Sediments* 19, no. 8 (2018): 3127–3138.

Gorovtsov, A., V. Rajput, M. Tatiana, M. Saglara, S. Svetlana, K. Igor, T. V. Grigoryeva et al. "The role of biochar-microbe interaction in alleviating heavy metal toxicity in Hordeum Vulgare L. grown in highly polluted soils." *Applied Geochemistry* 104 (2019): 93–101.

Gottesfeld, Perry, Faridah Hussein Were, Leslie Adogame, Semia Gharbi, Dalila San, Manti Michael Nota, and Gilbert Kuepouo. "Soil contamination from lead battery manufacturing and recycling in seven African countries." *Environmental Research* 161 (2018): 609–614.

Granados, R. R., and B. A. Federici. *The Biology of Baculoviruses.* CRC Press: Boca Raton, FL, USA, 1986.

Grewal, P. S., R. U. Ehlers, and D. I. Shapiro Ilan. *Nematodes as Biocontrol Agents.* CABI, New York, 2005.

Groom, Martha J., Gary K. Meffe, and Carl Ronald Carroll. *Principles of Conservation Biology.* No. Sirsi, i9780878935185. Sinauer Associates, Sunderland, 2006.

Gryglewicz, Stanislaw. "Rapeseed oil methyl esters preparation using heterogeneous catalysts." *Bioresource Technology* 70, no. 3 (1999): 249–253.

Guerra, L., Riccardo Guidi., I. Slot, S. Callegari, R. Sompallae, C. L. Pickett, S. Åström et al. "Bacterial genotoxin triggers FEN1-dependent RhoA activation, cytoskeleton remodeling and cell survival." *Journal of Cell Science* 124, no. 16 (2011): 2735–2742.

Gupta, D. K., and C. Walther, eds. *Radionuclide Contamination and Remediation through Plants.* Springer International Publishing, Switzerland, 2014.

Gupta, M., P. Sharma, N. B. Sarin, and A. K. Sinha. "Differential response of arsenic stress in two varieties of Brassica juncea L." *Chemosphere* 74, no. 9 (2009): 1201–1208.

Gupta, Neha, Krishna Kumar Yadav, and Vinit Kumar. "A review on current status of munici-
    pal solid waste management in India." *Journal of Environmental Sciences* 37 (2015):
    206–217.
Gupta, S., and A. K. Dikshit. "Biopesticides: An ecofriendly approach for pest control."
    Journal of Biopesticides 1 (2010): 186–188.
Gupta, S., and A. K. Dikshit. "Biopesticides: An ecofriendly approach for pest control." *J
    Biopest* 3 (2010): 186–188.
Gustin, J. L., M. E. Loureiro, D. Kim, G. Na, M. Tikhonova, and D. E. Salt. "MTP1-dependent
    Zn sequestration into shoot vacuoles suggests dual roles in Zn tolerance and accumula-
    tion in Zn-hyper accumulating plants." *The Plant Journal* 57, no. 6 (2009): 1116–1127.
Gutiérrez-Alcalá, Gloria, Cecilia Gotor, Andreas J. Meyer, Mark Fricker, José M. Vega, and
    Luis C. Romero. "Glutathione biosynthesis in Arabidopsis trichome cells." *Proceedings
    of the National Academy of Sciences* 97, no. 20 (2000): 11108–11113.
Miedema, H. M. E., and C. G. M. Oudshoorn. "Annoyance from transportation noise:
    Relationships with exposure metrics DNL and DENL and their confidence intervals."
    Environmental Health Perspectives 109 (2001): 409–416.
Hadacek, F. "Secondary metabolites as plant traits: Current assessment and future perspec-
    tives." *Critical Reviews in Plant Sciences* 21, no. 4 (2002): 273–322.
Hagerman, I., G. Rasmanis, V. Blomkvist, R. S. Ulrich, C. A. Eriksen, and T. Theorell.
    "Influence of coronary intensive care acoustics on the quality of care and physiological
    states of patients." *International Journal of Cardiology* 98 (2005): 267–270.
Hahn, M., S. Willscher, and G. Straube. "Copper leaching from industrial wastes by hetero-
    trophic microorganisms." *Biohydrometallurgical Technologies* 1 (1993): 99–108.
Hajek, A. E., and R. J. St. Leger. "Interactions between fungal pathogens and insect hosts."
    Annual Review of Entomology 39, no. 1 (1994): 293–322.
Hakansson, Jennie. *Genetic Aspects of Ex Situ Conservation*. Department of Biology, IFM.
    Linkoping University, Linkoping (SE), 2004.
Hallenbeck, Patrick C., and Dipankar Ghosh. "Advances in fermentative biohydrogen pro-
    duction: The way forward?" *Trends in Biotechnology* 27, no. 5 (2009): 287–297.
Hallman, William K. Consumer Concerns About Biotechnology: I Nternational Perspective.
    No. 1327-2016-103635, 2000.
Hamer, Geoffrey. "Solid waste treatment and disposal: Effects on public health and environ-
    mental safety." *Biotechnology Advances* 22, no. 1–2 (2003): 71–79.
Hanaki, Keisuke, and Joana Portugal-Pereira. "The effect of biofuel production on green-
    house gas emission reductions." In *Biofuels and Sustainability*. Springer, Tokyo, 2018,
    pp. 53–71.
Hannigan, J. A. "Nature, ecology and environmentalism: Constructing environmental knowl-
    edge (chapter 6)." *Environmental Sociology: A Social Constructionist Perspective*
    (1995): 107–127.
Hanski, Ilkka. "Habitat loss, the dynamics of biodiversity, and a perspective on conserva-
    tion." *Ambio* 40, no. 3 (2011): 248–255.
Hanski, Ilkka. *The Shrinking World: Ecological Consequences of Habitat Loss*. Vol. 14.
    International Ecology Institute, Oldendorf/Luhe, 2005.
Haque, Sk Moquammel, and Biswajit Ghosh. "High-frequency somatic embryogenesis and
    artificial seeds for mass production of true-to-type plants in *Ledebouria revoluta*: An
    important cardioprotective plant." *Plant Cell, Tissue and Organ Culture (PCTOC)* 127,
    no. 1 (2016): 71–83.
Harada, Masazumi, Shigeharu Nakachi, Taketo Cheu, Hirotaka Hamada, Yuko Ono,
    Toshihide Tsuda, Kohichi Yanagida, Takako Kizaki, and Hideki Ohno. "Monitoring
    of mercury pollution in Tanzania: Relation between head hair mercury and health."
    Science of the Total Environment 227, no. 2–3 (1999): 249–256.

Harada, T., M. Takeda, S. Kojima, and N. Tomiyama. "Toxicity and Carcinogenicity of Dichlorodiphenyltrichloroethane (DDT)." *Toxicological Research* 32 (2016): 21.

Haraguchi, Masahiko, Afreen Siddiqi, and Venkatesh Narayanamurti. "Stochastic cost-benefit analysis of urban waste-to-energy systems." *Journal of Cleaner Production* (2019).

Haritash, A. K., and C. P. Kaushik. "Biodegradation aspects of polycyclic aromatic hydrocarbons (PAHs): A review." *Journal of Hazardous Materials* 169, no. 1–3 (2009): 1–15.

Harms, H., D. Schlosser, and L. Y. Wick. "Untapped potential: Exploiting fungi in bioremediation of hazardous chemicals." *Nature Reviews Microbiology* 9, no. 3 (2011): 177–192.

Harms, H. H. "Bioaccumulation and metabolic fate of sewage sludge derived organic xenobiotics in plants." *Science of the Total Environment* 185, no. 1–3 (1996): 83–92.

Hassanshahian, M., G. Emtiazi, G. Caruso, and S. Cappello. "Bioremediation (bioaugmentation/biostimulation) trials of oil polluted seawater: A mesocosm simulation study." *Marine Environmental Research* 95 (2014): 28–38.

Hatra, G. "Radioactive pollution: An overview." *The Holistic Approach to Environment* 8, no. 2 (2018): 48–65.

Hawkins, Keith. "Environment and enforcement: Regulation and the social definition of pollution." (1984).

Haydon, M. J., and C. S. Cobbett. "Transporters of ligands for essential metal ions in plants." *New Phytologist* 174, no. 3 (2007): 499–506.

Hayes, W. J. *Pesticides Studied in Man*. Williams and Wilkins, Baltimore, MD, USA., 1982. ISBN: 13-9780683038965, p. 672.

He, S. A., and Y. Gu. "The challenge for the 21st Century for Chinese botanic gardens." Conservation into the 21st Century. Proceedings of the 4th International Botanic Gardens Conservation Congress (Perth, 1995), Kings Park and Botanic Garden, Perth, Australia, 1997, pp. 21–27.

Hein, M., S. Rotter, M. Schmitt-Jansen, P. C. von der Ohe, W. Brack, E. de Deckere, . . . and I. Muñoz. "Modelkey." *Environmental Sciences Europe* 22, no. 3 (2010): 217–228.

Herat, Sunil. "Electronic waste: An emerging issue in solid waste management in Australia." *Int. J. Environ. Waste Manag* 3, no. 1/2 (2009): 120–134.

Hernández, F., J. V. Sancho, M. Ibáñez, E. Abad, T. Portolés, and L. Mattioli. "Current use of high-resolution mass spectrometry in the environmental sciences." *Analytical and Bioanalytical Chemistry* 403, no. 5 (2012): 1251–1264.

Herniou, E. A., B. M. Arif, J. J. Becnel, G. W. Blissard, B. Bonning, R. Harrison et al. "Baculoviridae." In: King, A. M. Q., Adams, M. J., Carstens, E. B., and Lefkowitz, E. J. (eds.), *Virus Taxonomy: Classification and Nomenclature of Viruses: Ninth Report of the International Committee on Taxonomy of Viruses*. Elsevier Academic Press, San Diego, CA, USA, 2012, pp. 163–173.

Hidalgo-Ruz, V., L. Gutow, R. C. Thompson, and M. Thiel. "Microplastics in the marine environment: A review of the methods used for identification and quantification." *Environmental Science & Technology* 46, no. 6 (2012): 3060–3075.

Holder, D. J., and N. O. Keyhani. "Adhesion of the entomopathogenic fungus *Beauveria (Cordyceps) bassiana* to substrata." *Applied and Environmental Microbiology* 71, no. 9 (2005): 5260–5266.

Holoubek, I., P. Kořínek, Z. Šeda, E. Schneiderová, I. Holoubková, A. Pacl, J. Tříska, P. Cudlın, and J. Čáslavský. "The use of mosses and pine needles to detect persistent organic pollutants at local and regional scales." *Environmental Pollution* 109, no. 2 (2000): 283–292.

Hommen, U., J. M. Baveco, N. Galic, and P. J. van den Brink. "Potential application of ecological models in the European environmental risk assessment of chemicals I: Review of protection goals in EU directives and regulations." *Integrated Environmental Assessment and Management* 6, no. 3 (2010): 325–337.

Hoornweg, Daniel, and Perinaz Bhada-Tata. *What a Waste: A Global Review of Solid Waste Management.* Vol. 15. World Bank, Washington, DC, 2012.

Hossain, M. A., K. M. ALsabari, A. M. Weli, and Q. Al-Riyami. "Gas chro matography-mass spectrometry analysis and total phenolic contents of various crude extracts from the fruits of *Datura metel* L." *Journal of Taibah University for Science* 7 (2013): 209–215.

http://www.noiseoff.org/document/cenyc.noise.report.14.pdf

http://www.pubmedcentral.nih.gov/articlerender.fcgi?article 1253720

https://osha.europa.eu/en/publications/literature_reviews/combined-exposure-to-noise-and-ototoxic-substances

Huang, Z. M., Y. Z. Zhang, M. Kotakic, and S. Ramakrishna. "A review on polymer nanofibers by electrospinning and their applications in nanocomposites." *Compos Sci Technol* 63 (2003): 2223–2253.

Huelster, A., J. F. Mueller, and H. Marschner. "Soil-plant transfer of polychlorinated dibenzo-p-dioxins and dibenzofurans to vegetables of the cucumber family (Cucurbitaceae)." *Environmental Science and Technology* 28, no. 6 (1994): 1110–1115.

Hussain, S., M. Devers-Lamrani, N. El-Azahari, and F. Martin-Laurent. "Isolation and characterization of an isoproturon mineralizing Sphingomonas sp. strain SH from a French agricultural soil." *Biodegradation* 22, no. 3 (2011): 637–650.

Ibrahim, S. I., M. D. Abdel Lateef, H. M. S. Khalifa, and A. E. Abdel Monem. "Phytoremediation of atrazine-contaminated soil using *Zea mays* (maize)." *Ann Agric Sci* 58 (2013): 69–75.

Ijaz, K. J. I. Wattoo, B. Zeshan, T. Majeed, T. Riaz, S. Khalid, S. Baig, and M. A. Saleem. "Potential impact of microbial consortia in biomining and bioleaching of commercial metals." *Advancements in Life Sciences* 5, no. 1 (2017): 13–18.

Immunogenicity of biologically-derived therapeutics: Assessment and interpretation of non-clinical safety studies.

Isman, M. B. "Botanical insecticides, deterrents and repellents in modern agriculture and an increasingly regulated world." *Annu. Rev. Entomol.* 51 (2006): 45–66.

Itai, T., M. Otsuka, K. Ansong Asante, M. Muto, Y. Opoku-Ankomah, O. D. Ansa-Asare, and Sh. Tanabe. "Variation and distribution of metals and metalloids in soil/ash mixtures from Agbogbloshie e-waste recycling site in Accra, Ghana." *Science of the Total Environment* 470 (2014): 707–716.

IUCN,. Retrieved December 2003.

Iwata.). "Biodegradable and bio-based polymers: Future prospects of eco-friendly plastics." *Angew Chem Int Ed.* 54 (2015): 3210–3215.

IWMI.) "Water for food, water for life: A comprehensive assessment of water management in agriculture." In: *Earthscan and Colombo.* International Water Management Institute, London, 2007.

Jackson, Digger B., and Rhys Green. "The importance of the introduced hedgehog (Erinaceus *europaeus*) as a predator of the eggs of waders (Charadrii) on machair in South Uist, Scotland." *Biological Conservation* 93, no. 3 (2000): 333–348.

Jackson, Digger B., Robert J. Fuller, and Steve T. Campbell. "Long-term population changes among breeding shorebirds in the Outer Hebrides, Scotland, in relation to introduced hedgehogs (*Erinaceus europaeus*)." *Biological Conservation* 117, no. 2 (2004): 151–166.

Jacobsen, Carsten Suhr, and Mathis Hjort Hjelms. "Agricultural soils, pesticides and microbial diversity." *Current Opinion in Biotechnology* 27 (2014): 15–20.

Jain, P. K., V. K. Gupta, R. K. Gaur, M. Lowry, D. P. Jaroli, and U. K. Chauhan. "Bioremediation of petroleum oil contaminated soil and water." *Res J Environ Toxicol* 5, no. 1 (2011): 1–26.

Jambo, Siti Azmah, Rahmath Abdulla, Siti Hajar Mohd Azhar, Hartinie Marbawi, Jualang Azlan Gansau, and Pogaku Ravindra. "A review on third generation bioethanol feedstock." *Renewable and Sustainable Energy Reviews* 65 (2016): 756–769.

James, H. F. *Islands: Biological Diversity and Ecosystem Functioning*. Vitousek, P. M., Loope, L. L., and Adsersen, H. (eds.). Springer, Heidelberg, 1995, pp. 88–102.

Jansen, E. J. M., H. W. Helleman, W. A. Dreschler, and J. A. de Laat. "Noise-induced hearing loss and other hearing complaints among musicians of symphony orchestras." *Int Arch Occup Environ Health* 82 (2009): 153–164. [PubMed: 18404276].

Janssens, L., and R. Stoks. "Chlorpyrifos-induced oxidative damage is reduced under warming and predation risk: Explaining antagonistic interactions with a pesticide." *Environmental Pollution* 226 (2017): 79–88.

Jayaprakash, Sachin, H. Lohit, and B. Abhilash. "Design and development of compost bin for Indian kitchen." *Int J Waste Resour* 8 (2018).

Jebapriya, G. Roseline, and J. Joel Gnanadoss. "Bioremediation of textile dye using white rot fungi: A review." *International Journal of Current Research and Review* 5, no. 3 (2013): 1.

Jeffery, Simon, and Wim H. Van der Putten. "Soil borne human diseases." *Luxembourg: Publications Office of the European Union* 49, no. 10.2788 (2011): 37199.

Jha, Arvind K., S. K. Singh, G. P. Singh, and Prabhat K. Gupta. "Sustainable municipal solid waste management in low income group of cities: A review." *Tropical Ecology* 52, no. 1 (2011): 123–131.

Jha, P., J. Panwar, and P. N. Jha. "Secondary plant metabolites and root exudates: Guiding tools for polychlorinated biphenyl biodegradation." *International Journal of Environmental Science and Technology* 12, no. 2 (2015): 789–802.

Jha, Y., and R. B. Subramanian. "Regulation of plant physiology and antioxidant enzymes for alleviating salinity stress by potassium-mobilizing bacteria." In: Meena, V. S., Maurya, B. R., Verma, J. P., and Meena, R. S. (eds.), *Potassium Solubilizing Microorganisms for Sustainable Agriculture*. Springer, New Delhi, 2016, pp. 149–162.

Jiang, N. J., R. Liu, Y. J. Du, and Y. Z. Bi. "Microbial induced carbonate precipitation for immobilizing Pb contaminants: Toxic effects on bacterial activity and immobilization efficiency." *Science of the Total Environment* 672 (2019): 722–731.

Joel, F. R. *Polymer Science & Technology: Introduction to Polymer Science*. 3rd ed. Pub: Prentice Hall PTR Inc., Upper Saddle River, NJ, pp. 4–9.

Johnson, A.-C., and T. C. Morata. *The Nordic Expert Group for Criteria Documentation of Health Risks from Chemicals: Occupational Exposure to Chemicals and Hearing Impairment*. Gothenburg, Sweden, 2010.

Johnson, D. L., K. L. Maguire, D. R. Anderson, and S. P. McGrath. "Enhanced dissipation of chrysene in planted soil: The impact of a rhizobial inoculum." *Soil Biology and Biochemistry* 36, no. 1 (2004): 33–38.

Johnson, D. Barrie. "Biodiversity and ecology of acidophilic microorganisms." *FEMS Microbiology Ecology* 27, no. 4 (1998): 307–317.

Joshi, Rajkumar, and Sirajuddin Ahmed. "Status and challenges of municipal solid waste management in India: A review." *Cogent Environmental Science* 2, no. 1 (2016): 1139434.

Juhasz, A. L., and Ravendra Naidu. "Bioremediation of high molecular weight polycyclic aromatic hydrocarbons: A review of the microbial degradation of benzo[a]pyrene." *International Biodeterioration and Biodegradation* 45, no. 1–2 (2000): 57–88.

Kabaluk, J. T., A. M. Svircev, M. S. Goette, and S. G. Woo. "The use and regulation of microbial pesticides in representative jurisdictions worldwide." *IOBC Global* 99 (2010).

Kabra, A. N., R. V. Khandare, T. R. Waghmode, and S. P. Govindwar. "Differential fate of metabolism of a sulfonated azo dye Remazol Orange 3R by plants *Aster amellus* Linn., *Glandularia pulchella* (Sweet) Tronc. and their consortium." *Journal of Hazardous Materials* 190, no. 1–3 (2011): 424–431.

Kacálková, L., and P. Tlustoš. "The uptake of persistent organic pollutants by plants." *Central European Journal of Biology* 6, no. 2 (2011): 223–235.

Kachhawa, D. "Microorganisms as a biopesticides." *Journal of Entomology and Zoology Studies* 3 (2017): 468–473.

Kachieng'a, L. O., and M. N. B. Momba. "Biodegradation of fats and oils in domestic wastewater by selected protozoan isolates." *Water, Air, & Soil Pollution* 226, no. 5 (2015).

Kachieng'a, L., and M. N. B. Momba. "Kinetics of petroleum oil biodegradation by a consortium of three protozoan isolates (*Aspidisca* sp., *Trachelophyllum* sp. and *Peranema* sp.)." *Biotechnology Reports* 15 (2017): 125–131.

Kachienga, L., and M. N. B. Momba. "Biodegradation of hydrocarbon chains of crude oil byproducts by selected protozoan isolates in polluted wastewaters." An Interdisciplinary Response to Mine Water Challenges, Xuzhou, 2014, 743–756.

Kahar, P., J. Agus, Y. Kikkawa, K. Taguchi, Y. Doi, and T. Tsuge. "Effective production and kinetic characterization of ultra-high-molecular-weight poly (R)-3-hydroxybutyrate in recombinant *Escherichia coli*." *Polym Degrad Stab* 87 (2005): 161–169.

Kalia, Anu, and Ram Prakash Gupta. "Conservation and utilization of microbial diversity." *NBA Scientific Bulletin* 9, no. 7 (2005).

Kalyani, Khanjan Ajaybhai, and Krishan K. Pandey. "Waste to energy status in India: A short review." *Renewable and Sustainable Energy Reviews* 31 (2014): 113–120.

Kamran, M. A., S. A. M. A. S. Eqani, S. Bibi, R. K. Xu, M. F. H. Monis, A. Katsoyiannis, . . . and H. J. Chaudhary. "Bioaccumulation of nickel by *E. sativa* and role of plant growth promoting rhizobacteria (PGPRs) under nickel stress." *Ecotoxicology and Environmental Safety* 126 (2016): 256–263.

Kang, B. G., W. T. Kim, H. S. Yun, and S. C. Chang. "Use of plant growth-promoting rhizobacteria to control stress responses of plant roots." *Plant Biotechnology Reports* 4, no. 3 (2010): 179–183.

Kannan, K., S. Battula, B. G. Loganathan, C. S. Hong, W. H. Lam, D. L. Villeneuve, K. Sajwan, J. P. Giesy, and K. M. Aldous. "Trace organic contaminants, including toxaphene and trifluralin, in cotton field soils from Georgia and South Carolina, USA." *Archives of Environmental Contamination and Toxicology* 45, no. 1 (2003): 30–36.

Kapo, K. E., C. M. Holmes, S. D. Dyer, D. de Zwart, and L. Posthuma. "Developing a foundation for eco-epidemiological assessment of aquatic ecological status over large geographic regions utilizing existing data resources and models." *Environmental Toxicology and Chemistry* 33, no. 7 (2014): 1665–1677.

Karlsson, M. V., L. J. Carter, A. Agatz, and A. B. Boxall. "Novel approach for characterizing pH-dependent uptake of ionizable chemicals in aquatic organisms." *Environmental Science & Technology* 51, no. 12 (2017): 6965–6971.

Kartha, K. K. "Cryopreservation of secondary metabolite-producing plant cell cultures." In: Vasil, I. K. (ed.), Cell Culture and Somatic Cell Genetics of Plants. Vol. 4. Academic Press, Orlando, Florida, USA, 1987, pp. 217–227.

Kasso, Mohammed, and Mundanthra Balakrishnan. "Ex situ conservation of biodiversity with particular emphasis to ethiopia." *ISRN Biodiversity* 2013 (2013).

Kątska-Książkiewicz, Lucyna K. "Recent achievements in in vitro culture and preservation of ovarian follicles in mammals." *Reprod. Biol.* 6 (2006): 3–16.

Kaya, H. K., and Gaugler, R. "Entomopathogenic nematodes." *Annual Review of Entomology* 38 (1993): 181–206.

Kaza, Silpa, Lisa Yao, Perinaz Bhada-Tata, and Frank Van Woerden. *What a waste 2.0: A Global Snapshot of Solid Waste Management to 2050*. World Bank Publications, 2018.

Kekuda, P. T. R., S. Akarsh, S. A. N. Nawaz, M. C. Ranjitha, S. M. Darshini, and P. Vidya. "In vitro antifungal activity of some plants against bipolaris sarokiniana (Sacc.) shoem." *International Journal of Current Microbiology and Applied Sciences* 6 (2016): 331–337. https://doi.org/10.20546/ijcmas.2016.506.037

Kennedy, I. R., A. T. M. A. Choudhury, and M. L. Kecske´s. "Non-symbiotic bacterial diazo-trophs in crop-farming systems: Can their potential for plant growth promotion be bet-ter exploited?" *Soil Biol Biochem* 36, no. 8 (2004): 1229–1244.

Kennon-Lacy, Molly. "Water quality monitoring at impaired sites on the Cache River, AR, USA." M.tech diss., Arkansas State University, 2016.

Khalid, Azeem, Muhammad Arshad, Muzammil Anjum, Tariq Mahmood, and Lorna Dawson. "The anaerobic digestion of solid organic waste." *Waste Management* 31, no. 8 (2011): 1737–1744.

Khalid, S., M. Shahid, N. K. Niazi, B. Murtaza, I. Bibi, and C. Dumat. "A comparison of tech-nologies for remediation of heavy metal contaminated soils." *Journal of Geochemical Exploration* 182 (2017): 247–268.

Khan, S., Q. Cao, Y. M. Zheng, Y. Z. Huang, and Y. G. Zhu. "Health risks of heavy met-als in contaminated soils and food crops irrigated with wastewater in Beijing, China." *Environmental Pollution* 152, no. 3 (2008): 686–692.

Khan, Suliman, Ghulam Nabi, Muhammad Wajid Ullah, Muhammad Yousaf, Sehrish Manan, Rabeea Siddique, and Hongwei Hou. "Overview on the role of advance genom-ics in conservation biology of endangered species." *International Journal of Genomics* 2016 (2016).

Khare, E., and N. K. Arora. "Effects of soil environment on field efficacy of microbial inocu-lants." In: Arora, N. K. (eds.), *Plant Microbes Symbiosis: Applied Facets*. Springer, India, 2015, pp. 37–75.

Khater, H. F., and D. F. Khater. "The insecticidal activity of four medicinal plants against the blowfly *Lucilia sericata* (Diptera: Calliphoridae)." *Int. J. Dermatol* 48 (2009): 492–497.

Khater, H. F., A. Hanafy, A. D. Abdel-Mageed, M. Y. Ramadan, and R. S. El-Madawy. "Control of the myiasis-producing fly, *Lucilia sericata*, with Egyptian essential oils." *Int. J. Dermatol* 50 (2010): 187–194.

Khater, H. F. "Ecosmart biorational insecticides: Alternative insect control strategies." In: Perveen, F. (ed.), *Insecticides*. InTech, Rijeka, Croatia, 2011, ISBN 979-953-307-667-5.

Kim, M. G., S. M. Hong, H. J. Shim, Y. D. Kim, C. I. Cha, and S. G. Yeo. "Hearing threshold of Korean adolescents associated with the use of personal music players." *Yonsei Med J* 50 (2009): 771–776.

Kim, Y. H., J. P. Freeman, J. D. Moody, K. H. Engesser, and C. E. Cerniglia. "Effects of pH on the degradation of phenanthrene and pyrene by *Mycobacterium vanbaalenii* PYR-1." *Applied Microbiology and Biotechnology* 67, no. 2 (2005): 275–285.

Kinch, Michael S. "An overview of FDA-approved biologics medicines." *Drug Discovery Today* 20, no. 4 (2015): 393–398.

Klingemann, Hans-G., Andrew P. Grigg, Karen Wilkie-Boyd, Michael J. Barnett, Allen C. Eaves, D. E. Reece, J. D. Shepherd, and G. L. Phillips. "Treatment with recombinant interferon (alpha-2b) early after bone marrow transplantation in patients at high risk for relapse [corrected][published erratum appears in Blood 1992 Jun 15; 79 (12): 3397]." *Blood* 78, no. 12 (1991): 3306–3311.

Klotz, S., N. Kaufmann, A. Kuenz, and U. Prube. "Biotechnological production of enantio-merically pure d-lactic acid." *Appl Microbiol Biotechnol* 100 (2016): 9423–9437.

Klotz, S. "Drivers and pressures on biodiversity in analytical frameworks." *Biodiversity under Threat*, 25 (2007): 252.

Kmentova, E. "Response of plant to fluoranthene in environment." PhD diss., Ph.D Thesis, Masaryk University, Brno, Czech Republic, 2003.

Knejzlík, Z., J. Káš, and T. Ruml. "Mechanismus vstupu xenobiotik do organismu a jejich detoxikace." *Chemicke. Listy* 94 (2000): 913–918.

Knight, Andrew J. "Perceptions, knowledge and ethical concerns with GM foods and the GM process." *Public Understanding of Science* 18, no. 2 (2009): 177–188.

Knutson, K. L. "Sociodemographic and cultural determinants of sleep deficiency: Implications for cardiometabolic disease risk." *Soc Sci Med* 79 (2013): 7–15.

Knutson, K. L. "Sleep duration and cardiometabolic risk: Are view of the epidemiologic evidence." *Best Pract Res Clin Endocrinol Metab* 24, no. 5 (2010): 731–743.

Knutson, A., and J. Ruberson. "Field guide to predators, parasites and pathogens attacking insect and mite pests of cotton." In: Smith, E. M. (ed.), *Texas Cooperative Extension*, TX Publication, 2015, p. 136.

Kobayashi, Audrey. "Geographies of peace and armed conflict: Introduction." (2009): 819–826.

Kochian, L. V., O. A. Hoekenga, and M. A. Pineros. "How do crop plants tolerate acid soils? Mechanisms of aluminum tolerance and phosphorous efficiency." *Annual Review of Plant Biology* 55 (2004): 459–493.

Kogbara, R. B. "Ranking agro-technical methods and environmental parameters in the biodegradation of petroleum-contaminated soils in Nigeria." *Electronic Journal of Biotechnology* 11, no. 1 (2008): 0–0. doi: 10.2225/vol11-issue1-fulltext-4

Kokorīte, I., M. Kļaviņš, J. Šīre, O. Purmalis, and A. Zučika. "Soil pollution with trace elements in territories of military grounds in Latvia." Proceedings of the Latvian Academy of Sciences, Section B. Natural, Exact, and Applied Sciences, Versita. Vol. 62, no. 1–2, 2008, pp. 27–33.

Komárek, M., E. Čadková, V. Chrastný, F. Bordas, and J. C. Bollinger. "Contamination of vineyard soils with fungicides: A review of environmental and toxicological aspects." *Environment International* 36, no. 1 (2010): 138–151.

Korte, F., G. Kvesitadze, D. Ugrekhelidze, M. Gordeziani, G. Khatisashvili, O. Buadze, G. Zaalishvili, and F. Coulston. "Organic toxicants and plants." *Ecotoxicology and Environmental Safety* 47, no. 1 (2000): 1–26.

Kortenkamp, A., and M. Faust. "Combined exposures to anti-androgenic chemicals: Steps towards cumulative risk assessment." *International Journal of Andrology* 33, no. 2 (2010): 463–474.

Kotoky, Rhitu, L. Paikhomba Singha, and Piyush Pandey. "Draft genome sequence of polyaromatic hydrocarbon-degrading bacterium Bacillus subtilis SR1, which has plant growth-promoting attributes." *Genome Announc* 5, no. 49 (2017): e01339–17.

Kotterman, M. J. J., E. H. Vis, and J. A. Field. "Successive mineralization and detoxification of benzo[a]pyrene by the white rot fungusBjerkandera sp. strain BOS55 and indigenous microflora." *Applied Environmental Microbiology* 64, no. 8 (1998): 2853–2858.

Krauss, M., H. Singer, and J. Hollender. "LC-high resolution MS in environmental analysis: From target screening to the identification of unknowns." *Analytical and Bioanalytical Chemistry* 397, no. 3 (2010): 943–951.

Krystal, A. D. "Psychiatric disorders and sleep." *Neurol Clin* 30, no. 4 (2012): 1389–1413.

Kumar, A., J. S. Patel, I. Bahadur, and V. S. Meena. "The molecular mechanisms of KSMs for enhancement of crop production under organic farming." In: Meena, V. S., Maurya, B. R., Verma, J. P., and Meena, R. S. (eds.), *Potassium Solubilizing Microorganisms for Sustainable Agriculture*. Springer, New Delhi, 2016, pp. 61–75.

Kumar, Atul, and Sukha Ranjan Samadder. "A review on technological options of waste to energy for effective management of municipal solid waste." *Waste Management* 69 (2017): 407–422.

Kumar, M., R. Prasad, P. Goyal, P. Teotia, N. Tuteja, A. Varma, and V. Kumar. "Environmental biodegradation of xenobiotics: Role of potential microflora." In *Xenobiotics in the Soil Environment*. Springer, Cham, 2017, pp. 319–334.

Kumar, P. B. A. Nanda, Viatcheslav Dushenkov, Harry Motto, and Ilya Raskin. "Phytoextraction: The use of plants to remove heavy metals from soils." *Environmental Science and Technology* 29, no. 5 (1995): 1232–1238.

Kumar, Sunil, J. K. Bhattacharyya, A. N. Vaidya, Tapan Chakrabarti, Sukumar Devotta, and A. B. Akolkar. "Assessment of the status of municipal solid waste management in metro cities, state capitals, class I cities, and class II towns in India: An insight." *Waste Management* 29, no. 2 (2009): 883–895.

Kumar, Vaneet, and Navin Chand Kothiyal. "Distribution behavior of polycyclic aromatic hydrocarbons in roadside soil at traffic intercepts within developing cities." *International Journal of Environmental Science and Technology* 8, no. 1 (2011): 63–72.

Kumar, Vaneet, Navin Chand Kothiyal, and Saruchi. "Analysis of polycyclic aromatic hydrocarbon, toxic equivalency factor and related carcinogenic potencies in roadside soil within a developing city of Northern India." *Polycyclic Aromatic Compounds* 36, no. 4 (2016): 506–526.

Kumar, Vineet, S. K. Shahi, and Simranjeet Singh. "Bioremediation: An eco-sustainable approach for restoration of contaminated sites." In *Microbial Bioprospecting for Sustainable Development*. Springer, Singapore, 2018, pp. 115–136.

Kumar, D. Shiva, S. Srikantaswamy, M. R. Abhilash, and A. Nagaraju. "A comparative study of aerobic and anaerobic wastewater treatment." *International Journal for Research in Applied Science & Engineering Technology* 3 (2015): 994–1004.

Kumari, Aloka, Ponnusamy Baskaran, and Johannes Van Staden. "In vitro propagation via organogenesis and embryogenesis of Cyrtanthus mackenii: A valuable threatened medicinal plant." *Plant Cell, Tissue and Organ Culture (PCTOC)* 131, no. 3 (2017): 407–415.

Kunimi, Y. "Current status and prospects on microbial control in Japan." *J Invertebr Pathol* 95 (2007): 181–186.

Kuppusamy, S., P. Thavamani, K. Venkateswarlu, Y. B. Lee, R. Naidu, and M. Megharaj. "Remediation approaches for polycyclic aromatic hydrocarbons (PAHs) contaminated soils: Technological constraints, emerging trends and future directions." *Chemosphere* 168 (2017): 944–968.

Lakshmikanthan, P., P. Sughosh, and G. L. Sivakumar Babu. "Studies on characterization of mechanically biologically treated waste from Bangalore City." *Indian Geotechnical Journal* 48, no. 2 (2018): 293–304.

Lampinen, J. "Trends in bioplastics and biocomposites." In: Harlin, A., and Vikman, M. (eds.), *Developments in Advanced Biocomposites*. VTT Tiedotteita-Research Notes, 2010.

Lande, Russell. "Genetics and demography in biological conservation." *Science* 241, no. 4872 (1988): 1455–1460.

Landis, W. G., K. K. Ayre, A. F. Johns, H. M. Summers, J. Stinson, M. J. Harris, . . . and A. J. Markiewicz. "The multiple stressor ecological risk assessment for the mercury-contaminated South River and upper Shenandoah River using the Bayesian network-relative risk model." *Integrated Environmental Assessment and Management* 13, no. 1 (2017): 85–99.

Lanza, Robert Paul, Jose B. Cibelli, Francisca Diaz, Carlos T. Moraes, Peter W. Farin, Charlotte E. Farin, Carolyn J. Hammer, Michael D. West, and Philip Damiani. "Cloning of an endangered species (Bos gaurus) using interspecies nuclear transfer." *Cloning* 2, no. 2 (2000): 79–90.

Lapsley Miller, J. A., L. Marshall, L. M. Heller, and L. M. Hughes. "Low-level otoacoustic emissions may predict susceptibility to noise-induced hearing loss." *J AcoustSoc Am* 120 (2006): 280–296. [PubMed: 16875225].

Larson, S. L., A. J. Bednar, J. H. Ballard, M. G. Shettlemore, D. B. Gent, Christos Christodoulatos, R. Manis, J. C. Morgan, and M. P. Fields. "Characterization of a military training site containing 232Thorium." *Chemosphere* 59, no. 7 (2005): 1015–1022.

Laureysens, I., R. Blust, L. De Temmerman, C. Lemmens, and R. Ceulemans. "Clonal variation in heavy metal accumulation and biomass production in a poplar coppice culture: I. Seasonal variation in leaf, wood and bark concentrations." *Environmental Pollution* 131, no. 3 (2004): 485–494.

Le Prell, C. G., D. F. Dolan, D. C. Bennett, and P. A. Boxer. "Nutrient plasma levels achieved during treatment that reduces noise-induced hearing loss." *Transl Res* 158 (2011): 54–70. [PubMed: 21708356].

Le Prell, C. G., and C. Spankovich. "Healthy diets, and dietary supplements: Recent changes in how we might think about hearing conservation." In: Griefahn, B. (ed.), *10th International Congress on Noise as a.*

Lee, K. Y., S. E. Strand, and S. L. Doty. "Phytoremediation of chlorpyrifos by populus and salix." *International Journal of Phytoremediation* 14, no. 1 (2012): 48–61.

Lee, S. H., W. S. Lee, Ch. H. Lee, and J. G. Kim. "Degradation of phenanthrene and pyrene in rhizosphere of grasses and legumes." *Journal of Hazardous Materials* 153, no. 1–2 (2008): 892–898.

Leger, D., V. Bayon, M. M. Ohayon, P. Philip, P. Ement, A. Metlaine et al. "Insomnia and accidents: A cross-sectional study (EQUINOX) on sleep-related home, work and car accidents in 5293 subjects with insomnia from 10 countries." *J Sleep Res* 23, no. 2 (2014): 143–152.

Lemoigne, M. "Products of dehydration and of polymerization of β-hydroxybutyric acid." *Bull Soc Chim Biol* 8 (1926): 770–782.

Lemtiri, A., G. Colinet, T. Alabi, D. Cluzeau, L. Zirbes, É. Haubruge, and F. Francis. "Impacts of earthworms on soil components and dynamics: A review." *Biotechnologie, Agronomie, Société et Environnement* 18 (2014).

Lepper, P. "Towards the derivation of quality standards for priority substances in the context of the Water Framework Directive." Final Report, Fraunhofer Institute, Germany, 2002.

Lev, Efraim, and Zohar Amar. "Ethnopharmacological survey of traditional drugs sold in the Kingdom of Jordan." *Journal of Ethnopharmacology* 82, no. 2–3 (2002): 131–145.

Leventhal, H. G. "Low-frequency noise and annoyance." *Noise Health* 6 (2004): 59–72.

Li, W. C., H. F. Tse, and L. Fok. "Plastic waste in the marine environment: A review of sources, occurrence and effects." *Sci Total Environ* 566 (2016): 333–349.

Li, Fei. "Heavy metal in urban soil: Health risk assessment and management." *Heavy Metals* (2018): 337.

Li, M., X. Cheng, and H. Guo. "Heavy metal removal by biomineralization of urease producing bacteria isolated from soil." *Int Biodeterioration Biodegrad* 76 (2013): 81–85.

Li, P., X. Feng, G. Qiu, L. Shang, Sh. Wang, and B. Meng. "Atmospheric mercury emission from artisanal mercury mining in Guizhou Province, Southwestern China." *Atmospheric Environment* 43, no. 14 (2009): 2247–2251.

Li, X., R. Zheng, X. Zhang, Z. Liu, R. Zhu, X. Zhang, and D. Gao. "A novel exoelectrogen from microbial fuel cell: Bioremediation of marine petroleum hydrocarbon pollutants." *Journal of Environmental Management* 235 (2019): 70–76.

Liang, M. J., F. Zhao, D. French, and Y. Q. Zheng. "Characteristics of noise-canceling headphones to reduce the hearing hazard for MP3 users." *J Acoust Soc Am* 131 (2012): 4526–4534. [PubMed: 22712926]

Lidder, Preetmoninder, and Andrea Sonnino. "Biotechnologies for the management of genetic resources for food and agriculture." In *Advances in Genetics*. Vol. 78. Academic Press, 2012, pp. 1–167.

Liste, H. H., and M. Alexander. "Plant-promoted pyrene degradation in soil." *Chemosphere* 40, no. 1 (2000): 7–10.

Liu, X., W. Zhang, Y. Hu, E. Hu, X. Xie, L. Wang, and H. Cheng. "Arsenic pollution of agricultural soils by concentrated animal feeding operations (CAFOs)." *Chemosphere* 119 (2015): 273–281.

Liu, Z., H. Wang, P. L. Carmichael, E. J. Deag, R. Duarte-Davidson, H. Li, . . . and R. F. Shore. "China begins to position for leadership on responsible risk-based global chemicals management." *Environmental Pollution (Barking, Essex: 1987)* 165 (2012): 170.

Locasciulli, Anna, William Arcese, Franco Locatelli, Eros Di Bona, and Andrea Bacigalupo. "Treatment of aplastic anaemia with granulocyte-colony stimulating factor and risk of malignancy." *The Lancet* 357, no. 9249 (2001): 43–44.

Lodge, M. W. "International seabed authority's regulations on prospecting and exploration for polymetallic nodules in the area." *Journal of Energy & Natural Resources Law* 20, no. 3 (2002): 270–295.

Lohmann, R., K. Booij, F. Smedes, and B. Vrana. "Use of passive sampling devices for monitoring and compliance checking of POP concentrations in water." *Environmental Science and Pollution Research* 19, no. 6 (2012): 1885–1895.

Loi, Pasqualino, Grazyna Ptak, Barbara Barboni, Josef Fulka Jr, Pietro Cappai, and Michael Clinton. "Genetic rescue of an endangered mammal by cross-species nuclear transfer using post-mortem somatic cells." *Nature Biotechnology* 19, no. 10 (2001): 962.

Lomolino, Mark V., Brett R. Riddle, Robert J. Whittaker, and James H. Brown. "Biogeography." Sinauer: Sunderland, MA (2010).

Londono-Zuluaga, C., H. Jameel, R. W. Gonzalez, and L. Lucia. "Crustacean shell-based biosorption water remediation platforms: Status and perspectives." *Journal of Environmental Management* 231 (2019): 757–762.

Lovett, Gary M. "Atmospheric deposition of nutrients and pollutants in North America: An ecological perspective." *Ecological Applications* 4, no. 4 (1994): 629–650.

Lowe, Philip, Judy Clark, Susanne Seymour, and Neil Ward. *Moralizing the Environment: The Countryside Change, Farming and Pollution.* UCL Press Limited, 1997.

Lucas, N., C. Bienaime, C. Belloy, M. Queneudec, F. Silvestre, and J. E. Nava-Saucedo. "Polymer biodegradation: Mechanisms and estimation techniques: A review." *Chemosphere* 73 (2008): 429–442.

Lunney, A. I., B. A. Zeeb, and K. J. Reimer. "Uptake of weathered DDT in vascular plants: Potential for phytoremediation." *Environmental Science & Technology* 38, no. 22 (2004): 6147–6154.

Lutman, M. E., and Hall, A. J. *Novel Methods for Early Identification of Noise-Induced Hearing Loss.* HSE Books, Sudbury, 2000.

Ma, X., A. R. Richter, S. Albers, and J. G. Burken. "Phytoremediation of MTBE with hybrid poplar trees." *International Journal of Phytoremediation* 6, no. 2 (2004): 157–167.

Ma, X., and J. G. Burken. "TCE diffusion to the atmosphere in phytoremediation applications." *Environmental Science & Technology* 37, no. 11 (2003): 2534–2539.

Mace, G. M., R. S. Hails, P. Cryle, J. Harlow, and S. J. Clarke. "Towards a risk register for natural capital." *Journal of Applied Ecology* 52, no. 3 (2015): 641–653.

Macek, T., M. Mackova, and J. Káš. "Exploitation of plants for the removal of organics in environmental remediation." *Biotechnology Advances* 18, no. 1 (2000): 23–34.

MacGregor, J. T. "Genetic toxicity assessment of microbial pesticides: Needs and recommended approaches." *Intern Assoc Environ Mutagen Soc* (2006): 1–17.

Machado, Antonio E. H., Aline M. Furuyama, Sandra Z. Falone, Reinaldo Ruggiero, Denilson da Silva Perez, and Alain Castellan. "Photocatalytic degradation of lignin and lignin models, using titanium dioxide: The role of the hydroxyl radical." *Chemosphere* 40, no. 1 (2000): 115–124.

MacLeod, M., J. West, I. Busch-Vishniac, and J. Dunn. "Quieting weinberg 5C: A case study in reducing hospital noise on a patient ward." *Journal of the Acoustical Society of America* 119, no. 5 (2006): 3327.

Madison, L. L., and G. W. Huisman. "Metabolic engineering of poly (3-hydroalkanoates): From DNA to plastic." *Microbiol Mol Biol Rev* 63 (1999): 21–53.

Mahajan, A., R. D. Gupta, and R. Sharma. "Bio-fertilizers-a way to sustainable agriculture." *Agrobios Newsl* 6, no. 9 (2008): 36–37.

Mahar, Amanullah, Ping Wang, Amjad Ali, Mukesh Kumar Awasthi, Altaf Hussain Lahori, Quan Wang, Ronghua Li, and Zengqiang Zhang. "Challenges and opportunities in the phytoremediation of heavy metals contaminated soils: A review." *Ecotoxicology and Environmental Safety* 126 (2016): 111–121.

Mahdi, S. S., G. I. Hassan, S. A. Samoon, H. A. Rather, S. A. Dar, and B. Zehra. "Bio-fertilizers in organic agriculture." *Journal of Phytology* 2, no. 10 (2010): 42–54.

Malcova, R., M. Vosatka, and M. Gryndler. "Effects of inoculation with Glomusintraradices on lead uptake by *Zea mays* L. and *Agrostis capillaris* L." *Applied Soil Ecology* 23 (2003): 255–267.

Malik, F. R., S. Ahmed, and Y. M. Rizki. "Utilization of lignocellulosic waste for the preparation of nitrogenous biofertilizer." *Pakistan Journal of Biological Sciences* 4 (2001): 1217–1220.

Malla, Muneer Ahmad, Anamika Dubey, Shweta Yadav, Ashwani Kumar, Abeer Hashem, and Elsayed Fathi Abd_Allah. "Understanding and designing the strategies for the microbe-mediated remediation of environmental contaminants using omics approaches." *Frontiers in Microbiology* 9 (2018): 1132.

Maron, M., M. G. Mitchell, R. K. Runting, J. R. Rhodes, G. M. Mace, D. A. Keith, and J. E. Watson. "Towards a threat assessment framework for ecosystem services." *Trends in Ecology & Evolution* 32, no. 4 (2017): 240–248.

Marques, A. P. G. C., A. O. S. S. Rangel, and P. M. L. Castro. "Remediation of heavy metal contaminated soils: Phytoremediation as a potentially promising clean-up technology." *Crit. Rev. Environ. Sci. Technol* 39 (2009): 622–654.

Martín, José Antonio Rodríguez, Carmen Gutiérrez, Miguel Escuer, Ma Teresa García-González, Raquel Campos-Herrera, and Nancy Águila. "Effect of mine tailing on the spatial variability of soil nematodes from lead pollution in La Union (Spain)." *Science of the Total Environment* 473 (2014): 518–529.

Martinez, A., N. R. Erdman, Z. L. Rodenburg, P. M. Eastling, and K. C. Hornbuckle. "Spatial distribution of chlordanes and PCB congeners in soil in Cedar Rapids, Iowa, USA." *Environmental Pollution* 161 (2012): 222–228.

Mathias Basner, M. D. "Auditory and non-auditory effects of noise on health." *Lancet* 383, no. 9925 (2014, April 12): 1325–1332.

Mathur, S. C. "Future of Indian pesticides industry in next millennium." *Pesticide Information* 24, no. 4 (1999): 9–23.

Matthew, M., B. A. Robbins, A. R. Vaccaro, and L. Madigan. "The use of bioabsorbable implants in spine surgery, Neurosurg." *Focus* 16 (2004): 3.

Mattina, Mary Jane Incorvia, Jason White, Brian Eitzer, and William Iannucci-Berger. "Cycling of weathered chlordane residues in the environment: Compositional and chiral profiles in contiguous soil, vegetation, and air compartments." *Environmental Toxicology and Chemistry: An International Journal* 21, no. 2 (2002): 281–288.

Maxted, Nigel, and Shelagh Kell. "Establishment of a global network for the in situ conservation of crop wild relatives: Status and needs." *FAO Commission on Genetic Resources for Food and Agriculture, Rome* 266 (2009): 509.

May, N., D. E. Ralph, and G. S. Hansford. "Dynamic redox potential measurement for determining the ferric leach kinetics of pyrite." *Minerals Engineering* 10 (1997): 1279–1290.

Mazid, M., and T. A. Khan. "Future of bio-fertilizers in Indian agriculture: An overview." *International Journal of Agricultural and Food Research* 3, no. 3 (2014): 10–23.

Mazid, M., T. A. Khan, and F. Mohammad. "Cytokinins, a classical multifaceted hormone in plant system." *J Stress Physiol Biochem* 7, no. 4 (2011): 347–368.

Mc Farlane, C., and S. Trapp. *Plant Contamination: Modeling and Simulation of Organic Chemical Processes.* CRC Press, 1994.

McGenity, T. J. "Hydrocarbon biodegradation in intertidal wetland sediments." *Current Opinion in Biotechnology* 27 (2014): 46–54.

McGlynn, S. E., and R. J. Livingston. "The distribution of polynuclear aromatic hydrocarbons between aquatic plants and sediments." *International Journal of Quantum Chemistry* 64, no. 3 (1997): 271–283.

Meager, Anthony, ed. *The Interferons: Characterization and Application.* John Wiley & Sons, 2006.

Meghani, Z. "Regulations of consumer products." In *Consumer Perception of Product Risks and Benefits.* Springer, Cham, 2017, pp. 495–513.

Mehlhorn, H., K. A. S. Al-Rasheid, and F. Abdel-Ghaffar. "The neem tree story: Extracts that really work." In: Mehlhorn, H. (ed.), *Nature Helps, Parasitology Research Monographs 1.* Springer-Verlag, Berlin, 2011, pp. 77–108.

Meng, W., Z. Wang, B. Hu, Z. Wang, H. Li, and R. Cole Goodman. "Heavy metals in soil and plants after long-term sewage irrigation at Tianjin China: A case study assessment." *Agricultural Water Management* 171 (2016): 153–161.

Mesjasz-Przybyłowicz, J. O. L. A. N. T. A., M. I. R. O. S. Ł. A. W. Nakonieczny, P. A. W. E. Ł. Migula, M. A. R. I. A. Augustyniak, M. O. N. I. K. A. Tarnawska, W. U. Reimold, . . . and E. L. Ż. B. I. E. T. A. Głowacka. "Uptake of cadmium, lead nickel and zinc from soil and water solutions by the nickel hyperaccumulator Berkheya coddii." *Acta Biol. Cracov. Bot* 46 (2004): 75–85.

Meudec, A., J. Dussauze, E. Deslandes, and N. Poupart. "Evidence for bioaccumulation of PAHs within internal shoot tissues by a halophytic plant artificially exposed to petroleum-polluted sediments." *Chemosphere* 65, no. 3 (2006): 474–481.

Miller, Daniel J., Jennifer R. Simpson, and Brian Silver. "Safety of thrombolysis in acute ischemic stroke: A review of complications, risk factors, and newer technologies." *The Neurohospitalist* 1, no. 3 (2011): 138–147.

Miller, E. K., J. A. Boydston, and W. E. Dyer. "Mechanisms of pentachlorophenol phytoremediation in soil." Pacific Northwest Regional Meeting of the ASAE, Fairmont Hot Springs, Montana, USA, 10–12 Sepember 1998, p. 7.

Milner-Gulland, Eleanor J., Elizabeth L. Bennett, Katharine Anne Abernethy, Mohamed Bakarr, Richard Bodmer, Justin S. Brashares et al. "Wild meat: The bigger picture." *Trends in Ecology & Evolution* 18, no. 7 (2003): 351–357.

Minkina, T. M., D. G. Nevidomskaya, V. A. Shuvaeva, A. V. Soldatov, V. S. Tsitsuashvili, Y. V. Zubavichus, V. D. Rajput, and M. V. Burachevskaya. "Studying the transformation of Cu2+ ions in soils and mineral phases by the Xrd, Xanes, and xequential fractionation methods." *Journal of Geochemical Exploration* 184 (2018): 365–371.

Mirsal, A. I. "Soil pollution: Origin, monitoring & remediation." *Springer Science and Business Media* (2004): 310.

Mishra, J., and N. K. Arora. "Bioformulations for plant growth promotion and combating phytopathogens: A sustainable approach." In: Arora, N. K., Mehnaz, S., and Balestrini, R. (eds.), *Bioformulations: For Sustainable Agriculture.* Springer, India, 2016, pp. 3–33. doi: 10.1007/978-81-322-2779-3_1

Mishra, V., R. Lal, and Srinivasan. "Enzymes and operons mediating xenobiotic degradation in bacteria." *Critical Reviews in Microbiology* 27, no. 2 (2001): 133–166.

Mizubuti, G. S. E., V. L. Junior, and G. A. Forbes. "Management of late blight with alternative products." *Pest Technology* 2 (2007): 106–116.

MNRE (Ministry of New and Renewable Energy). Government of India, New Delhi, India, 2018. http://www.mnre.gov.in

Moerman, Daniel E. "Native North American food and medicinal plants: Epistemological considerations." *Plants for Food and Medicine* (1998): 69–74.

Moore, C. J., S. L. Moore, M. K. Leecaster, and S. B. A. Weisberg. "Comparison of plastic and plankton in the north Pacific central gyre." *Mar. Pollut. Bull.* 42 (2001): 1297–1300.

Moore, N. F., L. A. King, and R. D. Possee. "Viruses of insects." *Insect Sci. Appl.* 3 (1987): 275–289.

Moorhouse, E. R., A. T. Gillespie, E. K. Sellers, and A. K. Charnley. "Influence of fungicides and insecticides on the entomogenous fungus *Metarhizium anisopliae* a pathogen of the vine weevil, *Otiorhynchus sulcatus*." *Biocontrol Science and Technology* 2, no. 1 (1992): 49–58.

Mordue, A. J., and A. Blackwell. "Azadirachtin: An update." *J. Insect Physiol* 39 (1993): 903–924.

Morgan, Richard K. "Environmental impact assessment: The state of the art." *Impact Assessment and Project Appraisal* 30, no. 1 (2012): 5–14.

Morillo, E., A. S. Romero, L. Madrid, J. Villaverde, and C. Maqueda. "Characterization and sources of PAHs and potentially toxic metals in urban environments of Sevilla (Southern Spain)." *Water, Air, and Soil Pollution* 187, no. 1–4 (2008): 41–51.

Morrissey, W. A. "Methyl bromide and stratospheric ozone depletion." CRS Report for Congress, 2006, pp. 1–6.

Mortazavi, A., Z. Abbasloob, L. Ebrahimi, A. Keshavarz, and A. Masoomi. "Geotechnical investigation and design of leaching heap No. 2, Meydook copper mine, Iran." *Minerals Engineering* 79 (2015): 185–195.

Mostafalou, S., and M. Abdollahi. "Concerns of environmental persistence of pesticides and human chronic diseases." *Clinical and Experimental Pharmacology* S5 (2012): e002.

Motelay-Massei, A., D. Ollivon, B. Garban, M. J. Teil, M. Blanchard, and M. Chevreuil. "Distribution and spatial trends of PAHs and PCBs in soils in the Seine River basin, France." *Chemosphere* 55, no. 4 (2004): 555–565.

Motghare, H., and R. Gauraha. "Biofertilizers-types and their application." *KrishiSewa*, 2012. http://www. krishisewa.com/articles/organic-agriculture/115-biofertilizers.html

Mouhamadou, Bello, Mathieu Faure, Lucile Sage, Johanna Marçais, Florence Souard, and Roberto A. Geremia. "Potential of autochthonous fungal strains isolated from contaminated soils for degradation of polychlorinated biphenyls." *Fungal Biology* 117, no. 4 (2013): 268–274.

Moya, Diego, Clay Aldás, Germánico López, and Prasad Kaparaju. "Municipal solid waste as a valuable renewable energy resource: A worldwide opportunity of energy recovery by using Waste-To-Energy Technologies." *Energy Procedia* 134 (2017): 286–295.

Muenmee, S., and W. Chiemchaisri. "Enhancement of biodegradation of plastic wastes via methane oxidation in semi-aerobic landfill." *Int Biodeterior Biodegr* 113 (2016): 244–255.

Muhr, P., and U. Rosenhall. "The influence of military service on auditory health and the efficacy of a Hearing Conservation Program." *Noise Health* 13 (2011): 320–327. [PubMed: 21768736].

Müller, F., and B. Burkhard. "The indicator side of ecosystem services." *Ecosystem Services* 1, no. 1 (2012): 26–30.

Müller, F., M. Bergmann, R. Dannowski, J. W. Dippner, A. Gnauck, P. Haase, . . . and M. Küster. "Assessing resilience in long-term ecological data sets." *Ecological Indicators* 65 (2016): 10–43.

Mullin, Barbara H. "Invasive plant species." *Council for Agricultural Science and Technology* no. 13 (2000).

Muraleedharan, H., S. Seshadri, and K. Perumal. *Biofertilizer (Phosphobacteria)*, Booklet published by Shri AMM Murugappa Chettiar Research Centre, Taramani, Chennai-600113, 2010.

Muratova, A., T. Hübner, S. Tischer, O. Turkovskaya, M. Möder, and P. Kuschk. "Plant-rhizosphere-microflora association during phytoremediation of PAH-contaminated soil." *International Journal of Phytoremediation* 5, no. 2 (2003): 137–151.

Murphy, Jerry D., and Eamon McKeogh. "Technical, economic and environmental analysis of energy production from municipal solid waste." *Renewable Energy* 29, no. 7 (2004): 1043–1057.

Myresiotis, C., Z. Vryzas, and E. Papadopoulou-Mourkidou. "Biodegradation of soil-applied pesticides by selected strains of plant growth-promoting rhizobacteria (PGPR) and their effects on bacterial growth." *Biodegradation* 23 (2012): 297–310.

NAAS.) *Efficient Utilization of Phosphorus.* Vol 68, Policy Paper. National Academy of Agricultural Sciences, New Delhi, 2014, p. 16.

Nadal, M., V. Kumar, M. Schuhmacher, and J. L. Domingo. "Definition and GIS-based characterization of an integral risk index applied to a chemical/petrochemical area." *Chemosphere* 64, no. 9 (2006): 1526–1535.

NAE (National Academy of Engineering). *Technology for a Quieter America.* National Academies Press, Washington, DC. Office for Europe, Copenhagen, Denmark, 2010.

Nagarajan, Rajkumar, Subramani Thirumalaisamy, and Elango Lakshumanan. "Impact of leachate on groundwater pollution due to non-engineered municipal solid waste land-fill sites of erode city, Tamil Nadu, India." *Iranian Journal of Environmental Health Science & Engineering* 9, no. 1 (2012): 35.

Nagpal, S., D. Donald, and O. Timothy. "Effect of carbon dioxide concentration on the bio-leaching of a pyrite arsenopyrite ore concentrates." *Biotechnology and Bioengineering* 41 (1993): 459–464.

Naing, W. K., M. Anees, H. X. Nyugen, S. Y. Lee, W. S. Jeon, Y. S. Kim, H. M. Kim, and Y. K. Kim. "Biocontrol of late blight diseases (*Phytophthora capsici*) of pepper and the plant growth promotion by Paenibacillus chimensis KWNJ8." *Journal of Phytopathology* 2 (2013): 164–165.

Narayan, R. "Biodegradable and biobased plastics: An overview." In: Malinconico, M. (ed.), *Soil Degradable Bioplastics for a Sustainable Modern Agriculture, Green Chemistry and Sustainable Technology.* Springer-Verlag GmbH, Germany, 2017.

National Centre of Organic Farming, Department of Agriculture & Cooperation (DAC), Indian Fertilizer Scenario 2013, Department of Fertilizers, Ministry of Chemicals and Fertilizers, Government of India.

National Centre of Organic Farming, Department of Agriculture & Cooperation (DAC), Indian Fertilizer Scenario 2014, Department of Fertilizers, Ministry of Chemicals and Fertilizers, Government of India.

National Centre of Organic Farming, Department of Agriculture & Cooperation (DAC), Indian Fertilizer Scenario 2015, Department of Fertilizers, Ministry of Chemicals and Fertilizers, Government of India.

National Research Council, & National Research Council. "Committee on improving risk analysis approaches used by the US EPA: Science and decisions: Advancing risk assessment." (2009).

Natarajan, K. A. "Biomineralization and biobeneficiation of bauxite." *Transactions of the Indian Institute of Metals* 69, no. 1 (2016): 15–21.

Navarro, M. C., C. Pérez-Sirvent, M. J. Martínez-Sánchez, J. Vidal, P. J. Tovar, and J. Bech. "Abandoned mine sites as a source of contamination by heavy metals: A case study in a semi-arid zone." *Journal of Geochemical Exploration* 96, no. 2–3 (2008): 183–193.

Naveen, B. P., J. Sumalatha, and R. K. Malik. "A study on contamination of ground and surface water bodies by leachate leakage from a landfill in Bangalore, India." *International Journal of Geo-Engineering* 9, no. 1 (2018): 27.

Nedelkoska, T. V., and P. M. Doran. "Characteristics of heavy metal uptake by plant species with potential for phytoremediation and phytomining." *Minerals Engineering* 13, no. 5 (2000): 549–561.

Nefzi, A., B. A. R. Abdallah, H. Jabnoun-Khiareddine, S. Saidiana-Medimagh, R. Haouala, and M. Danmi-Remadi. "Antifungal activity of aqueous and organic extracts from *Withania somnifera* L. against *Fusarium oxysporum* f. sp. *Radicis-Lycopersici*." *Journal of Microbial and Biochemical Technology* 8 (2016): 144–150.

Newell, Peter. "Globalization and the Governance of Biotechnology." *Global Environmental Politics* 3, no. 2 (2003): 56–71.

Nguyen, D. B., M. T. Rose, T. J. Rose, S. G. Morris, and L. Van Zwieten. "Impact of glyphosate on soil microbial biomass and respiration: A meta-analysis." *Soil Biology and Biochemistry* 92 (2016): 50–57.

Nikel, P. I., M. J. Pettinari, M. A. Galvagno, and B. S. Mendez. "Poly(3-hydroxybutyrate) synthesis by recombinant *Escherichia coli* arcA mutants in microaerobiosis." *Appl Environ Microbiol* 72 (2006): 2614–2620.

Nill, Kimball R. *Glossary of Biotechnology Terms.* 3rd ed. CRC Press LLC, 2002.

Noss, Reed F. "Indicators for monitoring biodiversity: A hierarchical approach." *Conservation Biology* 4, no. 4 (1990): 355–364.

Oerke, E. C. "Crop losses to pests." *J Agr Sci* 144 (2005): 31–43. of the United Nations.

Ogundele, D. T., A. A. Adio, and O. E. Oludele. "Heavy metal concentrations in plants and soil along heavy traffic roads in North Central Nigeria." *Journal of Environmental & Analytical Toxicology* 5, no. 6 (2015): 1.

Ogundele, L. T., O. K. Owoade, P. K. Hopke, and F. S. Olise. "Heavy metals in industrially emitted particulate matter in Ile-Ife, Nigeria." *Environmental Research* 156 (2017): 320–325.

Ohtake, Y., T. Kobayashi, H. Asabe, and N. Murakami. "Studies on biodegradation of LDPE-observation of LDPE films scattered in agricultural fields or in garden soil." *Polym Degrad Stabil* 60 (1998): 79–84.

Okang'Odumo, Benjamin, Gregoria Carbonell, Hudson Kalambuka Angeyo, Jayanti Purshottam Patel, Manuel Torrijos, and José Antonio Rodríguez Martín. "Impact of gold mining associated with mercury contamination in soil, biota sediments and tailings in Kenya." *Environmental Science and Pollution Research* 21, no. 21 (2014): 12426–12435.

Okunlola, A. I., and O. Akinrinnola. "Effectiveness of botanical formulations in vegetable production and bio-diversity preservation in Ondo State, Nigeria." *Journal of Horticulture and Forestry* 1 (2014): 6–13.

Olaniran, N. S. "Environment and health: An introduction." In: *Environment and Health.* Micmillan Nig. Pub. Co for NCF, Lagos, 1995, pp. 34–151.

Oostergo, H. "Milieudata, Emissies van metalen en PAK door wegverkeer [Emission of metals and PACs by road traffic]." By order of the Ministry of VROM, The Hague (1997).

Ouda, O. K. M., S. A. Raza, A. S. Nizami, M. Rehan, R. Al-Waked, and N. E. Korres. "Waste to energy potential: A case study of Saudi Arabia." *Renewable and Sustainable Energy Reviews* 61 (2016): 328–340.

Ouda, Omar K. M., Syed, A. Raza, Rafat Al-Waked, Jawad F. Al-Asad, and Abdul-Sattar Nizami. "Waste-to-energy potential in the Western Province of Saudi Arabia." *Journal of King Saud University-Engineering Sciences* 29, no. 3 (2017): 212–220.

Owa, F. W. "Water pollution: Sources, effects, control and management." *International Letters of Natural Sciences* 8 (2014): 1–6.

Owili, M. A. "Assesment of impact of sewage effluents on coastal water quality in Hafnarfjordur, Iceland." The United Nations Fishery Training Program, Final Report, 2003.

Paliwal, Ritu. "EIA practice in India and its evaluation using SWOT analysis." *Environmental Impact Assessment Review* 26, no. 5 (2006): 492–510.

Panda, H. *Handbook on Organic Farming and Processing.* Asia Pacific Business Press Inc., 2013, pp. 149–152.

Panda, H. *Manufacture of Biofertilizer and Organic Farming.* Asia Pacific Business Press Inc., 2011, pp. 103–121.

Pandey, Pramod K., Philip H. Kass, Michelle L. Soupir, Sagor Biswas, and Vijay P. Singh. "Contamination of water resources by pathogenic bacteria." *Amb Express* 4, no. 1 (2014): 51.

Pan-Germany. "Pesticide and health hazards." *Facts and Figures*, 2012, 1–16. www.pangermany.org/download/Vergift_EN-201112-web.pdf

Pappu, Asokan, Mohini Saxena, and Shyam R. Asolekar. "Solid wastes generation in India and their recycling potential in building materials." *Building and Environment* 42, no. 6 (2007): 2311–2320.

Parrish, Z. D., M. K. Banks, and A. P. Schwab. "Assessment of contaminant lability during phytoremediation of polycyclic aromatic hydrocarbon impacted soil." *Environmental Pollution* 137, no. 2 (2005): 187–197.

Parte, S. G., A. D. Mohekar, and A. S. Kharat. "Microbial degradation of pesticide: A review." *African Journal of Microbiology Research* 24, no. 11 (2017): 992–1012.

Passarini, M. R., M. V. Rodrigues, M. da Silva, and L. D. Sette. "Marine-derived filamentous fungi and their potential application for polycyclic aromatic hydrocarbon bioremediation." *Marine Pollution Bulletin* 62, no. 2 (2011): 364–370.

Passatore, L., S. Rossetti, A. A. Juwarkar, and A. Massacci. "Phytoremediation and bioremediation of polychlorinated biphenyls (PCBs): State of knowledge and research perspectives." *Journal of Hazardous Materials* 278 (2014): 189–202.

Patel, Bhargav C., Devayani R. Tipre, and Shailesh R. Dave. "Development of Leptospirillum ferriphilum dominated consortium for ferric iron regeneration and metal bioleaching under extreme stresses." *Bioresource Technology* 118 (2012): 483–489.

Pathak, Malabika Roy, and Mohammad S. Abido. "The role of biotechnology in the conservation of biodiversity." *Journal of Experimental Biology* 2 (2014): 4.

Patial, Vanit, Madhu Sharma, and Amita Bhattacharya. "Potential of thidiazuron in improved micropropagation of Picrorhiza kurroa-an endangered medicinal herb of alpine Himalaya." *Plant Biosystems: An International Journal Dealing with All Aspects of Plant Biology* 151, no. 4 (2017): 729–736.

Paul, D., G. Pandey, J. Pandey, and R. K. Jain. "Accessing microbial diversity for bioremediation and environmental restoration." *Trends in Biotechnology* 23, no. 3 (2005): 135–142.

Paunescu, Anca. "Biotechnology for endangered plant conservation: A critical overview." *Romanian Biotechnological Letters* 14, no. 1 (2009): 4095–4103.

Pavel, Lucian Vasile, and Maria Gavrilescu. "Overview of ex situ decontamination techniques for soil cleanup." *Environmental Engineering & Management Journal (EEMJ)* 7, no. 6 (2008).

Pawar, P. A., and A. H. Purwar. "Bioderadable polymers in food packaging." *Am J Eng Res* 2 (2013): 151–164.

Pawar, A. D., and B. Singh. "Prospects of botanicals and biopesticides." In: Parmar, B. S., Dev Kumar, C. P. (eds.), *Botanical and Biopesticide*, 1993, pp. 188–196.

Pedrero, F., I. Kalavrouziotis, J. J. Alarcón, P. Koukoulakis, and A. Takashi. "Use of treated municipal wastewater in irrigated agriculture: Review of some practices in Spain and Greece." *Agricultural Water Management* 97, no. 9 (2010): 1233–1241.

Peelman, N., P. Ragaert, B. De Meulenaer, D. Adons, R. Peeters, L. Cardon, F. Van Impe, and F. Devlieghere. "Application of bioplastics for food packaging." *Trends Food Sci Technol* 32 (2013): 128–141.

Pei, Shengji. "Ethnobotany and modernisation of traditional Chinese medicine." Paper at a Workshop on Wise Practices and Experiential Learning in the Conservation and Management of Himalayan Medicinal Plants, Kathmandu, Nepal, 2002, pp. 15–20.

Peixoto, R., A. Vermelho, and A. S. Rosado. "Petroleum-degrading enzymes: Bioremediation and new prospects." *Enzyme Research* (2011): 1–7.

Pemberton, D., N. P. Brothers, and R. Kirkwood. "Entanglement of Australian fur seals in man-made debris in Tasmanian waters." *Wildl Res.* 19 (1992): 151–159.

Peng, J., Y. Zhang, J. Su, Q. Qiu, Z. Jia, and Y. G. Zhu. "Bacterial communities predominant in the degradation of 13C(4)-4,5,9,10-pyrene during composting." *Bioresour. Technol.* 143 (2013): 608–614.

Peng, W., X. Li, J. Song, W. Jiang, Y. Liu,&and W. Fa). "Bioremediation of cadmium-and zinc-contaminated soil using Rhodobacter sphaeroides." *Chemospher*, 197 (2018), 33–41.

Perdue, W. C., L. A. Stone, and L. O. Gostin. "The built environment and its relationship to the public's health: The legal framework." *Am J Public Health* 93 (2003): 1390–1394.

Pérez-de-Luque, Alejandro, and Diego Rubiales. "Nanotechnology for parasitic plant control." *Pest Management Science: Formerly Pesticide Science* 65, no. 5 (2009): 540–545.

Perkins, Devin N., Marie-Noel Brune Drisse, Tapiwa Nxele, and Peter D. Sly. "E-waste: A global hazard." *Annals of Global Health* 80, no. 4 (2014): 286–295.

Piccardo, Maria Teresa, Mauro Pala, Bruna Bonaccurso, Anna Stella, Anna Redaelli, Gaudenzio Paola, and Federico Valerio. "*Pinus nigra* and *Pinus pinaster* needles as passive samplers of polycyclic aromatic hydrocarbons." *Environmental Pollution* 133, no. 2 (2005): 293–301.

Pichersky, E., and J. Gershenzon. "The formation and function of plant volatiles: Perfumes for pollinator attraction and defense." *Curr. Opin. Plant Biol* 5 (2002): 237–243.

Pilla, S. "Engineering applications of bioplastics and biocomposites: An ovverview." In: Pilla, S. (ed.), *Handbook of Bioplastics and Biocomposites Engineering Applications*. Scrivener Publishing LLC, 2011, pp. 1–16.

Pilon-Smits, E. A. H., and J. L. Freeman. "Environmental cleanup using plants: Biotechnological advances and ecological considerations." *Frontiers in Ecology and the Environment* 4, no. 4 (2006): 203–210.

Pimm, Stuart L., and Peter Raven. "Biodiversity: Extinction by numbers." *Nature* 403, no. 6772 (2000): 843.

Pindi, P. K., and Satyanarayana, S. D. V. "Liquid microbial consortium: A potential tool for sustainable soil health." *Journal of Biofertilizers &Biopesticides* 3, no. 4 (2012): 124. doi: 10.4172/2155–6202.1000124

Planning Commission, Government of India. Report of the task Force on waste to energy (Volume I) in the context of integrated municipal solid waste management, 2014.

Platt, D. K. *The Starch-Based Biodegradable Polymer Market: Applications and Market.* iSmithers Rapra Publishing, Shawbury, UK, 2006.

Pollard, A. J., K. D. Powell, F. A. Harper, and J. Andrew C. Smith. "The genetic basis of metal hyperaccumulation in plants." *Critical Reviews in Plant Sciences* 21, no. 6 (2002): 539–566.

Ponce, Rafael, Leslie Abad, Lakshmi Amaravadi, Thomas Gelzleichter, Elizabeth Gore, James Green, Shalini Gupta et al. "Immunogenicity of biologically-derived therapeutics: Assessment and interpretation of nonclinical safety studies." *Regulatory Toxicology and Pharmacology* 54, no. 2 (2009): 164–182.

Popp, J. "Cost-benefit analysis of crop protection measures." *Journal of Consumer Protection and Food Safety* 6, Supplement 1 (2011): 105–112. doi: 10.1007/s00003-011-0677-4, Springer, May.

Popp, J., K. Pető, and J. Nagy. "Pesticide productivity and food security: A review." *Agronomy for Sustainable Development* 33, no. 1 (2013): 243–255.

Prasad, M. N. V. "Sunflower (*Helinathus annuus* L.): A potential crop for environmental industry." *Helia* 30 (2007): 167–174.

Prasad, Rajesh Kumar, Soumya Chatterjee, Pranab Behari Mazumder, Santosh Kumar Gupta, Sonika Sharma, Mohan Gunvant Vairale, Sibnarayan Datta, Sanjai Kumar Dwivedi, and Dharmendra Kumar Gupta. "Bioethanol production from waste lignocelluloses: A review on microbial degradation potential." *Chemosphere* (2019).

Pray, L., and K. Zhaurova. "Barbara McClintock and the discovery of jumping genes (transposons)." *Nature Education* 1, no. 1 (2008): 16.

Pretty, J. *The Pesticide Detox, towards a More Sustainable Agriculture.* Earthscan, London, 2009.

PTV Vision: Visum 11.5 - Basics. PTV AG. 2010. Karlsruhe.

PTV Visum Overview. 2004. PTV America.

Public Health Problem of the International Commission on Biological Effects of Noise. London, UK, 2011. http://www.icben.org/proceedings.html

Puentes, Rodrigo, Agustín Furtado, Maria José Estradé, Rafael Aragunde, Nicolás Cazales, Gabriela Costa, Jorge Estévez, Fernando Cirilo, and Cecilia Galosi. "Tissue preservation and cell culture of Przewalskii's horse (*Equus przewalskii*): An endangered species." *Bioscience Journal* 27, no. 5 (2011).

Puglisi, E. "Response of microbial organisms (aquatic and terrestrial) to pesticides." *EFSA Supporting Publications* 9, no. 11 (2012): 359E.

Puglisi, E., F. Cappa, G. Fragoulis, M. Trevisan, and A. A. Del Re). "Bioavailability and degradation of phenanthrene in compost amended soils." *Chemosphere* 67 (2007): 548–556.

Pussente, Igor C., Guillaume ten Dam, Stefan van Leeuwen, and Rodinei Augusti. "PCDD/Fs and PCBs in soils: A study of case in the City of Belo Horizonte-MG." *Journal of the Brazilian Chemical Society* 28, no. 5 (2017): 858–867.

Qian, X., C. Fang, M. Huang, and V. Achal. "Characterization of fungal-mediated carbonate precipitation in the biomineralization of chromate and lead from an aqueous solution and soil." *Journal of Cleaner Production* 164 (2017): 198–208.

Qin, R., Y. Hirano, and I. Brunner. "Exudation of organic acid anions from poplar roots after exposure to Al, Cu and Zn." *Tree Physiology* 27, no. 2 (2007): 313–320.

Qishlaqi, A., F. Moore, and G. Forghani. "Impact of untreated wastewater irrigation on soils and crops in Shiraz suburban area, SW Iran." *Environmental Monitoring and Assessment* 141, no. 1–3 (2008): 257–273.

Qu, J., Y. Xu, G. M. Ai, Y. Liu, and Z. P. Liu. "Novel chryseobacterium sp. PYR2 degrades various organochlorine pesticides OCPs and achieves enhancing removal and complete degradation of DDT in highly contaminated soil." *Journal of Environmental Economics and Management* 161 (2015): 350–357.

Quillaguaman, J., O. Delgado, B. Mattiasson, and R. Hatti-Kaul. "Poly(β-hydroxybutyrate) production by a moderate halophile, *Halomonas boliviensis* LC1." *Enz Microb Technol* 38 (2006): 148–154.

Quillaguaman, J., M. Munoz, B. Mattiasson, and R. Hatti-Kaul. "Optimizing conditions for poly(β-hydroxybutyrate) production by *Halomonas boliviensis* LC1 in batch culture with sucrose as carbon source." *Appl Microbiol Biotechnol* 74 (2007): 981–986.

Quraishi, Afaque, Snigdha Mehar, Durga Sahu, and Shailesh Kumar Jadhav. "In vitro mid-term conservation of *Acorus calamus* L. via cold storage of encapsulated microrhizome." *Brazilian Archives of Biology and Technology* 60 (2017).

Radhika, Kanupriya, and V. Radhika. "Role of biotechnology in conservation and utilization of agricultural biodiversity." *Current Science (00113891)* 115, no. 11 (2018).

Raghukumar, Chandralata, C. Mohandass, Shilpa Kamat, and M. S. Shailaja. "Simultaneous detoxification and decolorization of molasses spent wash by the immobilized white-rot fungus Flavodon flavus isolated from a marine habitat." *Enzyme and Microbial Technology* 35, no. 2–3 (2004): 197–202.

Raiand, R., and I. Roy. "Polyhydroxyalkanoates: The emerging new green polymers of choice." In: Sharma, S. K. and Mudhoo, A. (eds.), *A Handbook of Applied Biopolymer Technology: Synthesis, Degradation and Applications*. Royal Society of Chemistry, London, 2011, pp. 79–101.

Raja, N. "Biopesticides and biofertilizers: Ecofriendly sources for sustainable agriculture." *J Biofertil Biopestici* 4, no. 1 (2013).

Rajasekhar, M., N. Venkat Rao, G. Chinna Rao, G. Priyadarshini, and N. Jeevan Kumar. "Energy generation from municipal solid waste by innovative technologies-plasma gasification." *Procedia Materials Science* 10 (2015): 513–518.

Rajput, V., T. Minkina, A. Fedorenko, S. Sushkova, S. Mandzhieva, V. Lysenko, N. Duplii et al. "Toxicity of copper oxide nanoparticles on spring barley (*Hordeum Sativum Distichum*)." *Science of the Total Environment* 645 (2018): 1103–1113.

Rajput, V., T. Minkina, B. Ahmed, S. Sushkova, R. Singh, M. Soldatov, B. Laratte et al. "Interaction of copper-based nanoparticles to soil, terrestrial, and aquatic systems: Critical review of the state of the science and future perspectives." *Reviews Environmental Contamination Toxicology* (2019): 1–46. https://doi.org/10.1007/398_2019_34

Rakosy-Tican, Elena, Barna Bors, and Ana-Maria Szatmari. "In vitro culture and medium-term conservation of the rare wild species Gladiolus imbricatus." *African Journal of Biotechnology* 11, no. 81 (2012): 14703–14712.

Ramakrishnan, N., and S. K. Kumar. *Biological Control of Insects by Pathogen and Nematodes*, Pesticides, 1976, pp. 32–47.

Ramdani, A., P. Dold, S. Déléris, D. Lamarre, A. Gadbois, and Y. Comeau. "Biodegradation of the endogenous residue of activated sludge." *Water Research* 44, no. 7 (2010): 2179–2188.

Rameshkumar, Ramakrishnan, M. J. V. Largia, Lakkakula Satish, Jayabalan Shilpha, and Manikandan Ramesh. "In vitro mass propagation and conservation of *Nilgirianthus ciliatus* through nodal explants: A globally endangered, high trade medicinal plant of Western Ghats." *Plant Biosystems: An International Journal Dealing with All Aspects of Plant Biology* 151, no. 2 (2017): 204–211.

Ranade, Pinak, and Geeta Bapat. "Estimation of power generation from solid waste generated in sub urban area using spatial techniques: A case study for Pune City, India." *International Journal of Geomatics and Geosciences* 2, no. 1 (2011): 179–187.

Rao, D. L. N. "Recent advances in biological nitrogen fixation in agricultural systems." *Proc Indian Natl Sci Acad* 80, no. 2 (2014): 359–378.

Rasal, R. M., A. V. Janorkar, and D. E. Hirt. "Poly(lactic acid) modifications." *Prog Polym Sci* 35 (2010): 338–356.

Rastegar, S. O., S. M. Mousavi, and S. A. Shojaosadati. "Cr and Ni recovery during bio-leaching of dewatered metal-plating sludge using Acidithiobacillus ferrooxidans." *Bioresource Technology* 167 (2014): 61–68.

Ravallion, M., and G. Datt. "How important to India's poor is the sectoral composition of economic growth?" *World Bank Econ Rev* 10 (1996): 1–25.

Rawat, Janhvi Mishra, Aakriti Bhandari, Susmita Mishra, Balwant Rawat, Ashok Kumar Dhakad, Ajay Thakur, and Anup Chandra. "Genetic stability and phytochemical pro-filing of the in vitro regenerated plants of *Angelica glauca* Edgew: An endangered medicinal plant of Himalaya." *Plant Cell, Tissue and Organ Culture (PCTOC)* 135, no. 1 (2018): 111–118.

Rawlings, D. E. "Heavy metal mining using microbes." *Annual Reviews in Microbiology* 56 (2002): 65–91.

Rea, P. A. "Plant ATP-binding cassette transporters." *Annu. Rev. Plant Biol.* 58 (2007): 347–375.

Redford, Kent H., and Brian D. Richter. "Conservation of biodiversity in a world of use." *Conservation Biology* 13, no. 6 (1999): 1246–1256.

Reed, Barbara M., Viswambharan Sarasan, Michael Kane, Eric Bunn, and Valerie C. Pence. "Biodiversity conservation and conservation biotechnology tools." *In Vitro Cellular & Developmental Biology-Plant* 47, no. 1 (2011): 1–4.

Rehmann, K., H. P. Noll, C. E.W Steinberg, and A. A. Kettrup. "Pyrene degradation by Mycobacterium sp. strain KR2." *Chemosphere* 36, no. 14 (1998): 2977–2992.

Reichenauer, T. G., and J. J. Germida. "Phytoremediation of organic contaminants in soil and groundwater." *ChemSusChem: Chemistry and Sustainability Energy and Materials* 1, no. 8–9 (2008): 708–717.

"Review article advancements in life sciences." *International Quarterly Journal of Biological Sciences* 5, no. 1 (November): 13–18.

Rey, Federico E., Erin K. Heiniger, and Caroline S. Harwood. "Redirection of metabolism for biological hydrogen production." *Appl. Environ. Microbiol.* 73, no. 5 (2007): 1665–1671.

Rezania, S., M. Ponraj, A. Talaiekhozani, S. E. Mohamad, M. F. M. Din, S. M. Taib, and F. M. Sairan. "Perspectives of phytoremediation using water hyacinth for removal of heavy metals, organic and inorganic pollutants in wastewater." *Journal of Environmental Management* 163 (2015): 125–133.

Rhim, J. W., S. I. Hong, and C. S. Ha. "Tensile, water vapour barrier and antimicrobial prop-erties of PLA/nanoclay composite films." *Food Sci Technol* 42 (2009): 612–617.

Rice, Kevin M., Ernest M. Walker Jr, Miaozong Wu, Chris Gillette, and Eric R. Blough. "Environmental mercury and its toxic effects." *Journal of Preventive Medicine and Public Health* 47, no. 2 (2014): 74–83.

Rich, N. "The lawyer who became DuPont's worst nightmare." *New York Times Magazine* 6 (2016).

Richardson, S. D., and T. A. Ternes. "Water analysis: Emerging contaminants and current issues." *Analytical Chemistry* 83, 12 (2011): 4614–4648.

Rico, A., P. J. Van den Brink, R. Gylstra, A. Focks, and T. C. Brock. "Developing ecologi-cal scenarios for the prospective aquatic risk assessment of pesticides." *Integrated Environmental Assessment and Management* 12, no. 3 (2016): 510–521.

Rissato, Sandra R., Mário S. Galhiane, Valdecir F. Ximenes, Rita M. B. De Andrade, Jandira L. B. Talamoni, Marcelo Libânio, Marcos V. De Almeida, Benhard M. Apon, and Aline A. Cavalari. "Organochlorine pesticides and polychlorinated biphenyls in soil and water samples in the Northeastern part of São Paulo State, Brazil." *Chemosphere* 65, no. 11 (2006): 1949–1958.

Rivard, C., L. Moens, K. Roberts, J. Brigham, and S. Kelley. "Starch esters as biodegradable plastics: Effects of ester group chain length and degree of substitution on anaerobic biodegradation." *Enz Microbial Tech* 17 (1995): 848–852.

Robertson, G. "State of the art biobased food packaging materials." In: Chiellini, E. (ed.), *Environmentally Compatible Food Packaging*. CRC Press, Boca Raton, FL, 2008, pp. 3–28.

Rockne, Karl J., Joanne C. Chee-Sanford, Robert A. Sanford, Brian P. Hedlund, James T. Staley, and Stuart E. Strand. "Anaerobic naphthalene degradation by microbial pure cultures under nitrate-reducing conditions." *Applied and Environmental Microbiology* 66, no. 4 (2000): 1595–1601.

Rodríguez-Eugenio, N., M. McLaughlin, and Daniel Pennock. *Soil Pollution: A Hidden Reality*. FAO, 2018.

Rohr, J. R., C. J. Salice, and R. M. Nisbet. "The pros and cons of ecological risk assessment based on data from different levels of biological organization." *Critical Reviews in Toxicology* 46, no. 9 (2016): 756–784.

Rohrbacher, F., and M. St-Arnaud. "Root exudation: The ecological driver of hydrocarbon rhizoremediation." *Agronomy* 6, no. 19 (2016): 1–27.

Rohrmann, G. F. *Baculovirus Molecular Biology: Third Edition (Internet)*. National Center for Biotechnology Information (US), Bethesda, MD, 2013. http://www.ncbi.nlm.nih.gov/books/NBK114593/

Rokhzadi, A., A. Asgharzade., F. Darvis., G. Nourmohammadi, and E. MajidE) "Influence of plant growth-promoting rhizobacteria on dry matter accumulation and yield of chickpea (Cicer arietinum L.) under field condition." *Am-Euras J Agric Environ Sci* 3, no. 2 (2008): 253–257.

Rome, Ros, M., I. Rodríguez, C. García, and T. Hernández. "Microbial communities involved in the bioremediation of an aged recalcitrant hydrocarbon polluted soil by using organic amendments." *Bioresource Technology* 101, no. 18 (2010): 6916–6923.

Rose, A., O. Kehinde, and A. Babajide. "The level of persistent, bioaccumulative, and toxic (PBT) organic micropollutants contamination of Lagos soils." *Journal of Environmental Chemistry and Ecotoxicology* 5, no. 2 (2013): 26–38.

Roy, A., B. Moktan, and P. K. Sarkar. "Characteristics of *Bacillus cereus* isolates from legume-based Indian fermented foods." *Food Contr* 18 (2007): 1555–1564.

Rugh, C. L., H. D. Wilde, N. M. Stack, D. M. Thompson, A. O. Summers, and R. B. Meagher. "Mercuric ion reduction and resistance in transgenic Arabidopsis thaliana plants expressing a modified bacterial merA gene." *Proceedings of the National Academy of Sciences* 93, no. 8 (1996): 3182–3187.

Rujnic-Sokele, M., and A. Pilipovic. "Challenges and opportunities of biodegradable plastics: A mini review." *Waste Manag Res* 35 (2017): 132–140.

Ryder, Oliver A. "Cloning advances and challenges for conservation." *Trends in Biotechnology* 20, no. 6 (2002): 231–232.

Ryder, Oliver Allison, and Manabu Onuma. "Viable cell culture banking for biodiversity characterization and conservation." *Annual Review of Animal Biosciences* 6 (2018): 83–98.

Keziah Merlyn, S., and C. Subathra Devi. "Essentials of conservation biotechnology: A mini review." *Materials Science and Engineering Conference Series* 263, no. 2 (2017): 022047.

Namba, S., S. Kuwano, and T. Okamoto. "Sleep disturbance caused by meaningful sounds and the effect of background noise." *Journal of Sound and Vibration* 277 (2004): 445–452.

Sadegh, Hamidreza, Gomaa A. M. Ali, Vinod Kumar Gupta, Abdel Salam Hamdy Makhlouf, Ramin Shahryari-ghoshekandi, Mallikarjuna N. Nadagouda, Mika Sillanpää, and Elżbieta Megiel. "The role of nanomaterials as effective adsorbents and their applications in wastewater treatment." *Journal of Nanostructure in Chemistry* 7, no. 1 (2017): 1–14.

Sadeq, Manal Ahmed, Malabika Roy Pathak, Ah med Ali Salih, Mohammed Abido, and Asma Abahussain. "Effect of plant growth regulators on regeneration of the endangered medicinal plant *Calligonum comosum* L. Henry in the Kingdom of Bahrain." *African Journal of Biotechnology* 13, no. 25 (2014).

Sahney, Sarda, Michael J. Benton, and Howard Falcon-Lang. "Rainforest collapse triggered Carboniferous tetrapod diversification in Euramerica." *Geology* 38, no. 12 (2010): 1079–1082.

Said, Omar, Khaled Khalil, Stephen Fulder, and Hassan Azaizeh. "Ethnopharmacological survey of medicinal herbs in Israel, the Golan Heights and the West Bank region." *Journal of Ethnopharmacology* 83, no. 3 (2002): 251–265.

Saikia, S. P., and V. Jain. "Biological nitrogen fixation with non-legumes: An achievable target or a dogma." *Curr Sci* 92, no. 3 (2007): 317–322.

Sajjadi, Baharak, Wei-Yin Chen, Abdul Aziz Abdul Raman, and Shaliza Ibrahim. "Microalgae lipid and biomass for biofuel production: A comprehensive review on lipid enhancement strategies and their effects on fatty acid composition." *Renewable and Sustainable Energy Reviews* 97 (2018): 200–232.

Sakaya, K., D. A. Salam, and P. Campo. "Assessment of crude oil bioremediation potential of seawater and sediments from the shore of Lebanon in laboratory microcosms." *Science of the Total Environment* 660 (2019): 227–235.

Salt, D. E., R. C. Prince, I. J. Pickering, and I. Raskin. "Mechanisms of cadmium mobility and accumulation in Indian mustard." *Plant Physiology* 109, no. 4 (1995): 1427–1433.

Salt, D. E., R. D. Smith, and I. Raskin. "Phytoremediation." *Annual Review of Plant Biology* 49, no. 1 (1998): 643–668.

Samardjieva, K. A., R. F. Gonçalves, P. Valentão, P. B. Andrade, J. Pissarra, S. Pereira, and F. Tavares. "Zinc accumulation and tolerance in *Solanum nigrum* are plant growth dependent." *International Journal of Phytoremediation* 17, no. 3 (2015): 272–279.

Saratale, R. G., G. D. Saratale, J. S. Chang, and S. P. Govindwar. "Bacterial decolorization and degradation of azo dyes: A review." *Journal of the Taiwan Institute of Chemical Engineers* 42, 1 (2011): 138–157.

Saravi, S. S. S., and M. Shokrzadeh. "Role of pesticides in human life in the modern age: A review." In: Stoytcheva, M. (ed.), *Pesticides in the Modern World-Risks and Benefits.* In-Tech, 2011, pp. 4–11.

Sax, Dov F., and Steven D. Gaines. "Species diversity: From global decreases to local increases." *Trends in Ecology & Evolution* 18, no. 11 (2003): 561–566.

Sax, Dov F., Steven D. Gaines, and James H. Brown. "Species invasions exceed extinctions on islands worldwide: A comparative study of plants and birds." *The American Naturalist* 160, no. 6 (2002): 766–783.

Saxena, R. C., D. K. Adhikari, and H. B. Goyal. "Biomass-based energy fuel through biochemical routes: A review." *Renewable and Sustainable Energy Reviews* 13, no. 1 (2009): 167–178.

Saxena, Shikha, R. K. Srivastava, and A. B. Samaddar. "Sustainable waste management issues in India." *IUP Journal of Soil & Water Sciences* 3, no. 1 (2010).

Sazakli, E., G. Siavalas, A. Fidaki, K. Christanis, H. K. Karapanagioti, and M. Leotsinidis. "Concentrations of persistent organic pollutants and organic matter characteristics as river sediment quality indices." *Toxicological and Environmental Chemistry* 98, no. 7 (2016): 787–799.

Sazima, I., O. B. F. Gadig, R. C. Namora, and F. S. Motta. "Plastic debris collars on juvenile carcharhinid sharks (*Rhizoprionodon lalandii*) in southwest Atlantic." *Mar. Pollut. Bull.* 44 (2002): 1147–1149.

Scheffer, M., J. Bascompte, W. A. Brock, V. Brovkin, S. R. Carpenter, V. Dakos, . . . and G. Sugihara. "Early-warning signals for critical transitions." *Nature* 461, no. 7260 (2009): 53.

Schippers, A., and W. Sand. "Bacterial leaching of metal sulfides proceeds by two indirect mechanisms via thiosulfate or via polysulfides and sulfur." *Applied Environmental Microbiology* 65 (1999): 319–321.

Schmidt, J. H., E. R. Pedersen, P. M. Juhl et al. "Sound exposure of symphony orchestra musicians." *Ann OccupHyg* 55 (2011): 893–905. [PubMed: 21841154].

Schneider, J., R. Grosser, K. Jayasimhulu, W. Xue, and D. Warshawsky. "Degradation of pyrene, benz[a]anthracene, and benzo[a]pyrene by *Mycobacterium* sp. strain RJGII-135, isolated from a former coal gasification site." *Applied and Environmental Microbiology* 62, no. 1 (1996): 13–19.

Schnoor, J. L., L. A. Licht, S. C. McCutcheon, N. L. Wolfe, and L. H. Carreira. "Phytoremediation of organic and nutrient contaminants." *Environmental Science and Technology* 29, no. 7 (1995): 318A–323A.

Schroder, P., and C. D. Collins. *Organic Xenobiotics and Plants*. Springer Press, University of Groningen, Groningen, Holand, 2011, p. 308.

Schröter, M., A. Bonn, S. Klotz, R. Seppelt, and C. Baessler, eds. *Atlas of Ecosystem Services: Drivers, Risks, and Societal Responses*. Springer, London, 2019.

Schröter, M., A. Bonn, S. Klotz, R. Seppelt, and C. Baessler. "Ecosystem services: Understanding drivers, opportunities, and risks to move towards sustainable land management and governance." In: *Atlas of Ecosystem Services*. Springer, Cham, 2019, pp. 401–403.

Schulze, T., A. Bahlmann, P. I. Bustos, C. Hug, M. Krauss, and K. H. Walz. "AutomatisiertegroßvolumigeFestphasenextraktion (LVSPE) für das effekt-basierte Monitoring von Oberflächen-, Grund-und Abwässern." *SETAC GLB Poster* 1663 (2012).

Scientific Committee on Emerging and Newly Identified Health Risks (SCENIHR) of the European Commission. "Potential health risks of exposure to noise from personal music players and mobile phones including a music playing function." 2008. http://ec.europa.eu/health/ph_risk/committees/04_scenihr/docs/scenihr_o_017.pdf accessed on 11 July 2013.

Secretariat of the Convention on Biological Diversity. "Sustaining life on earth: How the convention on biological diversity promotes nature and humal well-being." Secretariat of the Convention on Biological Diversity, 2000.

Secretariat, C. B. D. "Global biodiversity outlook 3 secretariat of the Convention on Biological Diversity (CBD) Montreal." (2010).

Seebens, Hanno, Tim M. Blackburn, Ellie E. Dyer, Piero Genovesi, Philip E. Hulme, Jonathan M. Jeschke, Shyama Pagad et al. "No saturation in the accumulation of alien species worldwide." *Nature Communications* 8 (2017): 14435.

Segner, H. "Moving beyond a descriptive aquatic toxicology: The value of biological process and trait information." *Aquatic Toxicology* 105, no. 3–4 (2011): 50–55.

Semeniuc, C. A., C. R. Pop, and A. M. Rotar. "Antibacterial activity and interactions of plant essential oil combinations gainst gram-positive and gram-negative bacteria." *Journal of Food and Drug Analysis* 25 (2017): 403–408.

Seo, J. S., Y. S. Keum, and Q. Li. "Bacterial degradation of aromatic compounds." *International Journal of Environmental Research and Public Health* 6, no. 1 (2009): 278–309.

Sesan, T. E., E. Enache, M. Iacomi, M. Oprea, F. Oancea, and C. Iacomi. "Antifungal activity of some plant extract against botrytis cinerea Pers. in the Blackcurrant Crop (Ribes nigrum L)." *Acta Scientiarum Polonorum Technologia Alimentaria* 1 (2015): 29–43.

Shah, K., and J. M. Nongkynrih. "Metal hyperaccumulation and bioremediation." *Biologia Plantarum* 51, no. 4 (2007): 618–634.

Shah, P. A., and M. S. Goettel. *Directory of Microbial Control Products and Services.* Microbial Control Division, Society for Invertebrate Pathology, Gainesville, FL, 1999, p. 31.

Shah, Syed Naseer, Amjad M. Husaini, and Fatima Shirin. "Micropropagation of the Indian birthwort arsitolochia indica L." *Int J Biotechnol Mol Biol Res* 4 (2013): 86–92.

Shann, J. R. "The role of plants and plant/microbial systems in the reduction of exposure." *Environmental Health Perspective* 103, no. 5 (1995): 13–15.

Shapiro, S. A. *The Dormant Noise Control Act and Options to Abate Noise Pollution.* Washington, DC: Noise Pollution Clearinghouse, 1991. http://www.nonoise.org/library/shapiro/shapiro.htm accessed on 16 November 2013.

Shapiro-Ilan, D. I., D. H. Gough, S. J. Piggott, and Fife Patterson J. "Application technology and environmental considerations for use of entomopathogenic nematodes in biological control." *Biological Control* 38 (2006): 124–133.

Sharholy, Mufeed, Kafeel Ahmad, Gauhar Mahmood, and R. C. Trivedi. "Municipal solid waste management in Indian cities: A review." *Waste Management* 28, no. 2 (2008): 459–467.

Sharma, B., A. K. Dangi, and P. Shukla. "Contemporary enzyme based technologies for bioremediation: A review." *Journal of Environmental Management* 210 (2018): 10–22.

Sharma, Dushyant Kumar, and Tripti Sharma. "Biotechnological approaches for biodiversity conservation." *Indian Journal of Scientific Research* 4, no. 1 (2013): 183.

Sharma, S., and P. Malik. "Biopestcides: Types and applications." *International Journal of Advances in Pharmacy, Biology and Chemistry* 1, no. 4 (2012): 508–515.

Sharma, S., B. Singh, and V. K. Manchanda. "Phytoremediation: Role of terrestrial plants and aquatic macrophytes in the remediation of radionuclides and heavy metal contaminated soil and water." *Environmental Science and Pollution Research* 22, no. 2 (2014): 946–962.

Shazwin, Taib Mat, and Nobukazu Nakagoshi. "Sustainable waste management through international cooperation: Review of comprehensive waste management technique and training course." *J. Int. Dev. Coop* 16, no. 1 (2010): 23–33.

Shekhawat, Mahipal S., M. Manokari, and C. P. Ravindran. "Micropropagation, micromorphological studies, and in vitro flowering in rungia pectinata L." *Scientifica* 2016 (2016).

Sheldon, Ian M. "Regulation of biotechnology: Will we ever 'freely' trade GMOs?" *European Review of Agricultural Economics* 29, no. 1 (2002): 155–176.

Sheoran, V., A. S. Sheoran, and P. Poonia. "Phytomining: A review." *Minerals Engineering* 22, no. 12 (2009): 1007–1019.

Sherburne, L. A., J. D. Shrout, and P. J. J. Alvarez. "Hexahydro-1,3,5-trinitro-1,3,5-triazine (RDX) degradation by Acetobacterium paludosum." *Biodegradation* 16, no. 6 (2005): 539–547.

Sherif, N. Ahamed, J. H. Franklin Benjamin, Thirupathi Senthil Kumar, and Mandali Venkateswara Rao. "Somatic embryogenesis, acclimatization and genetic homogeneity assessment of regenerated plantlets of Anoectochilus elatus Lindl., an endangered terrestrial jewel orchid." *Plant Cell, Tissue and Organ Culture (PCTOC)* 132, no. 2 (2018): 303–316.

Shi. Q., C. Chen, L. Gao, and L. Jiao. "Physical and degradation properties of binary or ternary blends composed of poly (lactic acid), thermoplastic starch and GMA grafted POE." *Polym Degrad Stab.* 96 (2011): 175–182.

Shimao, M. "Biodegradation of plastics." *Curr Opinion Biotechnol* 12 (2001): 242–247.

Shiva, Vandana. "Protecting our biological and intellectual heritage in the age of biopiracy." Paper Prepared for the Seminar on IPRs, Community Rights and Biodiversity: A New Partnership for National Sovereignty Held at New Delhi, 20 February 1996. Research Foundation for Science, Technology and Natural Resource Policy.

Shuba, Eyasu Shumbulo, and Demeke Kifle. "Microalgae to biofuels: 'Promising' alternative and renewable energy, review." *Renewable and Sustainable Energy Reviews* 81 (2018): 743–755.

Siddique, M. H., A. Kumar, K. K. Kesari, and J. M. Arif. "Biomining: A useful approach toward metal extraction." *American-Eurasian Journal of Agronomy* 2, no. 2 (2009): 84–88.

Siddiqui, Mohd Haris, Ashish Kumar, Kavindra Kumar Kesari, and Jamal M. Arif. "Biomining: A useful approach toward metal extraction." *American-Eurasian Journal of Agronomy* 2, no. 2 (2009): 84–88.

Siddiqui, Wequar Ahmad, and Rajiv Ranjan Sharma. "Assessment of the impact of industrial effluents on groundwater quality in Okhla industrial area, New Delhi, India." *Journal of Chemistry* 6, no. S1 (2009): S41–S46.

Sidle, Roy C., J. E. Hook, and L. T. Kardos. "Accumulation of heavy metals in soils from extended wastewater irrigation." *Journal (Water Pollution Control Federation)* (1977): 311–318.

Siegel, J. P., J. V. Maddox, and W. G. Ruesink. *Journal of Invertebrate Pathology* 48 (1986): 167–173.

Sigler, Michelle. "The effects of plastic pollution on aquatic wildlife: Current situations and future solutions." *Water, Air, & Soil Pollution* 225, no. 11 (2014): 2184.

Silva, Vera, Hans G. J. Mol, Paul Zomer, Marc Tienstra, Coen J. Ritsema, and Violette Geissen. "Pesticide residues in European agricultural soils: A hidden reality unfolded." *Science of the Total Environment* 653 (2019): 1532–1545.

Simberloff, Daniel. "Introduced species: The threat to biodiversity and what can be done." *Action Bioscience.* org original article (2000).

Simonich, S. L., and R. A. Hites. "Organic pollutant accumulation in vegetation." *Environmental Science and Technology* 29, no. 12 (1995): 2905–2914.

Sindhu, Raveendran, Parameswaran Binod, Ashok Pandey, Snehalata Ankaram, Yumin Duan, and Mukesh Kumar Awasthi. "Biofuel production from biomass: Toward sustainable development." In *Current Developments in Biotechnology and Bioengineering.* Elsevier, 2019, pp. 79–92.

Singh, G., A. Kumari, A. Mittal, A. Yadav, and N. K. Aggarwal. "Poly β-hydroxybutyrate production by Bacillus subtilis NG220 using sugar industry waste water." *Biomed Res Int* (2013): 1–10.

Singh, A., and V. K. Srivastava. "Toxic effect of synthetic pyrethroid permethrin on the enzyme system of the freshwater fish *Channa striatus.*" *Chemosphere* 39 (1999): 1951–1956.

Singh, B. K., A. Walker, J. A. W. Morgan, and D. J. Wright. "Biodegradation of chlorpyrifos by enterobacter strain B-14 and its use in bioremediation of contaminated soils." *Applied and Environmental Microbiology* 70, no. 8 (2004): 4855–4863.

Singh, Brajesh K., Allan Walker, J. Alun W. Morgan, and Denis J. Wright. "Biodegradation of chlorpyrifos by Enterobacter strain B-14 and its use in bioremediation of contaminated soils." *Appl. Environ. Microbiol.* 70, no. 8 (2004): 4855–4863.

Singh, D. K., S. K. Singh, A. K. Singh, and V. S. Meena. "Impact of long term cultivation of lemon grass (*Cymbopogon citratus*) on post-harvest electro-chemical properties of soil." *Ann Agri Bio Res* 19, no. 1 (2014): 45–48.

Singh, H., A. Verma, M. Kumar, R. Sharma, R. Gupta, M. Kaur, M. Negi, and S. K. Sharma. "Phytoremediation: A green technology to clean up the sites with low and moderate level of heavy metals." *Austin Biochemistry* 2, no. 2 (2017): 1012.

Singh, N. "Enhanced degradation of hexachlorocyclohexane isomers in rhizosphere soil of Kochia sp." *Bulletin of Environmental Contamination and Toxicology* 70, no. 4 (2003): 775–782.

Singh, O. V., and R. K. Jain. "Phytoremediation of toxic aromatic pollutants from soil." *Applied Microbiology and Biotechnology* 63, no. 2 (2003): 128–135.

Singh, R. B. "Biotechnology, biodiversity and sustainable agriculture: A contradiction." Regional Conference in Agricultural Biotechnology Proceedings: Biotechnology Research and Policy: Needs and Priorities in the Context of Southeast Asia's Agricultural Activities, SEARCA (SEAMEO)/FAO/APSA, Bangkok, 2000.

Singh, R. P., V. V. Tyagi, Tanu Allen, M. Hakimi Ibrahim, and Richa Kothari. "An overview for exploring the possibilities of energy generation from municipal solid waste (MSW) in Indian scenario." *Renewable and Sustainable Energy Reviews* 15, no. 9 (2011): 4797–4808.

Singh, S., B. K. Singh, S. M. Yadav, and A. K. Gupta. "Potential of biofertilizers in crop production in Indian agriculture." *Am J Plant Nutr Fert Technol* 4, no. 2 (2014): 33–40.

Singh, Vijai. "Conservation of microbial diversity to sustain primary and secondary productivity." *Journal of Horticulture and Forestry* 3, no. 3 (2011): 93–95.

Singkaravanit, S., H. Kinoshita, F. Ihara, and T. Nihira. "Cloning and functional analysis of the second geranylgeranyl diphosphate syntheses gene influencing helvolic acid biosynthesis in *Metarhizium anisopliae.*" *Applied Microbiology and Biotechnology* 87, 3, no. (2010): 1077–1088.

Sinha, Rajiv K., Dalsukh Valani, Shanu Sinha, Shweta Singh, and Sunil Herat. "Bioremediation of contaminated sites: A low-cost nature's biotechnology for environmental clean up by versatile microbes, plants & earthworms." *Solid Waste Management and Environmental Remediation* (2009): 978–1 .

Siracusa, V., P. Rocculi, S. Romani, and M. D. Rosa. "Biodegradable polymers for food packaging: A review." *Trends Food Sci Technol* 19 (2008): 634–643.

Skea, J. "What drives clean fuel technology?" *Proceedings of the Institution of Mechanical Engineers, Part A: Journal of Power and Energy* 211, no. 1 (1997): 1–10.

Sliwinska-Kowalska, M. "Contribution of genetic factors to noise-induced hearing loss." In: Griefahn, B. (ed.), *10th International Congress on Noise as a Public Health Problem of the International Commission on Biological Effects of Noise*, London, UK, 2011. http://www.icben.org/proceedings.html

Smahi, Driss, Ouafa El Hammoumi, and Ahmed Fekri. "Assessment of the impact of the landfill on groundwater quality: A case study of the Mediouna Site, Casablanca, Morocco." *Journal of Water Resource and Protection* 5, no. 4 (2013): 440.

Smedes, F., L. A. van Vliet, and K. Booij. "Multi-ratio equilibrium passive sampling method to estimate accessible and pore water concentrations of polycyclic aromatic hydrocarbons and polychlorinated biphenyls in sediment." *Environmental Science & Technology* 47, no. 1 (2012): 510–517.

Smičiklas, I., and M. Šljivić-Ivanović. "Radioactive contamination of the soil: Assessments of pollutants mobility with implication to remediation strategies." Environmental Sciences, Soil Contamination-Current Consequences and Further Solutions. *InTech, Rijeka* (2016): 253–276.

Smith, P. A., A. Davis, M. Ferguson, and M. E. Lutman. "The prevalence and type of social noise exposure in young adults in England." *Noise Health* 2 (2000): 41–56. [PubMed: 12689478]

Smith, W. M. *Manufacture of Plastic. USA: Technology and Engineering.* Reinhold Pub. Corp., 1964.

Smith, Martin D., Frank Asche, Atle G. Guttormsen, and Jonathan B. Wiener. "Genetically modified salmon and full impact assessment." *Science* 330, no. 6007 (2010): 1052–1053.

Smithsonian Ocean. "Gulf OIL Spill." 2018. https://ocean.si.edu/conservation/pollution/gulf-oil-spill

Snow, Ray W., Matthew L. Brien, Michael S. Cherkiss, Laurie Wilkins, and Frank J. Mazzotti. "Dietary habits of the Burmese python, Python molurus bivittatus, in Everglades National Park, Florida." *Herpetological Bulletin* 101 (2007): 5.

Sode, S., A. Bruhn, T. J. Balsby, M. M. Larsen, A. Gotfredsen, and M. B. Rasmussen. "Bioremediation of reject water from anaerobically digested waste water sludge with macroalgae (*Ulva lactuca, Chlorophyta*)." *Bioresource Technology* 146 (2013): 426–435.

Solter, L. F., and J. J. Becnel. "Entomopathogenic microsporodia: Field manual of technique in invertebrate pathology." In: Lacey, L. A., and Kaya, H. K. (eds.), *Application and Evaluation of Pathogens for Control of Insects and Other Invertebrate Pests.* Kluwer Academic, Dordrecht, 2000, pp. 231–254.

Song, Y., and Q. Zheng. "Improved tensile strength of glycerol-plasticized gluten bioplastic containing hydrophobic liquids." *Bioresour Technol.* 99 (2008): 7665–7671.

Song, B., G. Zeng, J. Gong, J. Liang, P. Xu, Z. Liu, . . . and S. Ye. "Evaluation methods for assessing effectiveness of in situ remediation of soil and sediment contaminated with organic pollutants and heavy metals." *Environment International* 105 (2017): 43–55.

Spiegel, K., E. Tasali, P. Penev, and E. Van Cauter. "Brief communication: Sleep curtailment in healthy young men is associated with decreased leptin levels, elevated ghrelin levels, and increased hunger and appetite." *Ann Intern Med* 141, no. 11 (2004): 846–850.

Spiegel, K., E. Tasali, R. Leproult, and E. Van Cauter. "Effects of poor and short sleep on glucose metabolism and obesity risk." *Nat Rev Endocrinol* 5, no. 5 (2009): 253–261.

Sponza, Delia Teresa, and Pınar Demirden. "Treatability of sulfamerazine in sequential upflow anaerobic sludge blanket reactor (UASB)/completely stirred tank reactor (CSTR) processes." *Separation and Purification Technology* 56, no. 1 (2007): 108–117.

Srivastava, K. P., and G. S. Dhaliwal. *A Textbook of Applied Entomology.* Kalyani Publishers, New Delhi, 2010, p. 113.

Srivastava, Anjana, and Ram Prasad. "Triglycerides-based diesel fuels." *Renewable and Sustainable Energy Reviews* 4, no. 2 (2000): 111–133.

Stansfeld, S. A., B. Berglund, C. Clark et al. "Aircraft and road traffic noise and children's cognition and health: Across national study." *Lancet* 365 (2005): 1942–1949.

Stansfeld, S. A., and M. P. Matheson. "Noise pollution: Non-auditory effects on health." *Br Med Bull* 10, no. 68 (2003): 243–257.

Stationery Office. "The Control of Noise at Work Regulations." 2005. http://www.legislation.gov.uk/uksi/2005/1643/contents/made accessed on 11 July 11, 2013.

Steadman, David W. *Extinction and Biogeography of Tropical Pacific Birds.* University of Chicago Press, 2006, 480 p.

Steger, Ulrich. "Environmental management systems: Empirical evidence and further perspectives." *European Management Journal* 18, no. 1 (2000): 23–37.

Sterrett, S. B., R. L. Chaney, C. H. Gifford, and H. W. Mielke. "Influence of fertilizer and sewage sludge compost on yield and heavy metal accumulation by lettuce grown in urban soils." *Environmental Geochemistry and Health* 18, no. 4 (1996): 135–142.

Stevens, E. S. *Green Plastics: An Introduction to the New Science of Biodegradable Plastics.* Princeton University Press, 2002.

Stoneman, B. "Challenges to commercialization of biopesticides." Proceedings Microbial Biocontrol of Arthropods, Weeds and Plant Pathogens: Risks, Benefits and Challenges. National Conservation Training Center, Shepherdstown, WV, 2010.

Stroud, J. L., G. I. Paton, and K. T. Semple. "Microbe-aliphatic hydrocarbon interactions in soil: Implications for biodegradation and bioremediation." *Journal of Applied Microbiology* 102 (2007): 1239–1253.

Subha, B., Y. C. Song, and J. H. Woo. "Bioremediation of contaminated coastal sediment: Optimization of slow release biostimulant ball using response surface methodology (RSM) and stabilization of metals from contaminated sediment." *Marine Pollution Bulletin* 114, no. 1 (2017): 285–295.

Sudesh, K., H. Abe, and Y. Doi. "Synthesis, structure and properties of polyhydroxyalkanoates: Biological polyesters." *Prog Polym Sci.* 25 (2000): 1503–1555.

Suja, F., F. Rahim, M. R. Taha, N. Hambali, M. R. Razali, A. Khalid, and A. Hamzah. "Effects of local microbial bioaugmentation and biostimulation on the bioremediation of total petroleum hydrocarbons (TPH) in crude oil contaminated soil based on laboratory and field observations." *International Biodeterioration & Biodegradation* 90 (2014): 115–122.

Sun, H., J. Xu, S. Yang, G. Liu, and S. Dai. "Plant uptake of aldicarb from contaminated soil and its enhanced degradation in the rhizosphere." *Chemosphere* 54, no. 4 (2004): 569–574.

Sun, Mingming, Dengqiang Fu, Ying Teng, Yuanyuan Shen, Yongming Luo, Zhengao Li, and Peter Christie. " In situ phytoremediation of PAH-contaminated soil by intercropping alfalfa (*Medicago sativa* L.) with tall fescue (*Festuca arundinacea* Schreb.) and associated soil microbial activity." *Journal of Soils and Sediments* 11, no. 6 (2011): 980–989.

Suresh, B., and G. A. Ravishankar. "Phytoremediation: A novel and promising approach for environmental clean-up." *Critical Reviews in Biotechnology* 24, nos. 2–3 (2004): 97–124.

Susarla, S., V. F. Medina, and S. C. McCutcheon. "Phytoremediation: An ecological solution to organic chemical contamination." *Ecological Engineering* 18, no. 5 (2002): 647–658.

Sushkova, S., I. Deryabkina, E. Antonenko, R. Kizilkaya, V. Rajput, and G. Vasilyeva. "Benzo[a]pyrene degradation and bioaccumulation in soil-plant system under artificial contamination." *Science of the Total Environment* 633 (2018): 1386–1391.

Sushkova, S., T. Minkina, I. Deryabkina, V. Rajput, E. Antonenko, O. Nazarenko, B. Kumar Yadav, E. Hakki, and D. Mohan. "Environmental pollution of soil with PAHs in energy producing plants zone." *Science of the Total Environment* (2018). https://doi.org/10.1016/j.scitotenv.2018.11.080

Sushkova, Svetlana N., Tatiana Minkina, Irina Deryabkina, Saglara Mandzhieva, Inna Zamulina, Tatiana Bauer.

Sutherland, I., C. Thickitt, N. Douillet, K. Freebairn, D. Johns, C. Mountain, . . . and D. Keay. "Scalable technology for the extraction of pharmaceutics: Outcomes from a 3 year collaborative industry/academia research programme." *Journal of Chromatography A* 1282 (2013): 84–94.

Suttinun, O., E. Luepromchai, and R. Müller. "Cometabolism of trichloroethylene: Concepts, limitations and available strategies for sustained biodegradation." *Reviews in Environmental Science and Bio/Technology* 12, no. 1 (2012): 99–114.

Swain, Dhaneswar, Smita Lenka, Tapasuni Hota, and Gyana Ranjan Rout. "Micropropagation of Hypericum gaitii Haines, an endangered medicinal plants: Assessment of genetic fidelity." *The Nucleus* 59, no. 1 (2016): 7–13.

Swartjes, Frank A., ed. *Dealing with Contaminated Sites: From Theory towards Practical Application.* Springer Science & Business Media, 2011.

Syed, Khajamohiddin, Harshavardhan Doddapaneni, Venkataramanan Subramanian, Ying Wai Lam, and Jagjit S. Yadav. "Genome-to-function characterization of novel fungal P450 monooxygenases oxidizing polycyclic aromatic hydrocarbons (PAHs)." *Biochemical and Biophysical Research Communications* 399, no. 4 (2010): 492–497.

Sylvain, B., M. H. Mikael, M. Florie, J. Emmanuel, S. Marilyne, B. Sylvain, and M. Domenico. "Phytostabilization of As, Sb and Pb by two willow species (*S. viminalis* and *S. purpurea*) on former mine technosols." *Catena*, 136 (2016): 44–52.

Tabashnik, B. E., J. B. J. Van Rensburg, and Y. Carrière. "Field-evolved insect resistance to Bt crops: Definition, theory, and data." *Journal of Economic Entomology* 102 (2009): 2011–2025.

Taheri, S., L. Lin, D. Austin, T. Young, and E. Mignot. "Short sleep duration is associated with reduced leptin, elevated ghrelin, and increased body mass index." *PLoS Med* 1, no. 3 (2004): e62.

Tait, Joyce, and Les Levidow. "Proactive and reactive approaches to risk regulation: The case of biotechnology." *Futures* 24, no. 3 (1992): 219–231.

Takáčová, A., M. Smolinská, J. Ryba, T. Mackuľak, J. Jokrllová, P. Hronec, and G. Čík. "Biodegradation of benzo[a]pyrene through the use of algae." *Central European Journal of Chemistry* 12, no. 11 (2014): 1133–1143.

Takagi, S. I., K. Nomoto, and T. Takemoto. "Physiological aspect of mugineic acid, a possible phytosiderophore of graminaceous plants." *Journal of Plant Nutrition* 7, nos. 1–5 (1984): 469–477.

Talyan, Vikash, R. P. Dahiya, and T. R. Sreekrishnan. "State of municipal solid waste management in Delhi, the capital of India." *Waste Management* 28, no. 7 (2008): 1276–1287.

Tan, Sieting, Haslenda Hashim, Chewtin Lee, Mohd Rozainee Taib, and Jinyue Yan. "Economical and environmental impact of waste-to-energy (WTE) alternatives for waste incineration, landfill and anaerobic digestion." *Energy procedia* 61 (2014): 704–708.

Tanase, C. E., and I. Spiridon. "PLA/chitosan/keratin composites for biomedical applications." *Mater Sci Eng C* 40 (2014): 242–247.

Tanda, Y., and H. K. Kaya. *Insect Pathology*. Academic Press, Harcourt Brace Jovanovich Publishers, San Diego, 1993.

Tang, Ch.-Sh., W. H. Sun, M. Toma, F. M. Robert, and R. K. Jones. "Evaluation of agriculture-based phytoremediation in Pacific island ecosystems using trisector planters." *International Journal of Phytoremediation* 6, no. 1 (2004): 17–33.

Tang, W. "Research progress of microbial degradation of organophosphorus pesticides." *Progress in Applied Microbiology* 1 (2018): 29–35.

Tansel, Berrin. "From electronic consumer products to e-wastes: Global outlook, waste quantities, recycling challenges." *Environment International* 98 (2017): 35–45.

Tao, Shu, X. C. Jiao, S. H. Chen, W. X. Liu, R. M. Coveney Jr., L. Z. Zhu, and Y. M. Luo. "Accumulation and distribution of polycyclic aromatic hydrocarbons in rice (Oryza sativa)." *Environmental Pollution* 140, no. 3 (2006): 406–415.

Tassy, Caroline, Catherine Feuillet, and Pierre Barret. "A method for the medium-term storage of plant tissue samples at room temperature and successive cycles of DNA extraction." *Plant Molecular Biology Reporter* 24, no. 2 (2006): 247–248.

Tchobanoglous, George, Hilary Theisen, Samuel A. Vigil, and Victor M. Alaniz. *Integrated Solid Waste Management: Engineering Principles and Management Issues*. Vol. 4. McGraw-Hill, New York, 1993.

terminology ." *Clinical Infectious Diseases* 46, no. 7 (2008): 1121–1122.

Terry, N., C. Carlson, T. K. Raab, and A. M. Zayed. "Rates of selenium volatilization among crop species." *Journal of Environmental Quality* 21, no. 3 (1992): 341–344.

Testiati, E., J. Parinet, C. Massiani, I. Laffont-Schwob, J. Rabier, H. R. Pfeifer, V. Lenoble, V. Masotti, and P. Prudent. "Trace metal and metalloid contamination levels in soils and in two native plant species of a former industrial site: Evaluation of the phytostabilization potential." *Journal of Hazardous Materials* 248 (2013): 131–141.

Thakare, Shweta, and Somnath Nandi. "Study on potential of gasification technology for municipal solid waste (MSW) in Pune city." *Energy Procedia* 90 (2016): 509–517.

Thakur, Indu Shekhar. "Xenobiotics: Pollutants and their degradation-methane, benzene, pesticides, bioabsorption of metals." (2008).

Thirtle, C., L. Lin, and J. Piesse. "The impact of research-led agricultural productivity growth on poverty reduction in Africa, Asia and Latin America." *World Dev* 31 (2003): 1959–1975.

Tippani, Radhika, Rama Swamy Nanna, Praveen Mamidala, and Christopher Thammidala. "Assessment of genetic stability in somatic embryo derived plantlets of Pterocarpus marsupium Roxb. using inter-simple sequence repeat analysis." *Physiology and Molecular Biology of Plants* 25, no. 2 (2019): 569–579.

Tiwari, K. K., N. K. Singh, M. P. Patel, M. R. Tiwari, and U. N. Rai. "Metal contamination of soil and translocation in vegetables growing under industrial wastewater irrigated agricultural field of Vadodara, Gujarat, India." *Ecotoxicology and Environmental Safety* 74, no. 6 (2011): 1670–1677.

Tofangsazi, N., S. P. Arthurs, and R. M. G. Davis. "Entomopathogenic Nematodes (Nematoda: Rhabditida: Families Steinernematidae and Heterorhabditidae)." One of a series of the Entomology and Nematology Department, UF/IFAS Extension 1–5.

Toledo, Victor Manuel. "New paradigms for a new ethnobotany: Reflections on the case of Mexico." *Ethnobotany: Evolution of a Discipline* (1995): 75–88.

Tomei, M. C., and A. J. Daugulis. "Ex situ bioremediation of contaminated soils: An overview of conventional and innovative technologies." *Critical Reviews in Environmental Science and Technology* 43, no. 20 (2013): 2107–2139.

Tonjes, D. J., and K. L. Greene. "A review of national municipal solid waste generation assessments in the USA." *Waste Manag Res.* 30 (2012): 758–771.

Trapido, M. "Polycyclic aromatic hydrocarbons in Estonian soil: Contamination and profiles." *Environmental Pollution* 105, no. 1 (1999): 67–74.

Triantafyllidis, Kostas, Angelos Lappas, and Michael Stöcker, eds. *The Role of Catalysis for the Sustainable Production of Bio-fuels and Bio-chemicals*. Newnes, 2013.

Tripathi, A. K., S. Upadhyay, M. Bhuiyan, and P. R. Bhattacharya. "A review on prospects of essential oils as biopesticides in insect pest management." *J. Pharmacog. Phytother* 1 (2009): 52–63.

Tursi, A., A. Beneduci, F. Chidichimo, N. De Vietro, and G. Chidichimo. "Remediation of hydrocarbons polluted water by hydrophobic functionalized cellulose." *Chemosphere* 201 (2018): 530–539.

U.S. EPA. "Air Quality Criteria for Ozone and Related Photochemical Oxidants." Volume I. EPA 600/R05/004aF. National Center for Environmental Assessment-RTP Office Office of Research and Development U.S. Environmental Protection Agency Research Triangle. Park, NC (2006).

U.S. Green Building Council. "Indoor Environmental Quality Prerequisite 3." Minimum Acoustical Performance (2010).

U.S. Green Building Council. 2013. LEED.

Uddin, M. N., M. Ali, M. F. Muhammad, N. Ahmad, J. Jamil, M. A. Kalsoom, . . . and A. Khan. "Characterizing microbial populations in petroleum-contaminated soils of Swat district, Pakistan." *Polish Journal of Environmental Studies* 25, no. 4 (2016).

Ulrich, R. S. "Research on building design and patient outcomes." In: Ulrich, R. S., Lawson, B., and Martinez, M. (eds.), *Exploring the Patient Environment: An NHS Estates Workshop*. London: The Stationery Office.

"Urban solid waste management in Indian cities." *PEARL*. www.citiesalliance.org/sites/citiesalliance.org/files/GPIN3%20SWM accessed on April, 2019.

US Department of Health and Human Services, Public Health Service. *Toxicological Profile for Mercury*. US Department of Health and Human Services, Atlanta, 1999, pp. 1–600.

Usman, A., K. M. Zia, M. Zuber, S. Tabasum, S. Rehman, and F. Zia. "Chitin and chitosan based polyurethanes: A review of recent advances and prospective biomedical applications." *Int J Biol Macromol.* 86 (2016): 630–645.

Utracki, L. A. "Role of polymer blends technology in polymer recycling." In: *Polymer Blends Handbook*. Springer, Netherlands.

Vaishampayan, A., R. P. Sinha, D. P. Hader, T. Dey, A. K. Gupta, U. Bhan, and A. L. Rao. "Cyanobacterial biofertilizers in rice agriculture." *Botanical Review* 67, no. 4 (2001): 453–516.

Van Bohemen, H. D., and W. H. Janssen Van de Laak. "The influence of road infrastructure and traffic on soil, water, and air quality." *Environmental Management* 31, no. 1 (2003): 50–68.

Van Cauter, E., K. Spiegel, E. Tasali, and R. Leproult. "Metabolic consequences of sleep and sleep loss." *Sleep Med* 9, Suppl. 1 (2008): S23–S28.

Van den Brink, P. J., A. B. Boxall, L. Maltby, B. W. Brooks, M. A. Rudd, T. Backhaus, . . . and S. E. Apitz. "Toward sustainable environmental quality: Priority research questions for Europe." *Environmental Toxicology and Chemistry* 37, no. 9 (2018).: 2281–2295.

Vandecasteele, B., E. Meers, P. Vervaeke, B. De Vos, P. Quataert, and F. M. Tack. "Growth and trace metal accumulation of two Salix clones on sediment-derived soils with increasing contamination levels." *Chemosphere* 58, no. 8 (2005): 995–1002.

Vane, C. H., A. W. Kim, D. J. Beriro, M. R. Cave, K. Knights, V. Moss-Hayes, and P. C. Nathanail. "Polycyclic aromatic hydrocarbons (PAH) and polychlorinated biphenyls (PCB) in urban soils of Greater London, UK." *Applied Geochemistry* 51 (2014): 303–314.

Váňová, Lucie. "The use of in vitro cultures for effect assessment of persistent organic pollutants on plants." PhD diss., Masarykova univerzita, Přírodovědecká fakulta, 2009.

Van-Thuoc, D., J. Quillanguaman, G. Mamo, and B. Mattiason. "Utilization of agricultural residues for poly(3-hydroxybutyrate) production by *Halomonas boliviensis* LC1." *J Appl Microbiol.* 104 (2008): 420–428.

Vasanthan, N., and O. Ly. "Effect of microstructure on hydrolytic degradation studies of poly (L-lactic acid) by FTIR spectroscopy and differential scanning calorimetry." *Polym Degrad Stabil.* 94 (2009): 1364–1372.

Venosa, A. D., M. T. Suidan, B. A. Wrenn, K. L. Strohmeier, J. R. Haines, B. L. Eberhart, . . . and E. Holder. "Bioremediation of an experimental oil spill on the shoreline of Delaware Bay." *Environmental Science & Technology* 30, no. 5 (1996): 1764–1775.

Venter, Oscar, Nathalie N. Brodeur, Leah Nemiroff, Brenna Belland, Ivan J. Dolinsek, and James W. A. Grant. "Threats to endangered species in Canada." *Bioscience* 56, no. 11 (2006): 903–910.

Venuti, A., L. Alfonsi, and A. Cavallo. "Anthropogenic pollutants on top soils along a section of the Salaria state road, central Italy." *Annals of Geophysics* 59, no. 5 (2016): 544.

Verbruggen, N., C. Hermans, and H. Schat. "Molecular mechanisms of metal hyperaccumulation in plants." *New Phytol* 181 (2009): 759–776.

Verlinden, R. A. J., D. J. Hill, M. A. Kenward, C. D. Williams, and I. Radecka. "Bacterial synthesis of biodegradable polyhydroxyalkanoates." *J Appl Microbiol.* 102 (2007): 1437–1449.

Vicente, Gemma, Mercedes Martınez, and Jose Aracil. "Integrated biodiesel production: A comparison of different homogeneous catalysts systems." *Bioresource technology* 92, no. 3 (2004): 297–305.

Vidali, M. "Bioremediation: An overview." *Pure and Applied Chemistry* 73, no.7 (2001): 1163–1172.

Vie, Jean-Christophe, Craig Hilton-Taylor, and Simon N. Stuart, eds. *Wildlife in a Changing World: An Analysis of the 2008 IUCN Red List of Threatened Species*. IUCN, 2009.

Vieira, Gabriela A. L., Mariana Juventina Magrini, Rafaella C. Bonugli-Santos, Marili V. N. Rodrigues, and Lara D. Sette. "Polycyclic aromatic hydrocarbons degradation by marine-derived basidiomycetes: Optimization of the degradation process." *Brazilian Journal of Microbiology* 49, no. 4 (2018): 749–756.

Vignais, P. M., and A. Colbeau. "Molecular biology of microbial hydrogenases." *Current Issues in Molecular Biology* 6, no. 2 (2004): 159–188.

Vikrant, K., B. S. Giri, N. Raza, K. Roy, K. H. Kim, B. N. Rai, and R. S. Singh. "Recent advancements in bioremediation of dye: Current status and challenges." *Bioresource Technology* 253 (2018): 355–367.

Vogel, I., J. Brug, C. P. Van der Ploeg, and H. Raat. "Adolescents' risky MP3-player listening and its psychosocial correlates." *Health Educ Res* 26 (2011): 254–264.

von Arnold, Sara, Izabela Sabala, Peter Bozhkov, Julia Dyachok, and Lada Filonova. "Developmental pathways of somatic embryogenesis." *Plant Cell, Tissue and Organ Culture* 69, no. 3 (2002): 233–249.

von der Ohe, P. C., V. Dulio, J. Slobodnik, E. De Deckere, R. Kühne, R. U. Ebert, . . . and W. Brack. "A new risk assessment approach for the prioritization of 500 classical and emerging organic microcontaminants as potential river basin specific pollutants under the European Water Framework Directive." *Science of the Total Environment* 409, no. 11 (2011): 2064–2077.

Waitz, I. A., R. J. Bernhard, and C. E. Hanson. "Challenges and promises in mitigating transportation noise." *Bridge* 37 (2007): 25–32.

Walker, T. S. "Root exudation and rhizosphere biology." *Plant Physiology* 132, no. 1 (2003): 44–51.

Walter, U., M. Beyer, J. Klein, and H.-J. Rehm. "Degradation of pyrene by Rhodococcus sp. UW1." *Applied Microbiology and Biotechnology* 34, no. 5 (1991): 671–676.

Wang, X. L., K. K. Yang, and Y. Z. Wang. "Properties of starch blends with biodegradable polymers." *J Macromol Sci Part C: Polym Rev.* 43 (2003): 385–409.

Wang, F., Z. Wang, C. Kou, Z. Ma, and D. Zhao. "Responses of wheat yield, macro-and micro-nutrients, and heavy metals in soil and wheat following the application of manure compost on the North China Plain." *PloS One* 11, no. 1 (2016): e0146453.

Wang, K., S. Pang, X. Mu, S. Qi, D. Li, F. Cui, and C. Wang. "Biological response of earthworm, Eisenia fetida, to five neonicotinoid insecticides." *Chemosphere* 132 (2015): 120–126.

Wang, X. T., Y. Miao, Y. Zhang, Y. C. Li, M. H. Wu, and G. Yu. "Polycyclic aromatic hydrocarbons (PAHs) in urban soils of the megacity Shanghai: Occurrence, source apportionment and potential human health risk." *Science of the Total Environment* 447 (2013): 80–89.

Wang, Y., C. L. Luo, J. Li, H. Yin, X. D. Li, and G. Zhang. "Characterization and risk assessment of polychlorinated biphenyls in soils and vegetation near an electronic waste recycling site, South China." *Chemosphere* 85, no. 3 (2011): 344–350.

Warren, G. F. "Spectacular Increases in Crop Yields in the United States in the Twentieth Century." *Weed Tech* 12 (1998): 752.

Warshawsky, D., M. Radike, K. Jayasimhulu, and T. Cody. "Metabolism of benzo[a]pyrene by a dioxygenase enzyme system of the freshwater green alga *selenastrumcapricornutum*." *Biochemical and Biophysical Research Communications* 152, no. 2 (1988): 540–544.

Warshawsky, D., T. Cody, M. Radike, R. Reilman, B. Schumann, K. LaDow, and J. Schneider. "Biotransformation of benzo[a]pyrene and other polycyclic aromatic hydrocarbons and heterocyclic analogs by several green algae and other algal species under gold and white light." *Chemico-Biological Interactions* 97, no. 2 (1995): 131–148.

Warunasinghe, W. A. A. I., and P. I. Yapa. "A survey on household solid waste management (SWM) with special reference to a peri-urban area (Kottawa) in Colombo." *Procedia Food Science* 6 (2016): 257–260.

Watkin, E. L. J., S. E. Keeling, F. A. Perrot, D. W. Shiers, M.-L. Palmer, and H. R. Watling. "Metals tolerance in moderately thermophilic isolates from a spent copper sulfide heap, closely related to *Acidithiobacillus caldus, Acidimicrobium ferrooxidans* and *Sulfobacillus thermosulfidooxidans*." *Journal of Industrial Microbiology & Biotechnology* 36, no. 3 (2009): 461.

Watt, J. "Hedgehogs and ferrets on the South Uist machairs: A pilot study of predation on ground nesting birds." Unpublished Report, Scottish Natural Heritage, Inverness, UK, 1995.

Wawer, M., T. Magiera, G. Ojha, E. Appel, G. Kusza, S. Hu, and N. Basavaiah. "Traffic-related pollutants in roadside soils of different countries in Europe and Asia." *Water, Air, and Soil Pollution* 226, no. 7 (2015): 216.

Weathers, Kathleen C., and Gene E. Likens. "Acid rain." In: Rom, W. (eds.), In *Environmental and Occupational Medicine*. 4th ed. 2006, pp. 1549–1561.

Webster, J. P. G., R. G. Bowles, and N. T. Williams. "Estimating the Economic Benefits of Alternative Pesticide Usage Scenarios: Wheat Production in the United Kingdom." *Crop Production* 18 (1999): 83.

Wendlandt, K. D., W. Geyer, G. Mirschel, and F. A. Hemidi. "Possibilities for controlling a PHB accumulation process using various analytical methods." *J Biotechnol.* 117 (2005): 119–129.

Westbrook, Adam W., Dragan Miscevic, Shane Kilpatrick, Mark R. Bruder, Murray Moo-Young, and C. Perry Chou. "Strain engineering for microbial production of value-added chemicals and fuels from glycerol." *Biotechnology Advances* 37 (2018): 538–568.

White, C., A. K. Shaman, and G. M. Gadd. "An integrated microbial process for the bioremediation of soil contaminated with toxic metals." *Nature Biotechnol* 16 (1998): 572.

White, P. M., D. C. Wolf, G. J. Thoma, and C. M. Reynolds. "Phytoremediation of alkylated polycyclic aromatic hydrocarbons in a crude oil-contaminated soil." *Water, Air, and Soil Pollution* 169, no. 1–4 (2006): 207–220.

Whiteley, Susan E., Eric Bunn, Akshay Menon, Ricardo L. Mancera, and Shane R. Turner. "Ex situ conservation of the endangered species Androcalva perlaria (Malvaceae) by micropropagation and cryopreservation." *Plant Cell, Tissue and Organ Culture (PCTOC)* 125, no. 2 (2016): 341–352.

Whitlock, Raj, Helen Hipperson, Des B. A. Thompson, Roger K. Butlin, and Terry Burke. "Consequences of in-situ strategies for the conservation of plant genetic diversity." *Biological Conservation* 203 (2016): 134–142.

WHO, Nitrates. "Nitrite in Drinking Water." *Background Document for Guidelines for Drinking Water Quality 2011*, 2004.

Whyte, C., and M. O'Brien. "Changes in the number of waders breeding on Machair Habitats on the Uists and Benbecula, 1983– 1995." Unpublished Report, Royal Society for the Protection of Birds, Edinburgh, 1995.

Wilcove, David S., David Rothstein, Jason Dubow, Ali Phillips, and Elizabeth Losos. "Quantifying threats to imperiled species in the United States." *BioScience* 48, no. 8 (1998): 607–615.

Wild, E., J. Dent, G. O. Thomas, and K. C. Jones. "Visualizing the air-to-leaf transfer and within-leaf movement and distribution of phenanthrene: Further studies utilizing two-photon excitation microscopy." *Environmental Science and Technology* 40, no. 3 (2006): 907–916.

Wildt, David Edvin "Genome resource banking for wildlife research, management, and conservation." *ILAR Journal* 41, no. 4 (2000): 228–234.

Wilkie, David, and Julia F. Carpenter. "Bushmeat hunting in the Congo Basin: An assessment of impacts and options for mitigation." *Biodiversity & Conservation* 8, no. 7 (1999): 927–955.

Wilson, David Curran, Ljiljana Rodic, Prasad Modak, Reka Soos, A. Carpintero, K. Velis, Mona Iyer, and Otto Simonett. *Global Waste Management Outlook*. UNEP, 2015.

Wilson, E. O. *The Diversity of Life*. Harvard University Press: Cambridge, Massachusetts, p. 424 (1992).

Witholt, B., and B. Kessler. "Perspectives of medium chain length poly(hydroxyalkanotes), a versatile set of bacterial bioplastics." *Curr Opin Biotechnol.* 10 (1999): 279–285.

Wolfe, Martin S. "Crop strength through diversity." *Nature* 406, no. 6797 (2000): 681–682.

Wolvekamp, Monique C. J., Michelle L. Cleary, Shae Lee Cox, Jill M. Shaw, Graham Jenkin, and Alan O. Trounson. "Follicular development in cryopreserved common wombat ovarian tissue xenografted to nude rats." *Animal Reproduction Science* 65, nos. 1–2 (2001): 135–147.

Wong, P. A. L., M. K. Cheung, W. H. Lo, H. Chua, and P. H. F. Yu. "Investigation of the effects of the types of food waste utilized as carbon source on the molecular weight distributions and thermal properties of polyhydroxy-butyrate produced by two strains of microorganisms." *e-Polymers* 31 (2004): 1–11.

Wong, M. H. "Heavy metal contamination of soils and crops from auto traffic, sewage sludge, pig manure and chemical fertilizer." *Agriculture, Ecosystems and Environment* 13, no. 2 (1985): 139–149.

World Bank. "What a waste a global review of solid waste management." Urban Development Series Knowledge Paper No. 15, Washington, DC, 2012.

World Economic Forum. "The new plastics economy: Rethinking the future of plastics." World Economic Forum, Ellen MacArthur Foundation and McKinsey & Company, 2016. http://www.ellenmacarthurfoundation.org/publications

World Health Organization (WHO). "Permissible limits of heavy metals in soil and plants." *Geneva, Switzerland: Author* (1996).

World Health Organization. "Night noise guidelines for Europe." *WHO Regional* (2009).

Wu, A., S. Yan, B. Yang, J. Wang, and G. Qiu. "Study on preferential flow in dump leaching of low-grade ores." *Hydrometallurgy* 87, nos. 3–4 (2007): 124–132.

Wu, K., K. Lin, J. Miao, and Y. Zhang. "Field abundances of insect predators and insect pests on $\delta\delta$-Endotoxin-producing transgenic cotton in Northern China." Second International Symposium on Biological Control of Arthropods, Davos, 2005, pp. 362–368.

Wu, S., K. Cheung, Y. Luo, and M. Wong. "Effects of inoculation of plant growthpromoting rhizobacteria on metal uptake by Brassica juncea." *Environ. Pollut.* 140 (2006): 124–135.

Wu, Sh., X. Xia, L. Yang, and H. Liu. "Distribution, source and risk assessment of polychlorinated biphenyls (PCBs) in urban soils of Beijing, China." *Chemosphere* 82, no. 5 (2011): 732–738.

Wu, Y., X. Jing, C. Gao, Q. Huang, and P. Cai. "Recent advances in microbial electrochemical system for soil bioremediation." *Chemosphere* (2018).

Xiao, PeiGen, and Peng Yong. "Ethnopharmacology and research on medicinal plants in China." (1998).

Xiao, Y., and D. J. Roberts. "A review of anaerobic treatment of saline wastewater." *Environmental Technology* 31, nos. 8–9 (2010): 1025–1043.

Yadav, B. K., M. S. Akhtar, and J. Panwar. "Rhizospheric plant-microbe interactions: Key factors to soil fertility and plant nutrition." In: *Plant Microbes Symbiosis: Applied Facets.* Springer, New Delhi, 2015, pp. 127–145.

Yadav, B. K., M. S. Akhtar, and J. Panwar. "Rhizospheric plant-microbe interactions: Key factors to soil fertility and plant nutrition." In: *Plant Microbes Symbiosis: Applied Facets.* Springer, New Delhi, 2015, pp. 127–145.

Yan, J., L. Wang, P. P. Fu, and H. Yu. "Photomutagenicity of 16 polycyclic aromatic hydrocarbons from the US EPA priority pollutant list." *Mutation Research/Genetic Toxicology and Environmental Mutagenesis* 557, no. 1 (2004): 99–108.

Yang, F., H. Liu, J. Qu, and J. Paul Chen. "Preparation and characterization of chitosan encapsulated Sargassum sp. biosorbent for nickel ions sorption." *Bioresource Technol* 102, no. 3 (2011): 2821–2828.

Yang, G. C., S. C. Huang, Y. S. Jen, and P. S. Tsai. "Remediation of phthalates in river sediment by integrated enhanced bioremediation and electrokinetic process." *Chemosphere* 150 (2016): 576–585.

Yang, Jinyue, Baohong Hou, Jingkang Wang, BeiqianTian, Jingtao Bi, Na Wang, Xin Li, and Xin Huang. "Nanomaterials for the removal of heavy metals from wastewater." *Nanomaterials* 9, no. 3 (2019): 424.

Yang, S., S. Liang, L. Yi, B. Xu, J. Cao, Y. Guo, and Y. Zhou. "Heavy metal accumulation and phytostabilization potential of dominant plant species growing on manganese mine tailings." *Frontiers of Environmental Science & Engineering* 8, no. 3 (2013): 394–404.

Yang, Y., Z. Lu, X. Lin, C. Xia, G. Sun, Y. Lian, and M. Xu. "Enhancing the bioremediation by harvesting electricity from the heavily contaminated sediments." *Bioresource Technology* 179 (2015): 615–618.

Yang, Z., and C. Chu. "Towards understanding plant response to heavy metal stress." *Abiotic Stress in Plants–Mechanisms and Adaptations* 10 (2011): 24204.

Yang, Zhiliang, and Zisheng Zhang. "Recent advances on production of 2, 3-butanediol using engineered microbes." *Biotechnology Advances* 37 (2019).: 569–578.

Yankaskas, K. "Prelude: Noise-induced tinnitus and hearing loss in the military." *Hear Res.* 295 (2013): 3–8. [PubMed: 22575206]

Yap, H. Y., and J. D. Nixon. "A multi-criteria analysis of options for energy recovery from municipal solid waste in India and the UK." *Waste Management* 46 (2015): 265–277.

Yaradoddi, J. S., S. Hugar, N. R. Banapurmath, A. M. Hunashyal, M. B. Sulochana, A. S. Shettar, and S. V. Ganachari. "Alternative and renewable bio-based and biodegradable plastics." In: Martinez, L. M. T., Kharissova, O. V., and Kharisov, B. I. (eds.), *Handbook of Ecomaterials.* Springer International Publishing, 2018, pp. 1–20.

Ye, D., M. Akmal Siddiqi, A. E. Maccubbin, S. Kumar, and H. C. Sikka. "Degradation of polynuclear aromatic hydrocarbons by *Sphingomonas paucimobilis.*" *Environmental Science and Technology* 30, no. 1 (1995): 136–142.

Ye, X., F. Dong, and X. Lei. "Microbial resources and ecology-microbial degradation of pesticides." *Natural Resources Conservation and Research* 1, no. 1 (2018).

Yew, G. H., A. M. M. Yusof, Z. A. Z. Ishak, and U. S. Ishiaku. "Natural weathering effects on mechanical properties of polylactic acid/rice starch composites." 8th Polymers for Advanced Technologies International Symposium, Budapest, Hungary, 2005.

Yu, Y. L., X. Chen, Y. M. Luo, X. D. Pan, Y. F. He, and M. H. Wong. "Rapid degradation of butachlor in wheat rhizosphere soil." *Chemosphere* 50, no. 6 (2003): 771–774.

Yuan, M., H. He, L. Xiao, T. Zhong, H. Liu, S. Li, . . . and Y. Jing. "Enhancement of Cd phytoextraction by two Amaranthus species with endophytic Rahnella sp. JN27." *Chemosphere* 103 (2014): 99–104.

Yunus, Ian Sofian, Harwin, Adi Kurniawan, Dendy Adityawarman, and Antonius Indarto. "Nanotechnologies in water and air pollution treatment." *Environmental Technology Reviews* 1, no. 1 (2012): 136–148.

Zabed, H., J. N. Sahu, A. Suely, A. N. Boyce, and G. Faruq. "Bioethanol production from renewable sources: Current perspectives and technological progress." *Renewable and Sustainable Energy Reviews* 71 (2017): 475–501.

Zaib-Un-Nisa, Samin Jan, Muhammad Anwar, Safdar Hussain Shah Sajad, Ghulam Farooq, and Hazrat Ali. "60. Micropropagation through apical shoot explants and morphogenic potential of different explants of Saussurea lappa: An endangered medicinal plant." *Pure and Applied Biology (PAB)* 8, no. 1 (2019): 585–592.

Zaidi, S., S. Usmani, B. R. Singh, and J. Musarrat. "Significance of Bacillus subtilis strain SJ-101 as a bioinoculant for concurrent plant growth promotion and nickel accumulation in Brassica juncea." *Chemosphere*, 64, no. 6 (2006): 991–997.

Zalesny, J. R., S. Ronald, Edmund O. Bauer, Richard B. Hall, Jill A. Zalesny, Joshua Kunzman, Chris J. Rog, and Don E. Riemenschneider. "Clonal variation in survival and growth of hybrid poplar and willow in an in situ trial on soils heavily contaminated with petroleum hydrocarbons." *International Journal of Phytoremediation* 7, no. 3 (2005): 177–197.

Zimmermann, Gisbert. "The entomopathogenic fungus Metarhizium anisopliae and its potential as a biocontrol agent." *Pesticide Science* 37, no. 4 (1993): 375–379.

Zeng, An-Ping. "New bioproduction systems for chemicals and fuels: Needs and new development." *Biotechnology Advances* 37 (2019): 508–518.

Zhang, H. B., Y. M. Luo, Ming Hung Wong, Q. G. Zhao, and Guang L. Zhang. "Distributions and concentrations of PAHs in Hong Kong soils." *Environmental Pollution* 141, no. 1 (2006): 107–114.

Zhang, H., D. Ma, R. Qiu, Y. Tang, and C. Du. "Non-thermal plasma technology for organic contaminated soil remediation: A review." *Chemical Engineering Journal* 313 (2017): 157–170.

Zhang, H., Z. Wang, Y. Zhang, M. Ding, and L. Li. "Identification of traffic-related metals and the effects of different environments on their enrichment in roadside soils along the Qinghai-Tibet highway." *Science of the Total Environment* 521 (2015): 160–172.

Zhang, J., Sh. Fan, X. Du, J. Yang, W. Wang, and H. Hou. "Accumulation, allocation, and risk assessment of polycyclic aromatic hydrocarbons (PAHs) in soil-*Brassica chinensis* system." *PloS One* 10, no. 2 (2015): e0115863.

Zhao, S., C. Fan, X. Hu, J. Chen, and H. Feng. "The microbial production of polyhydroxybutyrate from methanol." *Appl Biochem Biotechnol.* 39 (1993): 191–199.

Zhao, Q., L. Li, C. Li, M. Li, F. Amini, and H. Zhang. "Factors affecting improvement of engineering properties of MICP-treated soil catalyzed by bacteria and urease." *Journal of Materials in Civil Engineering* 26, no. 12 (2014): 04014094.

Zhao, Y., Y. Bai, Q. Guo, Z. Li, M. Qi, X. Ma, . . . and B. Liang. "Bioremediation of contaminated urban river sediment with methanol stimulation: Metabolic processes accompanied with microbial community changes." *Science of the Total Environment*, 653 (2019): 649–657.

Zhu, X., W. Li, L. Zhan, M. Huang, Q. Zhang, and V. Achal. "The large-scale process of microbial carbonate precipitation for nickel remediation from an industrial soil." *Environmental Pollution*, 219 (2016): 149–155.

Zhu, Y. G., and G. Shaw. "Soil contamination with radionuclides and potential remediation." *Chemosphere* 41, nos. 1–2 (2000): 121–128.

Zutshi, Ambika, Amrik S. Sohal, and Carol Adams. "Environmental management system adoption by government departments/agencies." *International Journal of Public Sector Management* 21, no. 5 (2008): 525–539.

http://www.nhm.ac.uk/research-curation/research/biodiversity/conventionbiodiversity/convention-faqs/index.html

https://www.cbd.int/convention/articles/?a=cbd-02

https://www.cbd.int/doc/publications/plant-conservation-report-en.pdf

https://www.bioplasticsmagazine.com/en/news/meldungen/20201202-Market-update-2020-global-bioplastics-market-set-to-grow-by-36-percent-over-the-next-5-years.php

Index

Printed in the United States
by Baker & Taylor Publisher Services